THIRD GENERATION PHOTOVOLTAICS

Edited by **Vasilis Fthenakis**

Third Generation Photovoltaics
Edited by Vasilis Fthenakis

Published by InTech
Janeza Trdine 9, 51000 Rijeka, Croatia

Publishing Process Manager Anja Filipovic

Technical Editor Teodora Smiljanic

Cover Designer InTech Design Team

Additional hard copies can be obtained from orders@intechopen.com

Third Generation Photovoltaics, Edited by Vasilis Fthenakis
p. cm.
ISBN 978-953-51-0304-2

For EU product safety concerns:
IN TECH d.o.o., Prolaz Marije Krucifikse Kozulić 3, 51000 Rijeka, Croatia,
info@intechopen.com or visit our website at intechopen.com.

Meet the editor

Professor Vasilis Fthenakis is the founder and Director of the Center for Life Cycle Analysis at Columbia University and a senior scientist at Brookhaven National Laboratory. Dr. Fthenakis is recognized worldwide for his research that provided the underpinning of PV technologies in terms of their lifecycle impacts. He identified potential barriers in PV commercialization and proactively conducted research that resolved concerns associated with rapid growth of the PV market. He leads international collaborations on PV sustainability under the auspices of the International Energy Agency, is a Fellow of both the International Energy Foundation and the American Institute of Chemical Engineers, a member of the editorial boards of Progress in Photovoltaics and the Journal of Loss Prevention, and author or coauthor of more than 300 publications on energy and environmental topics.

Contents

Preface

The quest for alternative sustainable energy sources has led us to the third generation of photovoltaics. The feasibility for PV to become a major contributor in the electricity mixture of the U.S. and other major energy consuming countries has been proven. The production and sales of photovoltaics have increased by an average of 45% per year over the last fifteen years. As with all energy-generation technologies, PV deployment is subsidized but the expectation is that, within a decade, the cost of PV electricity generation will be in parity with electricity from the grid, making additional subsidies unnecessary.

For PV grid cost parity we need high efficiencies and low production costs. So far we have made strides on either high performance as exemplified by Sunpower producing modules with 20% efficiencies at a production cost of about \$1.75 /Wp, or on low cost as mastered by First Solar producing modules at a cost of 75 ¢/Wp and efficiencies of 12.5%. Third generation technologies promise further advances in each pathway and, more importantly, bridging the two with technologies of both high efficiency and low cost. In this book we present notable advances of 3rd generation technologies, namely organic cells and nanostructures, dye sensitized cells and multijunction III/V cells. The underlining promise of most 3rd generation technologies is that the thermodynamic Shockley-Queisser limit of single junction solar cells will be surpassed. In 1st generation crystalline silicon and 2nd generation a-Si, CdTe and CIGS solar cells, only photons within a narrow wavelength, corresponding to the semiconductor band gap, are effectively absorbed. Photons with energy lower than the semiconductor band gap are not absorbed and their energy is not used for carrier generation, whereas photons with energy larger than the band gap are absorbed but excess energy is lost to heat, negatively affecting the voltage generation of the devices. Primarily for this reason, the thermodynamic Shockley-Queisser of single junction cells is only about 31%, with spectral losses being as large as 50%. Several approaches have been proposed to reduce or eliminate spectral losses, e.g., multiple stacked cells, intermediate band gaps, multiple exciton generation, quantum dot concentrators, down- and up-converters, and down-shifters. The first approach is already commercialized, with III/V solar cells of efficiencies as high as 42% being produced in the lab and 38% efficient cells being deployed in terrestrial concentrator systems. The others involve the generation of more than one electron from a highly energetic photon and the upshifting of low energy photons so that their energy is also utilized in electron production. Nanotechnology is essential in realizing these concepts. Nanotechnology enabled organic photovoltaics (OPV), in addition to enabling high photon conversion efficiencies, have the potential to further lower the production cost of photovoltaics by using inexpensive materials. Dye-sensitized solar cells seem especially suited for low irradiation regions and their efficiencies can also be augmented with nanostructures; III/V solar cells are commercially produced and started being deployed, but challenges remain in decreasing their system costs.

The book begins with an overview of approaches based on nanotechnology for surpassing the Shockley-Queisser limit. The principles and advances of tuning chemical and physical properties in nanosized materials and inducing downconversion, downshifting and upconversion of photon energies, are explained in terms understood by general physics and engineering readership. The first chapter also introduces the reader into modelling of spectral conversion, a field where further advances are necessary to improve the understanding of the plethora of combinations with interlinked influences on a solar cell system. The second chapter presents specific cases of multi-scale modeling of quantum chemical interactions at the molecular level of organic solar cells. Case-specific calculations are shown to offer a universal view of such systems, from local interactions and structure, up to morphology and phase behavior. This chapter shows that a realistic description of OPV requires examination at multiple scales and that defining appropriate time and length scales is a prerequisite to selecting or developing a model. The discussion on organic solar cells continues in the third chapter with a tutorial on manufacturing organic solar cells, which will be especially interesting to people studying the transition from lab-scale to commercialization. This transition requires adjusting and optimizing of process and system parameters, such as solvent types, deposition techniques, drying conditions, and dimensions of individual cells and modules. The chapter establishes a knowledge base concerning the influence of different parameters and process conditions on the performance, cost and lifetime of polymer solar cells.

It is imperative that new energy technologies are safe and friendly to the environment. While solar cells, as an alternative to fossil fuel energy production, prevent the pollution and climate change impacts of the later, proactive evaluations are needed to ensure that new environmental issues are not created. For this reason, life cycle assessments (LCA) are performed of emerging organic technologies to evaluate a priori the environmental impact of their production and to identify sustainable energy production pathways. In chapter 4 a comprehensive LCA methodology, and its application on emerging organic photovoltaics, are presented. A major difficulty for applying LCA to new technologies is the availability of data representing actual commercial production. Based on extrapolation of laboratory data, this chapter shows that fullerenes used as exciton acceptors are the most energy-intensive components of organic photovoltaics and that a trend towards using higher quantities of larger fullerenes in OPV will increase the cumulative energy demand associated with their life cycle. Nevertheless, the LCAs of OPV inherently carry large uncertainties and the need for further studies on the environnmental impacts and recyclability of these technologies is pointed out.

Chapter 5 gives an overview of the state of knowledge about dye-sensitized solar cells and probes into the type of developments needed to produce durable and reliable PV modules from these cells. As advances are made in more effective semiconductor morphologies, lower-cost, and higher-efficiency dyes, and more effective and long-lasting hole conduction materials, these devices could become commercially viable. The chapter describes the functions and challenges for each of the components of a Grätzel cell, and outlines research needed for innovative ways of separating and transporting charges within these devices.

In chapter 6, an overview of III/V solar cells used in 3rd generation high concentration photovoltaics (HCPV) is given, explaining how these cells enable efficiencies well above the Shockley-Queisser limits and illustrating associated developments. The focus is on the importance of theoretical models for the prediction of the structural and electronic properties of IV-doped III-V alloys used in tandem structures, showing that efficiencies of 60% could be obtained. Concentrator systems are commercialized and Amonix is the industry leader with well tested HCPV based on Fresnel lenses and III/V solar cells. These systems are described in chapter 7 and performance data are presented along with the life-cycle environmental profile of the Amonix systems. One way to bring down the cost of generating electricity from photovoltaics is to replace relatively expensive reflecting materials in concentrating systems with low cost polymeric wavelength selective mirror films. Chapter 8 describes such films made by 3M Corporation that have the potential to significantly reduce the cost of electricity derived from PV.

Reliability of photovoltaics deployed in the field is an essential element of sustainable growth of these technologies. Accelerated aging tests, prescribed by qualification standards, can minimize risks to reliability. Proving the reliability of 3rd generation solar cells is a demanding challenge as some use electrolyte solutions, other use polymeric materials that have less resistance to light and temperature cycles than silicon or glass, and light concentrating systems that function at higher voltages and generate higher currents than the previous generations' solar cells. Reliability is best confirmed by decades of field operation. Chapter 9 describes how PV systems perform in the sunny and dusty south European/Middle-eastern climate of Cyprus. This chapter presents useful information of the seasonal performance and the effects of temperature, soiling and power rating of contemporary PV systems that would also guide the field evaluation of the forthcoming new technologies.

I am grateful to the authors of the chapters for sharing with us the technical knowledge for which they are renowned experts. I have worked together with several of the contributors and fully appreciate that in their busy schedules they found the time and energy to be part of this project. I also want to extend special thanks to Anja Filipovic, InTech Publishing, for keeping all of us on track.

Prof. Dr. Vasilis M. Fthenakis
Brookhaven National Laboratory and Columbia University
USA

1

Solar Spectrum Conversion for Photovoltaics Using Nanoparticles

W.G.J.H.M. van Sark[1], A. Meijerink[2] and R.E.I. Schropp[3]
[1]*Utrecht University, Copernicus Institute, Science, Technology and Society, Utrecht*
[2]*Utrecht University, Debye Institute for NanoMaterials Science, Condensed Matter and Interfaces, Utrecht*
[3]*Utrecht University, Debye Institute for NanoMaterials Science, Nanophotonics – Physics of Devices, Utrecht*
The Netherlands

1. Introduction

The possibility to tune chemical and physical properties in nanosized materials has a strong impact on a variety of technologies, including photovoltaics. One of the prominent research areas of nanomaterials for photovoltaics involves spectral conversion. Conventional single-junction semiconductor solar cells only effectively convert photons of energy close to the semiconductor band gap (E_g) as a result of the mismatch between the incident solar spectrum and the spectral absorption properties of the material (Green 1982, Luque and Hegedus 2003). Photons with an energy E_{ph} smaller than the band gap are not absorbed and their energy is not used for carrier generation. Photons with energy E_{ph} larger than the band gap are absorbed, but the excess energy E_{ph} – E_g is lost due to thermalization of the generated electrons. These fundamental spectral losses in a single-junction silicon solar cell can be as large as 50% (Wolf 1971), while the detailed balance limit of conversion efficiency for such a cell was determined to be 31% (Shockley and Queisser 1961). Several routes have been proposed to address spectral losses, and all of these methods or concepts obviously concentrate on a better exploitation of the solar spectrum, e.g., multiple stacked cells (Law et al. 2010), intermediate band gaps (Luque and Martí 1997), multiple exciton generation (Klimov 2006, Klimov et al. 2007), quantum dot concentrators (Chatten et al. 2003a) and down- and up-converters (Trupke et al. 2002a, b), and down-shifters (Richards 2006a, Van Sark 2005). In general they are referred to as Third or Next Generation photovoltaics (PV) (Green 2003, Luque et al. 2005, Martí and Luque 2004). Nanotechnology is essential in realizing most of these concepts (Soga 2006, Tsakalakos 2008), and semiconductor nanocrystals have been recognized as 'building blocks' of nanotechnology for use in next generation solar cells (Kamat 2008). Being the most mature approach, it is not surprising that the current world record conversion efficiency is 43.5% for a GaInP/GaAs/GaInNAs solar cell (Green et al. 2011), although this is reached at a concentration of 418 times.

As single-junction solar cells optimally perform under monochromatic light at wavelength $\lambda_{opt} \sim 1240/E_g$ (with λ_{opt} in nm and E_g in eV), an approach "squeezing" the wide solar spectrum (300-2500 nm) to a single small band spectrum without too many losses would greatly enhance solar cell conversion efficiency. Such a quasi-monochromatic solar cell could in principle reach efficiencies over 80%, which is slightly dependent on band gap (Luque and Martí 2003). For (multi)crystalline silicon ((m)c-Si) solar cells λ_{opt} = 1100 nm (with E_g = 1.12 eV); for amorphous silicon (a-Si:H) the optimum wavelength is λ_{opt} = 700 nm (with E_g = 1.77 eV). However, as these cells only contain a thin absorber layer, the optimum spectrum response occurs at about 550 nm (Schropp and Zeman 1998, Van Sark 2002).

Modification of the spectrum by means of so-called down- and/or upconversion or -shifting is presently being pursued for single junction cells (Richards 2006a), as illustrated in Fig. 1, as a relatively easy and cost-effective means to enhance conversion efficiency. In addition, so-called luminescent solar converters (LSC) employ spectrum modification as well (Goetzberger 2008, Goetzberger and Greubel 1977). Downconverters or -shifters are located on top of solar cells, as they are designed to modify the spectrum such that UV and visible photons are converted leading to a more red-rich spectrum that is converted at higher efficiency by the solar cell. Upconverters modify the spectrum of photons that are not absorbed by the solar cell to effectively shift the IR part of the transmitted spectrum to the NIR or visible part; a back reflector usually is applied as well.

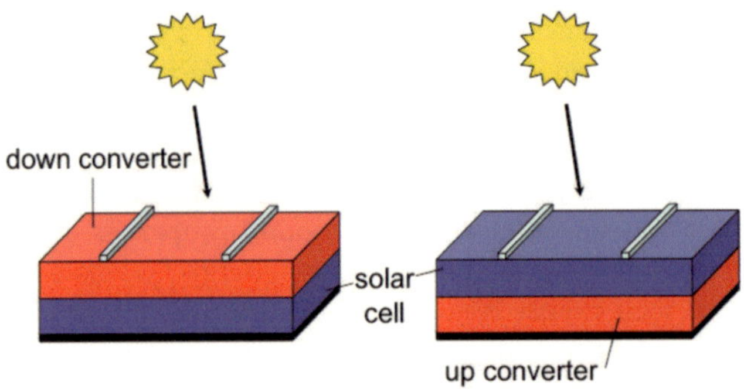

Fig. 1. Schematic drawings of a solar cell with down converter (or down shifter) layer on top (left) and a solar cell on top of an upconverter layer.

In case of downconversion (DC) an incident high-energy photon is converted into two or more lower energy photons which can lead to quantum efficiency of more than 100%, therefore it is also termed 'quantum cutting' (Timmerman et al. 2008, Wegh et al. 1999); for upconversion (UC) two or more low energy photons (sub band gap) are converted into one high-energy photon (Strümpel et al. 2007), see also Fig. 2. Downshifting (DS) is similar to downconversion where an important difference is that only one photon is emitted and that the quantum efficiency of the conversion process is lower than unity (Richards 2006a), although close to unity is preferred to minimize losses. Downshifting is also termed photoluminescence (Strümpel et al. 2007). DC, UC and DS layers only influence solar cell performance optically. As DC and DS both involve one incident photon per conversion, the intensity of converted or shifted emitted photons linearly scales with incident light intensity. UC involves two photons; therefore the intensity of converted light scales quadratically with incident light intensity.

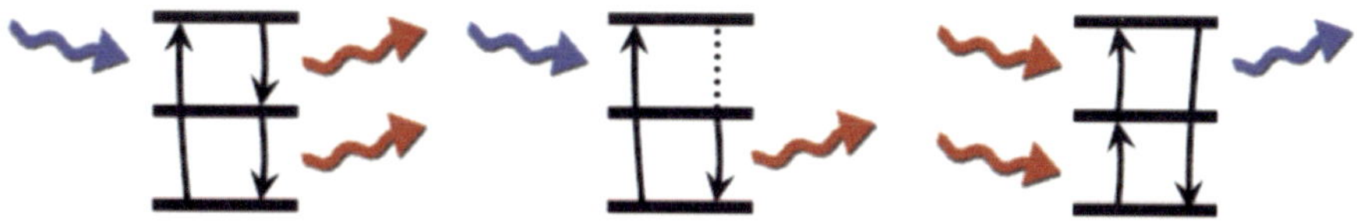

Fig. 2. Energy diagrams showing photon adsorption and subsequent downconversion, downshifting, and upconversion.

In this chapter a review is presented on the use of nanometer sized particles (including quantum dots) in solar spectrum conversion. Modification of the spectrum requires down- and/or upconversion or -shifting of the spectrum, meaning that the energy of photons is modified either to lower (down) or higher (up) energy. Nanostructures such as quantum dots, luminescent dye molecules, and lanthanide-doped glasses are capable of absorbing photons at a certain wavelength and emitting photons at a different (shorter or longer) wavelength. We will discuss down- and upconversion and -shifting by quantum dots, luminescent dyes, and lanthanide compounds, and assess their potential in contributing to ultimately lowering the cost per kWh of solar generated power.

2. Downconversion

2.1 Principles

Downconversion was theoretically suggested first by Dexter in the 1950s (Dexter 1953, 1957), and shown experimentally 20 years later using the lanthanide ion praseodymium Pr^{3+} in an yttrium fluoride YF_3 host (Piper et al. 1974, Sommerdijk et al. 1974). A VUV photon (185 nm) is absorbed in the host, and its energy is transferred into the $^1S_0$ state of the Pr^{3+} ion, from where two photons (408 and 620 nm) are emitted in a two-step process ($^1S_0 \rightarrow {}^3P_J$ at 408 nm followed by $^3P_J \rightarrow {}^3F_2$ at 620 nm). In this way a single absorbed high energy photon results in the emission of two visible photons and a higher-than-unity quantum efficiency is realized. Another frequently used ion is gadolinium Gd^{3+} (Wegh et al. 1997), either single or co-doped (Wegh et al. 1999). These lanthanide ions are characterized by a rich and well-separated energy level structure in the so-called Dieke energy diagrams (Dieke 1968), and have been identified as perfect "photon managers" (Meijerink et al. 2006). The energy levels arise from the interactions between electrons in the inner 4f shell. Trivalent lanthanides have as electronic configuration $[Xe]4f^n5s^25p^6$. Inside the filled 5s and 5p shells, there is a partially filled 4f shell where the number of electrons (n) can vary between 0 and 14. The number of possible arrangements for n electrons in 14 available f-orbitals is large (14 over n), which gives rise to a large number of different energy levels that are labelled by so-called term symbols. Transitions between the energy levels give rise to sharp absorption and emission lines. Energy transfer between neighbouring lanthanide ions is also possible and helps converting photons that are absorbed to photons of different energy. Based on their unique and rich energy level structure, lanthanide ions are promising candidates to realize efficient down conversion and recent research in this direction will be discussed below.

Downconversion in solar cells was theoretically shown to lead to a enlarged conversion efficiency of 36.5% (Trupke et al. 2002a) for nonconcentrated sunlight when applied in a single junction solar cell configuration, such as shown in Fig. 1. Note that detailed-balance calculations for single junction c-Si cells lead to a maximum efficiency of 31% (Shockley and Queisser 1961). These calculations were performed as a function of the band gap and refractive index of the solar cell material, for a 6000K blackbody spectrum. The efficiency limit is reached for a band gap of 1.05 eV, and asymptotically approaches 39.63% for very high refractive indices larger than 10. For c-Si, with refractive index of 3.6, the limit efficiency is 36.5%. Analysis of the energy content of the incident standard Air Mass 1.5 Global (AM1.5G) spectrum (ASTM 2003) and the potential gain DC can have shows that with a DC layer an extra amount of 32% is incident on a silicon solar cell (Richards 2006a), which can be converted at high internal quantum efficiency. Figure 3 illustrates the potential gains for DC and UC.

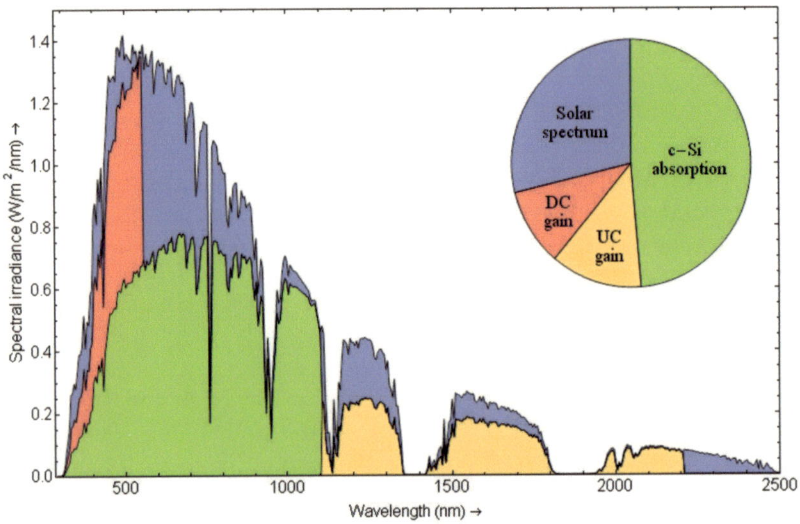

Fig. 3. Potential gain for down- and upconversion for a silicon solar cell. The green part reflects the energy conversion of the absorbed part of the solar spectrum for a c-Si solar cell, the red part reflects the extra energy conversion if every photon with an energy higher than $2E_g$ results in two NIR photons and the beige part reflects the energy gain if every pair of photons with an energy between $0.5E_g$ and E_g is converted to one NIR photon. Note that the figure considers no other losses than spectral mismatch losses (Courtesy of F. Rabouw, Utrecht University).

2.2 State of the art

The most promising systems for downconversion rely on lanthanide ions. The unique and rich energy level structures of these ions allow for efficient spectral conversion, including up- and downconversion processes mediated by resonant energy transfer between neighboring lanthanide ions (Auzel 2004, Wegh et al. 1999). Considering the energy levels of all lanthanides, as shown in the Dieke energy level diagram (Dieke 1968, Peijzel et al. 2005, Wegh et al. 2000) it is immediately evident that the energy level structure of Yb^{3+} is ideally suited to be used in down conversion for use in c-Si solar cells. The Yb^{3+} ion has a single excited state (denoted by the term symbol $^2F_{5/2}$) some 10,000 cm^{-1} above the $^2F_{7/2}$ ground state, corresponding to emission around 1000 nm. The absence of other energy levels allow Yb^{3+} to exclusively 'pick up' energy packages of 10,000 cm^{-1} from other lanthanide ions and emitting ~1000 nm photons that can be absorbed by c-Si. Efficient downconversion using Yb^{3+} as acceptor requires donor ions with an energy level around 20,000 cm^{-1} and an intermediate level around 10,000 cm^{-1}. From inspection of the Dieke diagram one finds that potential couples are (Er^{3+}, Yb^{3+}), (Nd^{3+}, Yb^{3+}) and (Pr^{3+}, Yb^{3+}) for a resonant two-step energy transfer process. Also, cooperative sensitization is possible where energy transfer occurs from a high excited state of the donor to two neighboring acceptor ions without an intermediate level.

The first report on efficient downconversion for solar cells was based on cooperative energy transfer from Tb^{3+} to two Yb^{3+} ions in $Yb_xY_{1-x}PO_4$:Tb^{3+} (Vergeer et al. 2005), as shown schematically in Fig. 4. The $^5D_4$ state of Tb^{3+} is around 480 nm (21,000 cm^{-1}) and from this state cooperative energy transfer to two Yb^{3+} neighbors occurs, both capable of emitting a 980 nm photon.

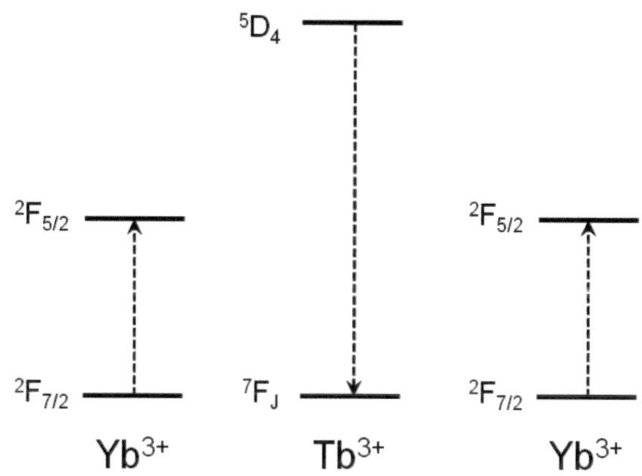

Fig. 4. Cooperative energy transfer from Tb to two Yb ions. From the $^5D_4$ state of one Tb ion two neighbouring Yb ions are excited to the $^7F_{5/2}$ state from where emission of 980 nm photons can occur (after Meijerink et al. 2006, Vergeer et al. 2005).

The same efficient down conversion process was observed in other host lattices: Zhang et al. (Zhang et al. 2007) observed cooperative quantum cutting in $(Yb,Gd)Al_3(BO_3)_4$:Tb^{3+} and quantum efficiencies up to 196% were reported. Cooperative downconversion for other couples of lanthanides was also claimed. For the couple (Pr, Yb) and (Tm, Yb) co-doped into borogermanate glasses a decrease of the emission from higher energy levels of Pr and Tm was observed (Liu et al. 2008). In case of Pr^{3+} the starting level for the cooperative down conversion is the $^3P_0$ level while for Tm^{3+} the $^1G_4$ is at twice the energy of the $^2F_{5/2}$ level of Yb^{3+}. The efficiency of the downconversion process was estimated from the decrease of the lifetime of the donor state. For the Tm^{3+} emission a decrease from 73 μs (for the sample doped with Tm only) to 45 μs (for a sample co-doped with 20% of Yb) was observed, giving a transfer efficiency of 38%. For the same glass co-doped with Pr and Yb a 65% decrease of the lifetime was observed, implying a 165% quantum yield. Similar results were found for the (Pr, Yb) couple in aluminosilicate glasses (Lakshminarayana et al. 2008). More recent work on the (Pr, Yb) couple confirmed the presence of downconversion, but the mechanism involved was pointed out to be a resonant two-step energy transfer rather than a cooperative energy transfer (Van der Ende et al. 2009a, b). For the (Pr, Yb) couple an intermediate level ($^1G_4$) is available around 10,000 cm^{-1}, which makes a two-step resonant energy transfer process possible. This

first-order process will have a much higher probability than the second-order (cooperative) transfer process. The results demonstrated efficient quantum cutting of one visible photon into two NIR photons in $SrF_2:Pr^{3+}$, Yb^{3+}. Comparison of absorption and excitation spectra provided direct evidence that the downconversion efficiency is close to 200%, in agreement with a two-step energy transfer process that can be expected based on the energy level diagrams of Pr^{3+} and Yb^{3+}. This first order energy transfer process is effective at relatively low Yb^{3+} concentrations (5%), where concentration quenching of the Yb^{3+} emission is limited. Comparison of emission spectra, corrected for the instrumental response, for $SrF_2:Pr^{3+}$ (0.1%) and $SrF_2:Pr^{3+}$ (0.1%), Yb^{3+} (5%) revealed an actual conversion efficiency of 140%.
An important aspect for downconversion is the incorporation in a solar cell. The solar spectrum has to be converted before entering the solar cell. A promising design is incorporation of downconverting material in a thin (10-100 nm) transparent layer on top of the solar cell. After downconversion the isotropically emitted 1000 nm need to be directed towards the solar cell. This can be achieved by a narrow-band reflective coating for 1000 nm NIR on top of the downconversion layer, thereby avoiding absorption in the other parts of the spectrum. The narrow band emission of the Yb^{3+} emission allows for a narrow band reflective coating. To realize a transparent downconversion layer, the downconverting lanthanide ion couples need to be incorporated in nanocrystals with dimensions smaller than ~50 nm to prevent light scattering. Chen et al. reported downconversion in transparent glass ceramics with embedded Pr^{3+}/Yb^{3+}:β-YF_3 nanocrystals (Chen et al. 2008a). Quantum cutting downconversion was shown in borate glasses using co-doping of Ce^{3+} and Yb^{3+} (Chen et al. 2008b). A UV photon (330 nm) excites a Ce^{3+} ion, and cooperative energy transfer between the Ce^{3+} and Yb^{3+} ion leads to the emission of two NIR photons (976 nm) with a 174% quantum efficiency, based on a 74% energy transfer efficiency calculated from the luminescence decay curves and assuming cooperative energy transfer. Recent work on the (Ce, Yb) couple suggests that the energy transfer mechanism is not cooperative energy transfer, but single step energy transfer via a metal-to-metal charge transfer state. If this is the case, the quantum efficiency is not above 100%. Efficient downconversion for other lanthanide ion couples has been reported, including the (Er, Yb) couple in low phonon hosts (Eilers et al. 2010). Research on new downconversion couples continues. It is not straightforward to establish the efficiency of the conversion to NIR photons and careful studies providing convincing evidence for the energy transfer mechanism and efficiency are required (Van Wijngaarden 2010).
Timmerman et al. (Timmerman et al. 2008) demonstrated so-called space-separated quantum cutting within SiO_2 matrices containing both silicon nanocrystals and Er^{3+}. Energy transfer from photo exited silicon nanocrystals to Er^{3+} had been observed earlier (Fujii et al. 1997), but Timmerman et al. show that upon absorption of a photon by a silicon nanocrystal a fraction of the photon energy is transferred generating an excited state within either an erbium ion or another silicon nanocrystal. As also the original silicon NC relaxes from a highly excited state towards the lowest-energy excited state, the net result is two electron-hole pairs for each photon absorbed. Recent quantum yield measurements as a function of excitation wavelength confirmed the occurrence of carrier multiplication in Si nanocrystals (Timmerman et al. 2011).

3. Downshifting

3.1 Principles

Downshifting (DS) or photoluminescence is a property of many materials, and is similar to downconversion, however, only one photon is emitted and energy is lost due to non-radiative relaxation, see Fig. 2. Therefore the quantum efficiency is lower than unity. DS can be employed to overcome poor blue response of solar cells (Hovel et al. 1979), due to, e.g., non-effective front surface passivation for silicon solar cells. Shifting the incident spectrum to wavelengths where the internal quantum efficiency of the solar cell is higher than in the blue can effectively enhance the overall conversion efficiency by ~10% (Van Sark et al. 2005). Improvement of front passivation may make down shifters obsolete, or at least less beneficial. Downshifting layers can also be used to circumvent absorption of high-energy photons in heterojunction window layers, e.g., CdS on CdTe cells (Hong and Kawano 2003). A recent review is presented by Klampaftis et al. (2010).

Downshifting was suggested in the 1970s to be used in so-called luminescent solar concentrators (LSCs) that were attached on to a solar cell (Garwin 1960, Goetzberger and Greubel 1977). In these concentrators, organic dye molecules absorb incident light and re-emit this at a red-shifted wavelength. Internal reflection ensures collection of all the re-emitted light in the underlying solar cells. As the spectral sensitivity of silicon is higher in the red than in the blue, an increase in solar cell efficiency was expected. Also, it was suggested to use a number of different organic dye molecules of which the re-emitted light was matched for optimal conversion by different solar cells. This is similar to using a stack of multiple solar cells, each sensitive to a different part of the solar spectrum. The expected high efficiency of ~30% (Smestad et al. 1990, Yablonovitch 1980) in practice was not reached as a result of not being able to meet the stringent requirements to the organic dye molecules, such as high quantum efficiency and stability, and the transparency of collector materials in which the dye molecules were dispersed (Garwin 1960, Goetzberger and Greubel 1977).

Nowadays, new organic dyes can have extremely high luminescence quantum efficiency (LQE) (near unity) and are available in a wide range of colors at better re-absorption properties that may provide necessary UV stability. Quantum dots (QDs) have been proposed for use in luminescent concentrators to replace organic dye molecules: the quantum dot concentrator (Barnham et al. 2000, Chatten et al. 2003a,b, Gallagher et al. 2007). Quantum dots are nanometer-sized semiconductor crystals of which the emission wavelength can be tuned by their size, as a result of quantum confinement (Alivisatos 1996, Gaponenko 1998). Recently, both QDs and new organic dyes have been evaluated for use in LSCs (Van Sark et al. 2008a). QDs have advantages over dyes in that: (i) their absorption spectra are far broader, extending into the UV, (ii), their absorption properties may be tuned simply by the choice of nanocrystal size, and (iii) they are inherently more stable than organic dyes (Bruchez Jr. et al. 1998). Moreover, (iv) there is a further advantage in that the red-shift between absorption and luminescence is quantitatively related to the spread of QD sizes, which may be determined during the growth process, providing an additional strategy for minimizing losses due to re-absorption (Barnham et al. 2000). However, as yet QDs can only provide reasonable LQE: a LQE > 0.8 has been reported for core-shell QDs (Peng et al. 1997). Performance in LSCs has been modeled using thermodynamic as well as ray-trace models (Burgers et al. 2005, Chatten et al. 2003b, 2004a, b, Chatten et al. 2005, Gallagher et al. 2004, Kennedy et al. 2008), and results are similar and also compare well with experimental values (Kennedy et al. 2008, Van Sark et al. 2008a). Calculated efficiencies vary between 2.4% for an LSC with mc-Si cell at certain mirror specifications to 9.1% for an LSC with InGaP cell for improved specifications.

Alternatives for dye molecules used in LSCs are luminescent ions. Traditionally, efficient luminescent materials rely on the efficient luminescence of transition metal ions and lanthanide ions. In case of transition metal ions intraconfigurational $3d^n$ transitions are responsible for the luminescence, while in case of lanthanide ions both intraconfigurational $4f^n$-$4f^n$ transitions and interconfigurational $4f^n$-$4f^{n-1}5d$ transitions are capable of efficient emission. In most applications efficient emission in the visible is required and emission from lanthanide ions and transition metal ions is responsible for almost all the light from artificial light sources (e.g. fluorescent tubes, displays (flat and cathode ray tube, CRT) and white light emitting diodes (LEDs) (Blasse and Grabmaier 1995). For LSCs to be used in combination with c-Si solar cells efficient emission in the NIR is needed. The optimum wavelength is between 700 and 1000 nm, which is close to the band gap of c-Si and in the spectral region where c-Si solar cells have their optimum conversion efficiency. Two types of schemes can be utilized to achieve efficient conversion of visible light into narrow band NIR emission. A single ion can be used if the ion shows a strong broad band absorption in the visible spectral range followed by relaxation to the lowest excited state from which efficient narrow band or line emission in the NIR occurs. Alternatively, a combination of two ions can be used where one ion (the sensitizer) absorbs the light and subsequently transfers the energy to a second ion (the activator), which emits efficiently in the NIR. Both concepts have been investigated for LSCs by incorporating luminescent lanthanides and transition metal ions in glass matrices. The stability of these systems is not a problem, in contrast to LSCs based on dye molecules, however, the quantum efficiency of luminescent ions in glasses appeared to be much lower than in crystalline compounds, especially in the infrared, thus hampering use for LSCs.

3.2 State of the art

Chung et al. (Chung et al. 2007) reported downshifting phosphor coatings consisting of $Y_2O_3:Eu^{3+}$ or $Y_2O_2S:Eu^{3+}$ dispersed in either polyvinyl alcohol or polymethylmethacrylate on top of mc-Si solar cells: an increase in conversion efficiency was found of a factor of 14 under UV illumination by converting the UV radiation (for which the response of c-Si is low) to 600 nm emission from the Eu^{3+} ion. The used solar cells used were encapsulated in an epoxy that absorbs photons with energy higher that ~3 eV. This protective coating can remain in place and down converters or shifters can easily be added. A 2.9% relative increase in efficiency has been reported for Eu^{3+} containing ethyline-vinyl-acetate layers (EVA) on top of c-Si cells; this is very relevant as EVA nowadays is the standard encapsulating matrix used for c-Si technology.

Svrcek et al. (Svrcek et al. 2004) demonstrated that silicon nanocrystals incorporated into spin-on-glass (SOG) on top of c-Si solar cells are successful as downshifter leading to a potential absolute efficiency enhancement of 1.2%, while they experimentally showed an enhancement of 0.4% (using nanocrystals of 7 nm diameter with a broad emission centered around 700 nm). McIntosh et al. (McIntosh et al. 2009) recently presented results on encapsulated c-Si solar cells, of which the PMMA encapsulant contained down-shifting molecules, i.e., Lumogen dyes. These results indicate a ~1% relative increase in the module efficiency, based on a 40% increase in external quantum efficiency for wavelengths < 400 nm. Mutlugun et al. (Mutlugun et al. 2008) claimed a two-fold increase in efficiency applying a so-called nanocrystal scintillator on top of a c-Si solar cell; it comprises a PMMA layer in which CdSe/ZnS core-shell quantum dots (emission wavelength 548 nm) are embedded. However, the quality of their bare c-Si cells is very poor. Yuan et al. (2011) presented a 14% improved internal quantum efficiency of a silicon oxide layer containing silicon nanocrystals on top of c-Si cells.

Stupca et al. (2007) demonstrated the integration of ultra thin films (2-10 nm) of monodisperse luminescent Si nanoparticles on polycrystalline Si solar cells. 1-nm sized blue emitting and 2.85-nm red emitting particles enhanced the conversion efficiency by 60% in the UV, and 3 to 10% in the visible for the red and blue emitting particles, respectively, similarly to what has been predicted (Van Sark et al. 2005). An 8% relative increase in conversion efficiency was reported (Maruyama and Kitamura 2001) for a CdS/CdTe solar cell, where the coating in which the fluorescent coloring agent was introduced increased the sensitivity in the blue; a maximum increase in efficiency was calculated to be 30-40%. Others showed results that indicate a 6% relative increase in conversion efficiency (Maruyama and Bandai 1999) upon coating a mc-Si solar cell. The employed luminescent species has an absorption band around 400 nm and a broad emission between 450 and 550 nm. As QDs have a much broader absorption it is expected that in potential the deployment of QDs in planar converters could lead to relative efficiency increases of 20-30%. Downshifting employing QDs in a polymer composite has been demonstrated in a light-emitting diode (LED), where a GaN LED was used as an excitation source for QDs emitting at 590 nm (Lee et al. 2000). Besides QDs, other materials have been suggested such as rare earth ions and dendrimers (Serin et al. 2002). A maximum increase of 22.8% was calculated for a thin film coating of $KMgF_3$ doped with Sm on top of a CdS/CdTe solar cell, while experimental results show an increase of 5% (Hong and Kawano 2003).

Recent efforts to surpass the historical 4% efficiency limit of LSCs (Goetzberger 2008, Wittwer et al. 1984, Zastrow 1994), albeit for smaller area size, have been successful. For example, Goldschmidt et al. (2009) showed for a stack of two plates with different dyes, to which four GaInP solar cells were placed at the sides, that the conversion efficiency is 6.7%; the plate was small (4 cm^2), and the concentration ratio was only 0.8. It was argued that the conversion efficiency was limited by the spectral range of the organic dyes used, and that if the same quantum efficiency as was reached for the 450-600 nm range could be realized for the range 650-1050 nm an efficiency of 13.5% could be within reach. They also discuss the benefits of a photonic structure on top of the plate, to reduce the escape cone loss (Goldschmidt et al. 2009). The proposed structure is a so-called rugate filter; this is characterized by a varying refractive index in contrast to standard Bragg reflectors, which suppresses the side loops that could lead to unwanted reflections. The use of these filters would increase the efficiency by ~20%, as was determined for an LSC consisting of one plate and dye. Slooff et al. (2008) presented results on 50x50x5 mm^3 PMMA plates in which both CRS040 and Red305 dyes were dispersed at 0.003 and 0.01 wt%, respectively. The plates were attached to either mc-Si, GaAs or InGaP cells, and a diffuse reflector (97% refection) was used at the rear side of the plate. The highest efficiency measured was 7.1% for 4 GaAs cells connected in parallel (7% if connected in series).

As stated above, quantum dots are potential candidates to replace organic dye molecules in an LSC, for their higher brightness, better stability, and wider absorption spectrum (Barnham et al. 2000, Chatten et al. 2003a, b, Gallagher et al. 2007). In fact, the properties and availability of QDs started renewed interest in LSCs around 2000 (Barnham et al. 2000), with the main focus on modeling, while more recently some experimental results have been presented. Schüler et al. (2007) proposed to make LSCs by coating transparent glass substrates with QD-containing composite films, using a potentially cheap sol-gel method. They reported on the successful fabrication of thin silicon oxide films that contain CdS QDs using a sol-gel dip-coating process, whereby the 1-2 nm sized CdS QDs are formed during thermal treatment after dip-coating. Depending on the anneal temperature, the colors of the LSC ranged from green for 250 °C to yellow for 350 °C and orange for 450 °C. Reda (2008) also prepared CdS QD concentrators, using sol-gel spin coating, followed by annealing. The annealing temperature was found to affect absorption and emission spectrum: luminescent intensity and Stokes' shift both decreased for 4 weeks outdoor exposure to sunlight, which probably was caused by aggregation and oxidation. It is known that oxidation leads to blue-shifts in emission (Van Sark et al. 2002). Blue-shifts have also been observed by Gallagher et al. (2007) who dispersed CdSe QDs in several types of resins (urethane, PMMA, epoxy), for fabrication of LSC plates.

Bomm et al. (2011) have addressed several problems regarding incorporation of QDs in an organic polymer matrix, viz. phase separation, agglomeration of particles leading to turbid plates, and luminescence quenching due to exciton energy transfer (Koole et al. 2006). They have synthesized QDSCs using CdSe core/multishell QDs (Koole et al. 2008) (QE = 60%) that were dispersed in laurylmethacrylate (LMA), see also Lee et al. (2000) and Walker et al. (2003). UV-polymerization was employed to yield transparent PLMA plates with QDs without any sign of agglomeration. To one side of this plate, a mc-Si solar cell was placed, and aluminum mirrors to all other sides. Compared to the bare cell (5x0.5 cm^2) that generated a current density of 40.28 mA/cm^2 at 1000 W AM1.5G spectrum, the best QDC made generated a current of 77.14 mA/cm^2, nearly twice as much. The QDC efficiency is 3.5%. In addition, exposure to a 1000 W sulphur lamp for 280 hours continuously showed very good stability: the current density decreased by 4% only, on average. However, re-absorption may still be a problem, as is demonstrated by a small red shift in the emission spectrum for long photon pathways; also absorption by the matrix is occurring. Besides QDs also nanorods (NRs) have been dispersed in PLMA, showing excellent transmittance of 93%; for long rods (aspect ratio of 6) a QE of 70% is observed, which is only slightly smaller than the QE in solution, implying that these rods are stable throughout the polymerization process. In addition, these NRs have also been dispersed in cellulosetriacetate (CTA), and a ~10 μm thin film on a glass substrate was made showing bright orange luminescence (Bomm et al. 2010).

Hyldahl et al. (2009) used commercially available CdSe/ZnS core/shell QDs with QE=57% in LSCs, both liquid (QDs dissolved in toluene, between two 6.2x6.2x0.3 cm glass plates) and solid (QDs dispersed in epoxy), and they obtained an efficiency of 3.98% and 1.97%, respectively. They also used the organic dye Lumogen F Red300, and obtained efficiency of 2.6% in toluene.

4. Upconversion

4.1 Principles

Upconversion was, like DC also suggested in the 1950s, by Bloembergen (1959), and was related to the development of IR detectors: IR photons would be detected through sequential absorption, as would be possible by the arrangement of energy levels of a solid. However, as Auzel (2004) pointed out the essential role of energy transfer was only recognized nearly 20 years later. Several types of upconversion mechanisms exist (Auzel 2004), of which the APTE (addition de photon par transferts d'energie) or, in English, ETU (energy transfer upconversion) mechanism is the most efficient; it involves energy transfer from an excited ion, named sensitizer, to a neighbouring ion, named activator (see Fig. 5).

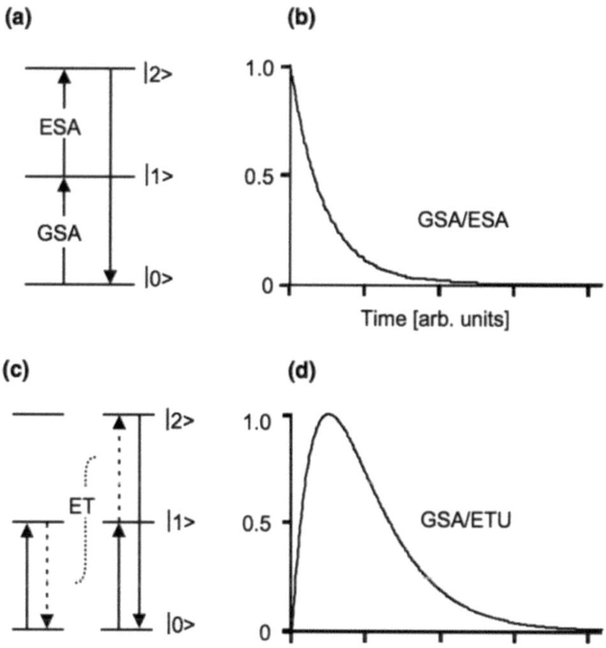

Fig. 5. Schematic representation of two upconversion processes and the characteristic time response of the up converted emission after a short excitation pulse. (a), (b) Ground state absorption (GSA) followed by excited state absorption (ESA). This is a single ion process and takes place during the excitation pulse. (c), (d) Upconversion is achieved by GSA followed energy transfer between ions and the delayed response is characteristic of the energy transfer up conversion (ETU) (from Suyver et al. 2005a).

Others are two-step absorption, being a ground state absorption (GSA) followed by an excited state absorption (ESA), and second harmonics generation (SHG). The latter mechanism requires extremely high intensities, of about 10^{10} times the sun's intensity on a sunny day, to take place (Strümpel et al. 2007). This may explain that research in this field with focus on enhancing solar cell efficiency was started only recently (Shalav et al. 2007). Upconverters usually combine an active ion, of which the energy level scheme is employed for absorption, and a host material, in which the active ion is embedded. The most efficient upconversion has been reported for the lanthanide ion couples (Yb, Er) and (Yb, Tm); the corresponding upconversion schemes are shown in Fig. 6. The first demonstration of such an UC layer on the back of solar cells comprised an ultra thin (3 μm) GaAs cell (band gap 1.43 eV) that was placed on a 100 μm thick vitroceramic containing Yb^{3+} and Er^{3+} (Gibart et al. 1996): it showed 2.5% efficiency upon excitation of 256 kW/m2 monochromatic sub-band-gap (1.391 eV) laser light (1 W on 0.039 cm^2 cell area), as well as a clear quadratic dependence on incident light intensity.

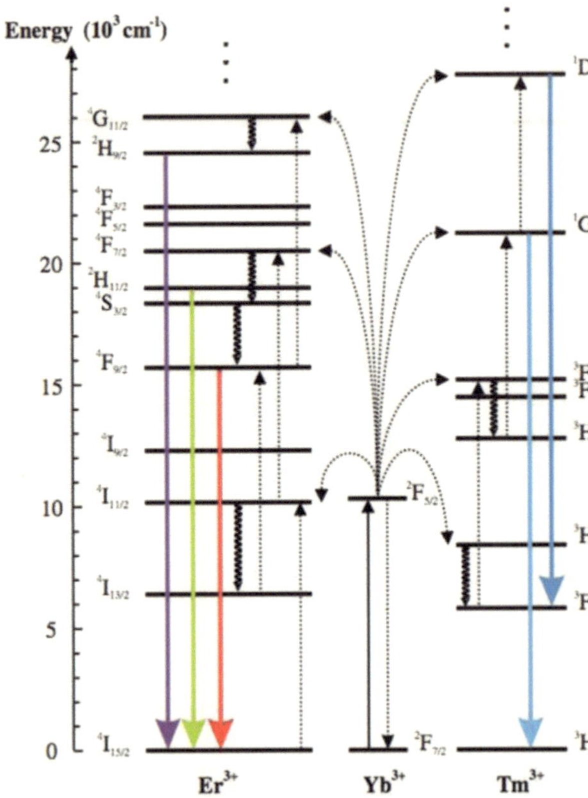

Fig. 6. Up conversion by energy transfer between Yb^{3+} and Er^{3+} (left) and Yb^{3+} and Tm^{3+} (right). Excitation is around 980 nm in the Yb^{3+} ion and by a two-step energy transfer or three-step energy transfer process higher excited states of Er^{3+} or Tm^{3+} are populated giving rise to visible (red, green or blue) emission. Full, dotted, and curly arrows indicate radiative, nonradiative energy transfer, and multiphonon relaxation processes, respectively (from Suyver et al. 2005a).

Upconversion in solar cells was calculated to potentially lead to a maximum conversion efficiency of 47.6 % (Trupke et al. 2002b) for nonconcentrated sunlight using a 6000K blackbody spectrum in detailed-balance calculations. This optimum is reached for a solar cell material of ~2 eV band gap. Applied on the back of silicon solar cells,the efficiency limit would be ~37% (Trupke et al. 2002b). The analysis of the energy content of the incident AM1.5G spectrum presented in Fig. 3 revealed that cells with an UC layer would benefit from an extra amount of 35% light incident in the silicon solar cell (Richards 2006a).

4.2 State of the art

Lanthanides have also been employed in upconverters attached to the back of bifacial silicon solar cells. Trivalent erbium is ideally suited for up conversion of NIR light due to its ladder of nearly equally spaced energy levels that are multiples of the $^4I_{15/2}$ to $^4I_{13/2}$ transition (1540 nm) (see also Fig. 3). Shalav et al. (2005) have demonstrated a 2.5% increase of external quantum efficiency due to up conversion using $NaYF_4$:20% Er^{3+}. By depicting luminescent emission intensity as a function of incident monochromatic (1523 nm) excitation power in a double-log plot, they showed that at low light intensities a two-step up conversion process ($^4I_{15/2} \rightarrow {}^4I_{13/2} \rightarrow {}^4I_{11/2}$) dominates, while at higher intensities a three-step up conversion process ($^4I_{15/2} \rightarrow {}^4I_{13/2} \rightarrow {}^4I_{11/2} \rightarrow {}^4S_{3/2}$ level) is involved.

Strümpel et al. have identified the materials of possible use in up- (and down-) conversion for solar cells (Strümpel et al. 2007). In addition to the $NaYF_4$:(Er,Yb) phosphor, they suggest the use of $BaCl_2$:(Er^{3+},Dy^{3+}) (Strümpel et al. 2005), as chlorides were thought to be a better compromise between having a low phonon energy and a high excitation spectrum, compared to the $NaYF_4$ (Gamelin and Güdel 2001, Ohwaki and Wang 1994, Shalav et al. 2007). These lower phonon energies lead to lower non-radiative losses. In addition, the emission spectrum of Dysprosium is similar to that of Erbium, but the content of Dy^{3+} should be <0.1%, to avoid quenching (Auzel 2004, Strümpel et al. 2007).

$NaYF_4$ co-doped with (Er^{3+}, Yb^{3+}) is to date the most efficient upconverter (Suyver et al. 2005a, Suyver et al. 2005b), with ~50% of all absorbed NIR photons up converted and emitted in the visible wavelength range. However, the (Yb, Er) couple is not considered beneficial for up conversion in c-Si cells, as silicon also absorbs in the 920-980 nm wavelength range. These phosphors can be useful for solar cells based on higher band gap materials such as the Grätzel-cell (O'Regan and Grätzel 1991), a-Si(Ge):H or CdTe. In that case, the $^2F_{5/2}$ level of Yb^{3+} would serve as an intermediate step for up conversion (Shalav et al. 2007) and IR-radiation between ~700 and 1000 nm that is not absorbed in these wider band gap solar cells can be converted into green (550 nm) light that can be absorbed. Nanocrystals of $NaYF_4$:Er^{3+}, Yb^{3+} also show upconversion. An advantage of using nanocrystals is that transparent solutions or transparent matrices with upconverting nanocrystals can be obtained. An excellent recent review on upconverting nanoparticles summarizes the status of a variety of UC materials that are presently available as nanocrystals, mostly phosphate and fluoride NCs (Haase et al., 2011). A problem with upconversion NCs is the lower upconversion efficiency (Boyer et al., 2010). There is a clear decrease in efficiency with decreasing size in the relevant size regime between 8 and 100 nm. The decrease in efficiency is probably related to surface effects and quenching by coupling with by high energy vibrations in molecules attached to the surface.

When applying an upconverter, it is advantageous to use it with a cell that with a rather high band gap, such as amorphous Si (1.7 eV). A typical external collection efficiency (ECE) graph of a standard single junction p-i-n a Si:H solar cell is shown in Fig. 7. These cells are manufactured on textured SnO_2:F coated glass substrates and routinely have >10% initial efficiency. Typically, the active Si layer in the cell has a thickness of 300 nm and the generated current is 14.0-14.5 mA/cm^2, depending on the light trapping properties of the textured metal oxide and the back reflector. After light-induced introduction of the stabilized defect density (Staebler-Wronski effect (Schropp and Zeman 1998)), the stabilized efficiency is 8.2-8.5%. From Fig. 7 it can be seen that the maximum ECE is 0.8 at ~550 nm, and the cut-off occurs at 700 nm, with a response tailing towards 800 nm. The response is shown with and without the use of a buffer layer at the front of the cell. The buffer layer causes already an improvement at the short wavelengths, and a down converter or shifter may not be beneficial. The purpose of an upconverter is to tune the energy of the emitted photons to the energy where the spectral response shows a maximum. If the energy of the emitted photons is too close to the absorption limit (the band gap edge), then the absorption coefficient is too low and the up converted light would not be fully used.

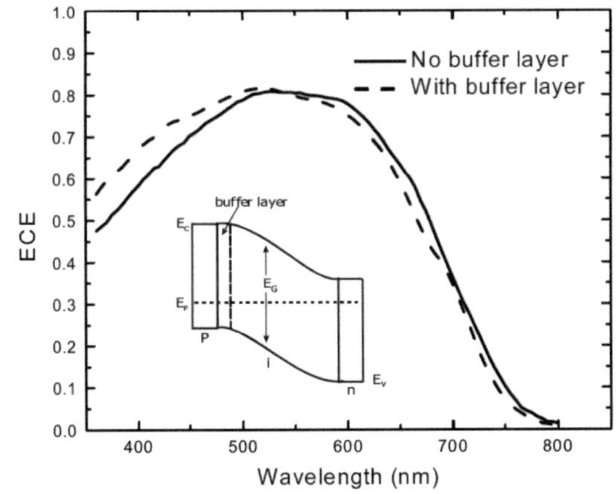

Fig. 7. The spectral response for cells with and without a buffer layer at the p/i interface in the as deposited state obtained at -1 V bias voltage. Inset: the p-i-n- structure showing the position of the buffer layer (from Munyeme et al. 2004).

The photogenerated current could be increased by 40% if the spectral response was sustained at high level up to the band gap cut off at 700 nm and by even more if light with wavelengths λ>700 nm could be more fully absorbed. These two effects can be achieved with the upconversion layer, combined with a highly reflecting back contact. While the upconversion layer converts sub-bandgap photons to super-bandgap photons that can thus be absorbed, a non-conductive reflector is a much better alternative than any metallic mirror, thus sending back both the unabsorbed super-bandgap photons as well as the "fresh" super-bandgap photons into the cell. It is estimated that the stabilized efficiency of the 8.2-8.5% cell can be enhanced to ~12 %. Besides a-Si, a material denoted as protocrystalline Si could be used; this is an amorphous material that is characterized by an enhanced medium range structural order and a higher stability against light-induced degradation compared to standard amorphous silicon. The performance stability of protocrystalline silicon is within 10% of the initial performance; its band gap is slightly higher (1.8 eV) than amorphous silicon.

De Wild et al. (2011) have demonstrated upconversion for a-Si cells with $NaYF_4$ co-doped with (Er^{3+}, Yb^{3+}) as upconverter. The upconverter shows absorption of 980 nm (by the Yb^{3+} ion) leading to efficient emission of 653 nm (red) and 520-540 nm (green) light (by the Er^{3+}) after a two-step energy transfer process. The narrow absorption band around 980 nm for Yb^{3+} limits the spectral range of the IR that can be used for up conversion. An external quantum efficiency of 0.02% at 980 nm laser irradiation was obtained. By using a third ion (for example Ti^{3+}) as a sensitizer the full spectral range between 700 and 980 nm can be efficiently absorbed and converted to red and green light by the Yb-Er couple. The resulting light emission in the green and red region is very well absorbed by the cell with very good quantum efficiency for electron-hole generation.

Upconversion systems consisting of lanthanide nanocrystals of $YbPO_4$ and $LuPO_4$ have been demonstrated to be visible by the naked eye in transparent solutions, however at efficiency lower than solid state up conversion phosphors (Suyver et al. 2005a). Other host lattices ($NaXF_4$, X=Y,Gd,La) have been used and codoping with Yb^{3+} and Er^{3+}, or Yb^{3+} and Tm^{3+} appeared successful, where Yb^{3+} acts as sensitizer. Nanocrystals of <30 nm in size, to prevent scattering in solution, have been prepared and they can be easily dissolved in organic solvents forming colloidal solutions, without agglomeration. Further efficiency increase is possible by growing a shell of undoped $NaYF_4$ around the nanocrystal; in addition, surface modification is needed to allow dissolution in water, for use in biological labeling.

Porous silicon layers are investigated for use as up converter layers as host for rare-earth ions, because these ions can easily penetrate the host due to the large surface area and porosity. A simple and low-cost dipping method has been reported (Díaz-Herrera et al. 2009), in which a porous silicon layer is dipped in to a nitrate solution of erbium and ytterbium in ethanol ($Er(NO_3)_3$:$Yb(NO_3)_3$:C_2H_5OH), which is followed by a spin-on procedure and a thermal activation process at 900 °C. Excitation of the sample at 980 nm revealed up-conversion processes as visible and NIR photoluminescence is observed; codoping of Yb with Er is essential, doping only with Er shows substantial quenching effects (González-Díaz et al. 2008).

Sensitized triplet-triplet annihilation (TTA) using highly photostable metal-organic chromophores in conjunction with energetically appropriate aromatic hydrocarbons has been shown to be another alternative up conversion system (Singh-Rachford et al. 2008). This mechanism was shown to take place under ambient laboratory conditions, i.e. low light intensity conditions, clearly of importance for outdoor operation. These chromophores (porphyins in this case) can be easily incorporated in a solid polymer so that the materials can be treated as thin film materials (Islangulov et al. 2007). A problem with TTA upconverters is the spectral range. No efficient upconversion of NIR radiation at wavelengths beyond 800 nm has been reported which limits the use to very wide bandgap solar cells (Singh-Rachford et al, 2010).

5. Modeling of spectral conversion

An extension to the models described above was presented in a study by Trupke et al. (2006), in which realistic spectra were used to calculate limiting efficiency values for up conversion systems. Using an AM1.5G spectrum leads to a somewhat higher efficiency of 50.69% for a cell with a band gap of 2.0 eV. For silicon, the limiting efficiency would be 40.2%, or nearly 10% larger than the value of 37% obtained for the 6000K blackbody spectrum (Trupke et al. 2002b). This increase was explained by the fact that absorption in the earths' atmosphere at energies lower than 1.5 eV (as evident in the AM1.5G spectrum) leads to a decrease in light intensity. Badescu and Badescu (2009) have presented an improved model, that according to them it appropriately takes into account the refractive index of solar cell and converter materials. Two configurations are studied: cell and rear converter (C-RC), the usual up converter application, and front converter and cell (FC-C), respectively. They confirm the earlier results of Trupke et al. (2002b) in that the limiting efficiency is larger than that of a cell alone, with higher efficiencies at high concentration. Also, the FC-C combination, i.e., up converter layer on top of the cell, does not improve the efficiency, which is obvious. Further, by studying the variation of refractive indexes of cell and converter separately, as opposed to Trupke et al. (2002b), it was found that the limiting efficiency increases with refractive index of both cell and up converter. In practice, a converter layer may have a lower refractive index (1.5, for a transparent polymer: polymethylmethacrylate (PMMA) (Richards and Shalav 2005)) , than that of a cell (3.4). Using a material with similar refractive index as the cell would improve the efficiency by about 10%.

In a series of papers the groups of Badescu, De Vos and co-workers (Badescu and De Vos 2007, Badescu et al. 2007, De Vos et al. 2009) have re-examined the model for down conversion as proposed by Trupke et al. (2002a), and have added the effects of non-radiative recombination and radiation transfer through interfaces. Analogous to the model for up converters they studied FC-C and C-RC configurations, with down conversion or shifting properties. First, neglecting non-radiative recombination, Badescu et al. (2007) qualitatively confirm the results presented by Trupke and co-workers (2002a). For both configurations, the efficiency of the combined system is larger than that of a single mono-facial cell, albeit that the efficiency is smaller (~26%) due to inclusion of front reflections. Second, including radiative recombination for both converter and cell does only increase the efficiency for high (near unity) radiative recombination efficiency values. Interestingly, they report that in this case the C-RC combination cell-rear converter yields a higher efficiency than the FC-C combination in high-quality solar cells, while for low quality solar cells, this is reversed. More realistic device values and allowing for different refractive indices in cell and converter was studied by (Badescu and De Vos 2007), leading to the conclusion that in reality down converters may not always be beneficial. However, extending the model once more, with the inclusion of anti-reflection coating and light trapping texture, showed a limiting efficiency of 39.9%, as reported by De Vos et al. (2009).

Del Cañizo et al. (2008) presented a Monte Carlo ray-tracing model, in which photon transport phenomena in the converter/solar cell system are coupled to non-linear rate equations that describe luminescence. The model was used to select candidate materials for up- and down-conversion, but was set-up for use with rare-earth ions. Results show that for both converters, the potential gain in short-circuit current is small, and may reach 6-7 mA/cm^2 at intensities as high as 1000 suns, in correspondence with earlier work by Shalav et al. (2007).

Modelling downshifting layers on solar cells was also extended for non-AM1.5G spectra, including varying air mass between 1 and 10, and diffuse and direct spectra (Van Sark 2005). Here, the PC1D model (Basore and Clugston 1996) was used to model quantum dots dispersed in a PMMA layer on top of a multi-crystalline silicon cell (mc-Si) as function of the concentration of quantum dots. Annual performance has been modelled by using modelled spectra from the model SEDES2 (Houshyani Hassanzadeh et al. 2007, Nann and Riordan 1991); these spectra can be considered realistic as actual irradiation data is used as input. It was found that the simulated short current enhancement, which varies between about 7% and 23%, is linearly related with the average photon energy (APE) of the spectra, based on hourly spectra of four typical and other randomly selected days throughout the year, and of monthly spectra. The annual short circuit increase was determined at 12.8% using the annual distribution of APE values and their linear relation (Van Sark 2007), which is to be compared with the 10% increase in case of the AM1.5G spectrum. For mc-Si cells with improved surface passivation and a concomitant improved blue response the relative short current increase has been calculated to be lower (Van Sark 2006).

Modeling large area LSCs has indicated the importance of top-surface losses that occur through the escape cone (Chatten et al. 2007) both through primary emission and through emission of luminescence that has been reabsorbed and might otherwise have been trapped via total internal reflection or by mirrors. As an example, for an idealized (perfectly transparent host, LQE = 1), mirrored (perfect mirrors on one short and two long edges, and the bottom surface), 40×5×0.5 cm LSC doped with CdSe/ZnS core-shell QDs with emission matched to a GaInP cell, 24% of the AM1.5G spectrum is absorbed. The photon concentration ratio, C, which is the ratio of the concentrated flux escaping the right-hand surface of the LSC to the flux incident on the top surface, is 4.18. However, since the concentrated flux escaping the right-hand surface is a narrow-band matched to the spectral response of the cell it can all be converted and the idealized LSC would produce 8 times the current compared to the cell alone exposed to AM1.5G. However, 78% of the luminescence is lost through the large top-surface area and only 22% may be collected at the right-hand surface. This lead to the use of wavelength-selective cholesteric liquid crystal coatings applied to the top surface in order to reduce the losses (Debije et al. 2006), as these coatings are transparent to incoming light but reflect the emitted light.

Ray-tracing for LSCs uses basic ray-tracing principles, which means that a ray representing light of a certain wavelength travelling in a certain direction, is traced until it leaves the system e.g. by absorption or reflection at the interface (Burgers et al. 2005, Gallagher et al. 2004). The main extension to the standard ray-tracing model is the handling of the absorption and emission by the luminescent species in the LSC. Ray-tracing has been used to perform parameter studies for a 5x5 cm^2 planar LSC to find attainable LSC efficiencies. The concentrator consists of a PMMA plate (refractive index n=1.49, absorption 1.5 m^{-1}) doped with two luminescent dyes, CRS040 from Bayer and Lumogen F Red 305 from BASF, with a FQE of 95% (Slooff et al. 2006). With the ray-tracing model the efficiency of this plate together with a mc-Si solar cell was determined to be 2.45%. Results for attaining efficiencies for other configurations are shown in Table 1 (Van Sark et al. 2008a). Replacing the mc-Si cell by a GaAs cell or an InGaP cell, will increase the efficiency from 3.8% to 6.5% and 9.1%, respectively (based on Voc (FF) values of 0.58 V (0.83), 1.00 V (0.83), and 1.38 V (0.84), for mc-Si, GaAs, InGaP, respectively). Thus, the use of GaAs or InGaP cells will result in higher efficiencies, but these cells are more expensive. A cost calculation must be performed to determine if the combination of the luminescent concentrator with this type of cells is an interesting alternative to mc-Si based solar technology.

mc-Si	GaAs	InGaP	parameters
2.4	4.2	5.9	fixed mirrors, 85% reflectivity, dyes with 95% LQE
2.9	5.1	7.1	97% reflectivity "air-gap mirrors" on sides, and 97% reflectivity Lambertian mirror at bottom
3.4	5.9	8.3	reduce background absorption of polymer matrix from 1.5 m^{-1} to 10^{-3} m^{-1}
3.8	6.5	9.1	increase of refractive index from 1.49 to 1.7

Table 1. Calculated efficiencies (in %) for an LSC for various optimized configurations and parameters (Van Sark et al. 2008a). A Lambertian mirror is one that reflects radiation isotropically.

Currie et al. (2008) projected conversion efficiencies as high as 6.8%, for a tandem LSC based on two single LSCs that consist of thin layer of deposited organic dye molecules onto a glass plate to which a GaAs cell was attached. Using CdTe or Cu(In,Ga)Se2 solar cells conversion efficiencies of 11.9% and 14.5%, respectively, were calculated.

Annual performance has been modeled using an LSC of which the properties and geometry resulted from a cost-per-unit-of-power optimization study by Bende et al. (2008). A square plate of 23.7x23.7x0.1 cm^3 was used and in the ray-trace model attached to four c-Si solar cells (18.59% efficiency) on all sides. Using a Lumogen F Red 305 (BASF) dye it was calculated that 46.5% of all photons in the wavelength range of 370-630 nm were collected, leading to an LSC efficiency of 4.24% (Van Sark et al. 2008b). The annual yield of this LSC was determined using realistic spectra representative for the Netherlands from Houshyani Hassanzadeh and collaborators (2007) and amounted to 41.3 kWh/m^2, and this is equivalent to an effective annual efficiency of 3.81%.

6. Discussion and outlook

Two issues remain to be solved before downconverters can be applied in solar cells: the absorption strength needs to be increased as the transitions involved for the trivalent lanthanides are sharp and weak (parity forbidden). A second issue is concentration quenching. High Yb-concentrations are needed for efficient energy transfer as every donor (Tb, Pr or Tm) needs to have two Yb-neighbours for energy transfer. For these high concentrations, energy migration over the Yb-sublattice occurs and trapping of the migrating excitation energy by quenching sites strongly reduces the emission output. At present research is conducted to resolve these issues. The limited absorption can be solved by the inclusion of a sensitizer for the $^3P_0$ level of Pr^{3+}, which is able to absorb efficiently over a broad wavelength range (300-500 nm) and subsequent energy transfer to the $^3P_0$ level of Pr^{3+}. In principle, such sensitization can be realized in an efficient and cost-effective manner by the inclusion of a sensitizer ion. The 4f-5d luminescence of Ce^{3+}, for example, is often used to sensitize the Tb^{3+} luminescence in phosphors for fluorescent tubes. Concentration quenching may be limited by optimization of the synthesis conditions (less quenching sites) while also the synthesis of nanocrystals may be beneficial. In nanocrystals the volume probed by energy migration is limited (due to the small size of the nanocrystal) and for defect-free nanocrystals high quantum yields can be expected, similar to the increase in quantum yield observed for quantum dots vs. bulk semiconductors.

For upconverters based on lanthanides absorption strengths need also to be increased and quenching decreased. In addition, upconversion could be useful for solar cells with band gap higher than that of crystalline silicon; thus, research is directed toward optimum matching of NIR absorption and visible emission with the band gap of the solar cell to which the up converter is attached.

Modelling studies on incorporation of conversion layers on top (down converter or down shifter) or at the bottom (upconverter) of single junction solar cells have shown that the conversion efficiency may increase by about 10% (Glaeser and Rau 2007, Richards 2006b, Strümpel et al. 2007, Trupke et al. 2006, Van Sark 2005, Van Sark et al. 2005). This is still an experimental challenge. Also, stability of converter materials is a critical issue, as their lifetime should be >20 years to comply with present solar module practice.

The usefulness of down- and upconversion and downshifting depends on the incident spectrum and intensity. While solar cells are designed and tested according to the ASTM standard (ASTM 2003), these conditions are rarely met outdoors, as discussed by Makrides and collaborators in Chapter 8 of this book. Spectral conditions for solar cells vary from AM0 (extraterrestrial) via AM1 (equator, summer and winter solstice) to AM10 (sunrise, sunset). The weighted average photon energy (APE) (Minemoto et al. 2007) can be used to parameterize this; the APE (using the range 300-1400 nm) of AM1.5G is 1.674 eV, while the APE of AM0 and AM10 are 1.697 eV and 1.307 eV, respectively. Further, overcast skies cause higher scattering leading to diffuse spectra, which are blue-rich, e.g., the APE of the AM1.5 diffuse spectrum is calculated to be 2.005 eV, indeed much larger than the APE of the AM1.5 direct spectrum of 1.610 eV. As DC and DS effectively red-shift spectra, the more blue an incident spectrum contains (high APE) the more gain can be expected (Van Sark 2005, 2008).

Application of DC layers will therefore be more beneficial for regions with high diffuse irradiation fraction, such as Northwestern Europe, where this fraction can be 50% or higher. Thus, luminescent solar concentrators are expected to be deployed successfully in such regions (Van Sark et al. 2008b). In contrast, solar cells with UC layers will be performing well in countries with high direct irradiation fractions or in early morning and evening due to the high air mass resulting in low APE, albeit that the non-linear response to intensity may be limiting.

The variation of the incident spectra are of particular concern for series-connected multiple junction cells, such as triple a-Si:H (Krishnan et al. 2009) and GaAs-based cells. Current-mismatch due to spectral differences with respect to the AM1.5G standard leads to lowering of the conversion efficiency, as the cell with the lowest current determines the total current. It has been shown that the calculated limiting efficiency of a GaAs-based triple cell is reduced to 32.6% at AM5 while its AM1.5G efficiency was 52.5% (Trupke et al. 2006). A single cell with an UC layer, having an AM1.5G efficiency of 50.7%, does not suffer that much from an increase in air mass: at AM5 the efficiency is lowered to 44.0% (Trupke et al. 2006). This so-called spectral robustness is due to the current matching constraints, which are much more relaxed in the cell/UC layer case.

Successfully optimizing absorption strength and quenching in lanthanides based down converters will bring the theoretical limits within reach. This also holds for up conversion, where, in addition triplet-triplet annihilation and mixed transition metal/lanthanide systems (Suyver et al. 2005a) constitute new material systems.

Future use of DS layers on top of solar cells may be limited as blue response of present cells will advance to higher levels (Van Sark 2006). On the other hand, these improvements might require additional expensive processing, while application of a DS layer is expected to be low-cost, as it only involves coating of a plastic with dispersed luminescent species.

LSC development will focus on material systems that: 1) absorb all photons with wavelength > 950 nm, and emit them red-shifted at ~1000 nm, for use with c-Si solar cells, 2) have as low as possible spectral overlap between absorption and emission spectra to minimize re-absorption losses; 3) have near unity luminescence quantum yield; 4) have low escape cone losses; 5) be stable outdoors for more than 10 years; 6) be easy to manufacture at low cost (Rowan et al. 2008). Much progress has occurred, which is illustrated by the recent efficiency record of 7.1% (Slooff et al. 2008) for organic dyes in PMMA. The present lack of NIR dyes will prohibit further increase of conversion efficiency towards the 30% limit. Here, quantum dots or nanorods may have to be used, as their broad absorption spectrum is very favourable. However, they should be emitting in the NIR at high quantum efficiency, larger than the present ~70-80%, and their Stokes' shift should be larger. The latter would be possible as the size distribution of a batch of QDs influences the Stokes' shift (Barnham et al. 2000). Another strategy could be the use of so-called type II QDs, as their Stokes' shift could be very large (~300 nm), however, but their stability and QE are not good enough yet. Stability could be improved using multishell QDs (Koole et al. 2008), while interfacial alloying can be optimized to obtain type II QDs with desired properties, i.e., a Stokes' shift of ~50-100 nm, without spectral overlap (Chin et al. 2007). An different approach was presented recently that employs resonance-shifting to circumvent reabsorption losses (Giebink et al. 2011). Also, Eu^{3+} has been employed succesfully to address self-absorption loss (Wang et al. 2011). Alternatively, the originally proposed three-plate stack (Goetzberger and Greubel 1977) could be further developed using perhaps a combination of organic dyes and nanocrystals, or even rare earth ions (Rowan et al. 2008), with optimized dedicated solar cells for each spectral region.

7. Conclusion

The possibility to tune chemical and physical properties in nanosized materials could have a transformational effect on the performance of solar cells and one of the prominent research areas of nanomaterials for photovoltaics involves spectral conversion. In this chapter we reviewed pathways of spectrum modification for application in solar cells by means of down- and upconversion, and downshifting. Nanoparticles, based on semiconductors or lanthanides, embedded in predominantly polymer layers on top or at the back of solar cells are an essential ingredient of spectral conversion mechanisms that may potentially lead to higher solar cell conversion efficiencies.

Increasing the conversion efficiency, while keeping manufacture cost as low as possible is the main R&D target in solar photovoltaic energy for the coming decades. Nanomaterial synthesis and nanotechnology will play a key role in this.

8. Acknowledgements

The authors gratefully acknowledge numerous colleagues at Utrecht University and elsewhere who contributed to the presented work. This work was financially supported by the European Commission as part of the Framework 6 integrated project FULLSPECTRUM (contract SES6-CT-2003-502620), SenterNovem as part of their Netherlands Nieuw Energie Onderzoek (New Energy Research) programme, Netherlands Foundation for Fundamental Research on Matter (FOM), and Netherlands Organisation for Scientific Research (NWO).

9. References

Alivisatos, A. P. 1996. Perspectives on the Physical Chemistry of Semiconductor Nanocrystals. Journal of Physical Chemistry 100: 13226-13239.

ASTM. 2003. Standard Tables for Reference Solar Spectral Irradiances: Direct Normal and Hemispherical on 37° Tilted Surface, Standard G173-03e1 West Conshohocken, PA, USA: American Society for Testing and Materials.

Auzel, F. 2004. Upconversion and Anti-Stokes Processes with f and d Ions in Solids. Chemical Reviews 104: 139-173.

Badescu, V. and A. M. Badescu. 2009. Improved model for solar cells with up-conversion of low-energy photons. Renewable Energy 34: 1538-1544.

Badescu, V. and A. De Vos. 2007. Influence of some design parameters on the efficiency of solar cells with down-conversion and down shifting of high-energy photons. Journal of Applied Physics 102: 073102-1 - 073102-7.

Badescu, V., A. De Vos, A. M. Badescu, and A. Szymanska. 2007. Improved model for solar cells with down-conversion and down-shifting of high-energy photons. Journal of Physics D: Applied Physics 40: 341-352.

Barnham, K., J. L. Marques, J. Hassard, and P. O'Brien. 2000. Quantum-dot concentrator and thermodynamic model for the global redshift. Applied Physics Letters 76: 1197-1199.

Basore, P. A. and D. A. Clugston. 1996. PC1D Version 4 for Windows: from Analysis to Design. In Proceedings of 25th IEEE Photovoltaic Specialists Conference, Eds. E. C. Boes, D. J. Flood, J. Schmid, and M. Yamaguchi, 377-381. IEEE.

Bende, E. E., A. R. Burgers, L. H. Slooff, W. G. J. H. M. Van Sark, and M. Kennedy. 2008. Cost and Efficiency Optimisation of the Fluorescent Solar Concentrator. In Proceedings of Twenty third European Photovoltaic Solar Energy Conference, Eds. G. Willeke, H. Ossenbrink, and P. Helm, 461-469. WIP, Munich, Germany.

Blasse, G. and B. C. Grabmaier. 1995. Luminescent Materials. Berlin, Germany: Springer.

Bloembergen, N. 1959. Solid state infrared quantum counters. Physical Review Letters 2: 84-85.

Bomm, J., A. Büchtemann, A. Fiore, et al. 2010. Fabrication and spectroscopic studies on highly luminescent CdSe nanorod polymer composites. Beilstein Journal of Nanotechnology 1: 94-100.

Bomm, J., A. Büchtemann, A.J. Chatten et al. 2011. Fabrication and full characterization of state-of-the-art CdSe quantum dot luminescent solar concentrators. Solar Energy Materials and Solar Cells 95: 2087-2094.

Boyer, C, F.C.J.M van Veggel (2010) Absolute Quantum Yield Measurements of Colloidal NaYF4: Er3+, Yb3+ Upconverting Nanoparticles, Nanoscale 2: 1417-1419

Bruchez Jr., M., M. Moronne, P. Gin, S. Weiss, and A. P. Alivisatos. 1998. Semiconductor Nanocrystals as Fluorescent Biological Labels. Science 281: 2013-2016.

Burgers, A. R., L. H. Slooff, R. Kinderman, and J. A. M. van Roosmalen. 2005. Modeling of luminescent concentrators by ray-tracing. In Proceedings of Twentieth European Photovoltaic Solar Energy Conference, Eds. W. Hoffmann, J.-L. Bal, H. Ossenbrink, W. Palz, and P. Helm, 394-397. WIP, Munich, Germany.

Chatten, A. J., K. W. J. Barnham, B. F. Buxton, N. J. Ekins-Daukes, and M. A. Malik. 2003a. A new approach to modelling quantum dot concentrators. Solar Energy Materials and Solar Cells 75: 363-371.

Chatten, A. J., K. W. J. Barnham, B. F. Buxton, N. J. Ekins-Daukes, and M. A. Malik. 2003b. The Quantum Dot Concentrator: Theory and Results. In Proceedings of Third World Congress on Photovoltaic Energy Conversion (WPEC-3), Eds. K. Kurokawa, L. Kazmerski, B. McNelis, M. Yamaguchi, C. Wronski, and W. C. Sinke, 2657-2660.

Chatten, A. J., K. W. J. Barnham, B. F. Buxton, N. J. Ekins-Daukes, and M. A. Malik. 2004a. Quantum Dot Solar Concentrators. Semiconductors 38: 909-917.

Chatten, A. J., K. W. J. Barnham, B. F. Buxton, N. J. Ekins-Daukes, and M. A. Malik. 2004b. Quantum Dot Solar Concentrators and Modules. In Proceedings of 19th European Photovoltaic Solar Energy Conference, Eds. W. Hoffmann, J.-L. Bal, H. Ossenbrink, W. Palz, and P. Helm, 109-112. WIP, Munich, Germany; ETA, Florence, Italy.

Chatten, A. J., D. J. Farrell, R. Bose, et al. 2007. Thermodynamic Modelling of Luminescent Solar Concentrators With Reduced Top Surface Losses. In Proceedings of Twenty Second European Photovoltaic Solar Energy Conference, Eds. G. Willeke, H. Ossenbrink, and P. Helm, 349-353. WIP, Munich, Germany.

Chatten, A. J., D. J. Farrell, C. M. Jermyn, et al. 2005. Thermodynamic Modelling of the Luminescent Solar Concentrator. In Proceedings of 31st IEEE Photovoltaic Specialists Conference, Eds. 82-85. IEEE.

Chen, D., Y. Wang, N. Yu, P. Huang, and F. Weng. 2008a. Near-infrared quantum cutting in transparent nanostructured glass ceramics. Optics Letters 33: 1884-1886.

Chen, D., Y. Wang, Y. Yu, P. Huang, and F. Weng. 2008b. Quantum cutting downconversion by cooperative energy transfer from Ce3+ to Yb3+ in borate glasses. Journal of Applied Physics 104: 116105-1 - 116105-3.

Chin, P. T. K., C. De Mello Donegá, S. S. Van Bavel, et al. 2007. Highly Luminescent CdTe/CdSe Colloidal Heteronanocrystals with Temperature-Dependent Emission Color. Journal of the American Chemical Society 129: 14880-14886.

Chung, P., H.-H. Chung, and P. H. Holloway. 2007. Phosphor coatings to enhance Si photovoltaic cell performance. Journal of Vacuum Science and Technology A 25: 61-66.

Currie, M. J., J. K. Mapel, T. D. Heidel, S. Goffri, and M. A. Baldo. 2008. High-Efficiency Organic Solar Concentrators for Photovoltaics. Science 321: 226-228.

De Vos, A., A. Szymanska, and V. Badescu. 2009. Modelling of solar cells with down-conversion of high energy photons, anti-reflection coatings and light trapping Energy Conversion and Management 50: 328-336.

De Wild, J., A. Meijerink, J.K. Rath, W.G.J.H.M. van Sark, R.E.I. Schropp. 2011. Upconverter solar cells: materials and applications. Energy & Environmental Science 4: 4835-4848

Debije, M. G., R. H. L. Van der Blom, D. J. Broer, and C. W. M. Bastiaansen. 2006. Using selectively-reflecting organic mirrors to improve light output from a luminescent solar concentrator. In Proceedings of World Renewable Energy Congress IX.

Del Cañizo, C., I. Tobias, J. Pérez-Bedmar, A. C. Pan, and A. Luque. 2008. Implementation of a Monte Carlo method to model photon conversion for solar cells. Thin Solid Films 516: 6757-6762.

Dexter, D. L. 1953. A theory of sensitized luminescence in solids. Journal of Chemical Physics 21: 836-850.

Dexter, D. L. 1957. Possibility of luminescent quantum yields greater than unity. Physical Review 108: 630-633.

Díaz-Herrera, B., B. González-Díaz, R. Guerrero-Lemus, et al. 2009. Photoluminescence of porous silicon stain etched and doped with erbium and ytterbium. Physica E 41: 525–528.

Dieke, G. H. 1968. Spectra and Energy Levels of Rare Earth Ions in Crystals. New York, NY, USA: Wiley Interscience.

Eilers J. J., Biner, D.; van Wijngaarden, J. T.; Kraemer, K.; Guedel, H.-U.; Meijerink, A.. 2010, Efficient visible to infrared quantum cutting through downconversion with the Er(3+)-Yb(3+) couple in Cs(3)Y(2)Br(9), Appl. Phys. Lett. 96: 151106.

Fujii, M., M. Yoshida, Y. Kanzawa, S. Hayashi, and K. Yamamoto. 1997. 1.54 μm photoluminescence of Er3+ doped into SiO2 films containing Si nanocrystals: Evidence for energy transfer from Si nanocrystals to Er3+. Applied Physics Letters 71: 1198-1200.

Gallagher, S. J., P. C. Eames, and B. Norton. 2004. Quantum dot solar concentrator behaviour predicted using a ray trace approach. Journal of Ambient Energy 25: 47-56.

Gallagher, S. J., B. C. Rowan, J. Doran, and B. Norton. 2007. Quantum dot solar concentrator: Device optimisation using spectroscopic techniques. Solar Energy 81: 540.

Gamelin, D. R. and H. U. Güdel. 2001. Upconversion Processes in Transition Metal and Rare Earth Metal Systems. Topics in Current Chemistry 214: 1-56.

Gaponenko, S. V. 1998. Optical Properties of Semiconductor Nanocrystals. Cambridge, U.K.: Cambridge University Press.

Garwin, R. L. 1960. The Collection of Light from Scintillation Counters. Review of Scientific Instruments 31: 1010-1011.

Gibart, P., F. Auzel, J.-C. Guillaume, and K. Zahraman. 1996. Below band-gap IR response of substrate-free GaAs solar cells using two-photon up-conversion. Japanese Journal of Applied Physics 351: 4401-4402.

Giebink, N.C., G.P. Wiederrecht, M.R. Wasielewski. 2011. Resonance-shifting to circumvent reabsorption loss in luminescent solar concentrators. Nature Photonics 5: 694-701.

Glaeser, G. C. and U. Rau. 2007. Improvement of photon collection in Cu(In,Ga)Se2 solar cells and modules by fluorescent frequency conversion. Thin Solid Films 515: 5964.

Goetzberger, A. 2008. Fluorescent Solar Energy Concentrators: Principle and Present State of Development. in High-Efficient Low-Cost Photovoltaics - Recent Developments, Eds. V. Petrova-Koch, R. Hezel, and A. Goetzberger, 159-176. Heidelberg, Germany: Springer.

Goetzberger, A. and W. Greubel. 1977. Solar Energy Conversion with Fluorescent Collectors. Applied Physics 14: 123-139.

Goldschmidt, J. C., M. Peters, A. Bösch, et al. 2009. Increasing the efficiency of fluorescent concentrator systems Solar Energy Materials and Solar Cells 93: 176-182.

González-Díaz, B., B. Díaz-Herrera, R. Guerrero-Lemus, et al. 2008. Erbium doped stain etched porous silicon. Materials Science and Engineering B 146: 171-174.

Green, M., K. Emery, Y. Hishikawa, W. Warta, E.D. Dunlop. 2011. Solar Cell Efficiency Tables (Version 38). Progress in Photovoltaics: Research and Applications 19: 565-572.

Green, M. A. 1982. Solar Cells; Operating Principles, Technology and Systems Application. Englewood Cliffs, NJ, USA: Prentice-Hall.

Green, M. A. 2003. Third Generation Photovoltaics, Advanced Solar Energy Conversion. Berlin, Germany: Springer Verlag.

Haase, M. And H. Schafer. 2011. Upconverting Nanoparticles, Angewandte Chemie, Int. Ed. 50: 5808-5829.

Hong, B.-C. and K. Kawano. 2003. PL and PLE studies of KMgF3:Sm crystal and the effect of its wavelength conversion on CdS/CdTe solar cell. Solar Energy Materials and Solar Cells 80: 417-432.

Houshyani Hassanzadeh, B., A. C. De Keizer, N. H. Reich, and W. G. J. H. M. Van Sark. 2007. The effect of a varying solar spectrum on the energy performance of solar cells. In Proceedings of 21st European Photovoltaic Solar Energy Conference, Eds. G. Willeke, H. Ossenbrink, and P. Helm, 2652 2658. WIP Renewable Energies, Munich, Germany.

Hovel, H. J., R. T. Hodgson, and J. M. Woodall. 1979. The effect of fluorescent wavelength shifting on solar cell spectral response. Solar Energy Materials 2: 19-29.

Hyldahl, M. G., S. T. Bailey, and B. P. Wittmershaus. 2009. Photo-stability and performance of CdSe/ZnS quantum dots in luminescent solar concentrators. Solar Energy 83: 566–573.

Islangulov, R. R., J. Lott, C. Weder, and F. N. Castellano. 2007. Noncoherent Low-Power Upconversion in Solid Polymer Films. Journal of the American Chemical Society 129: 12652-12653.

Kamat, P. V. 2008. Quantum dot solar cells. Semiconductor nanocrystals as light harvesters. Journal of Physical Chemistry C 112: 18737-18753.

Kennedy, M., A. J. Chatten, D. J. Farrell, et al. 2008. Luminescent solar concentrators: a comparison of thermodynamic modelling and ray-trace modelling predictions. In Proceedings of Twenty third European Photovoltaic Solar Energy Conference, Eds. G. Willeke, H. Ossenbrink, and P. Helm, 334-337. WIP, Munich, Germany.

Klampaftis, E., D. Ross, K.R. McIntosh, B.S. Richards. 2009. Enhancing the performance of solar cells via luminescent down-shifting of the incident spectrum: A review. Solar Energy Materials and Solar Cells 93: 1182-1194.

Klimov, V. I. 2006. Mechanisms for Photogeneration and Recombination of Multiexcitons in Semiconductor Nanocrystals: Implications for Lasing and Solar Energy Conversion. Journal of Physical Chemistry B 110: 16827-16845.

Klimov, V. I., S. A. Ivanov, J. Nanda, et al. 2007. Single-exciton optical gain in semiconductor nanocrystals. Nature 447: 441-446.

Koole, R., P. Liljeroth, C. De Mello Donegá, D. Vanmaekelbergh, and A. Meijerink. 2006. Electronic Coupling and Exciton Energy Transfer in CdTe Quantum-Dot Molecules. Journal of the American Chemical Society 128: 10436-10441.

Koole, R., M. Van Schooneveld, J. Hilhorst, et al. 2008. On the Incorporation Mechanism of Hydrophobic Quantum Dots in Silica Spheres by a Reverse Microemulsion Method. Chemistry of Materials 20: 2503–2512.

Krishnan, P., J. W. A. Schüttauf, C. H. M. Van der Werf, et al. 2009. Response to simulated typical daily outdoor irradiation conditions of thin-film silicon-based triple-band-gap, triple-junction solar cells. Solar Energy Materials and Solar Cells 93: 691-697.

Lakshminarayana, G., H. Yang, S. Ye, Y. Liu, and J. Qiu. 2008. Cooperative downconversion luminescence in Pr3+/Yb3+:SiO2–Al2O3–BaF2–GdF3 glasses. Journal of Materials Research 23: 3090-3095.

Law, D. C., R. R. King, H. Yoon, et al. 2010. Future technology pathways of terrestrial III–V multijunction solar cells for concentrator photovoltaic systems Solar Energy Materials and Solar Cells 94: 1314-1318.

Lee, J., V. C. Sundar, J. R. Heine, M. G. Bawendi, and K. F. Jensen. 2000. Full Color Emission from II-VI Semiconductor Quantum Dot-Polymer Composites. Advanced Materials 12: 1102-1105.

Liu, X., Y. Qiao, G. Dong, et al. 2008. Cooperative downconversion in Yb3+–RE3+ (RE = Tm or Pr) codoped lanthanum borogermanate glasses. Optics Letters 33: 2858-2860.

Luque, A. and S. Hegedus, Eds. Handbook of Photovoltaic Science and Engineering. 2003, Wiley: Chichester, U.K.

Luque, A. and A. Marti. 1997. Increasing the Efficiency of Ideal Solar Cells by Photon Induced Transitions at Intermediate Levels. Physical Review Letters 78: 5014-5017.

Luque, A. and A. Martí. 2003. Theoretical Limits of Photovoltaic Conversion. in Handbook of Photovoltaic Science and Engineering, Eds. A. Luque and S. Hegedus, 113-149. Chichester, U.K.: Wiley.

Luque, A., A. Martí, A. Bett, et al. 2005. FULLSPECTRUM: a new PV wave making more efficient use of the solar spectrum. Solar Energy Materials and Solar Cells 87: 467-479.

Martí, A. and A. Luque, eds. Next Generation Photovoltaics, High Efficiency through Full Spectrum Utilization. Series in Optics and Optoelectronics, ed. R. G. W. Brown and E. R. Pike. 2004, Institute of Physics Publishing: Bristol, UK.

Maruyama, T. and J. Bandai. 1999. Solar Cell Module Coated with Fluorescent Coloring Agent. Journal of the Electrochemical Society 146: 4406-4409.

Maruyama, T. and R. Kitamura. 2001. Transformations of the wavelength of the light incident upon CdS/CdTe solar cells. Solar Energy Materials and Solar Cells 69: 61-68.

McIntosh, K. R., G. Lau, J. N. Cotsell, K. Hanton, and D. L. Bätzner. 2009. Increase in External Quantum Efficiency of Encapsulated Silicon Solar Cells from a Luminescent Down-Shifting Layer. Progress in Photovoltaics: Research and Applications 17: 191-197.

Meijerink, A., R. Wegh, P. Vergeer, and T. Vlugt. 2006. Photon management with lanthanides Optical Materials 28: 575–581.

Minemoto, T., M. Toda, S. Nagae, et al. 2007. Effect of spectral irradiance distribution on the outdoor performance of amorphous Si//thin-film crystalline Si stacked photovoltaic modules. Solar Energy Materials and Solar Cells 91: 120-122.

Munyeme, G., M. Zeman, R. E. I. Schropp, and W. F. Van der Weg. 2004. Performance analysis of a-Si:H p–i–n solar cells with and without a buffer layer at the p/i interface. physica status solidi C 9: 2298-2303.

Mutlugun, E., I. M. Soganci, and H. V. Demir. 2008. Photovoltaic nanocrystal scintillators hybridized on Si solar cells for enhanced conversion efficiency in UV. Optics Express 16: 3537-3545.

Nann, S. and C. Riordan. 1991. Solar Spectral Irradiance under Clear and Cloudy Skies: Measurements and a Semiempirical Model. Journal of Applied Meteorology 30: 447-462.

O'Regan, B. and M. Grätzel. 1991. A low-cost, high-efficiency solar cell based on dye-sensitized colloidal TiO2 films. Nature 353: 737-740.

Ohwaki, J. and Y. Wang. 1994. Efficient 1.5 μm to Visible Upconversion in Er3+ Doped Halide Phosphors. Japanese Journal of Applied Physics 33: L334-L337.

Peijzel, P. S., A. Meijerink, R. T. Wegh, M. F. Reid, and G. W. Burdick. 2005. A complete 4fn energy level diagram for all trivalent lanthanide ions Journal of Solid State Chemistry 178: 448-453.

Peng, X., M. C. Schlamp, A. V. Kadavanich, and A. P. Alivisatos. 1997. Epitaxial Growth of Highly Luminescent CdSe/CdS Core/Shell Nanocrystals with Photostability and Electronic Accessibility. Journal of the American Chemical Society 119: 7019-7029.

Piper, W. W., J. A. DeLuca, and F. S. Ham. 1974. Cascade fluorescent decay in Pr3+-doped fluorides: achievement of a quantum yield greater than unity for emission of visible light. Journal of Luminescence 8: 344-348.

Reda, S. M. 2008. Synthesis and optical properties of CdS quantum dots embedded in silica matrix thin films and their applications as luminescent solar concentrators Acta Materialia 56: 259-264.

Richards, B. S. 2006a. Enhancing the performance of silicon solar cells via the application of passive luminescence conversion layers. Solar Energy Materials and Solar Cells 90: 2329-2337.

Richards, B. S. 2006b. Luminescent layers for enhanced silicon solar cell performance: Down-conversion. Solar Energy Materials and Solar Cells 90: 1189.

Richards, B. S. and A. Shalav. 2005. The role of polymers in the luminescence conversion of sunlight for enhanced solar cell performance. Synthetic Metals 154: 61-64.

Rowan, B. C., L. R. Wilson, and B. S. Richards. 2008. Advanced Material Concepts for Luminescent Solar Concentrators. IEEE Journal of Selected Topics in Quantum Electronics 14: 1312-1322.

Schropp, R. E. I. and M. Zeman. 1998. Amorphous and microcrystalline silicon solar cells: Modeling, Materials, and Device Technology. Boston, MA, USA: Kluwer Academic Publishers.

Schüler, A., M. Python, M. Valle del Olmo, and E. de Chambrier. 2007. Quantum dot containing nanocomposite thin films for photoluminescent solar concentrators. Solar Energy 81: 1159-1165.

Serin, J. M., D. W. Brousmiche, and J. M. J. Frechet. 2002. A FRET-Based Ultraviolet to Near-Infrared Frequency Convertor. Journal of the American Chemical Society 124: 11848-11849.

Shalav, A., B. S. Richards, and M. A. Green. 2007. Luminescent layers for enhanced silicon solar cell performance: up-conversion. Solar Energy Materials and Solar Cells 91: 829-842.

Shalav, A., B. S. Richards, T. Trupke, K. W. Krämer, and H. U. Güdel. 2005. Application of NaYF4:Er3+ up-converting phosphors for enhanced near-infrared silicon solar cell response. Applied Physics Letters 86: 013505-1 - 013505-3.

Shockley, W. and H. J. Queisser. 1961. Detailed Balance Limit of Efficiency of p-n Junction Solar Cells. Journal of Applied Physics 32: 510.

Singh-Rachford, T. N., A. Haefele, R. Ziessel, and F. N. Castellano. 2008. Boron Dipyrromethene Chromophores: Next Generation Triplet Acceptors/Annihilators for Low Power Upconversion Schemes. Journal of the American Chemical Society 130: 16164-16165.

Singh-Rachford, T.N., F.N. Castellano, Low Power Photon Upconversion based on Sensitized Triplet-Triplet Annihilation, Coordination Chemistry Review, 254: 2560-2573.

Slooff, L. H., E. E. Bende, A. R. Burgers, et al. 2008. A luminescent solar concentrator with 7.1% power conversion efficiency. Physica Status Solidi - Rapid Research Letters 2: 257-259.

Slooff, L. H., R. Kinderman, A. R. Burgers, et al. 2006. The luminescent concentrator illuminated. Proceedings of SPIE 6197: 61970k1-8

Smestad, G., H. Ries, R. Winston, and E. Yablonovitch. 1990. The thermodynamic limits of light concentrators. Solar Energy Materials 21: 99-111.

Soga, T., Ed. Nanostructured Materials for Solar Energy Conversion. 2006, Elsevier: Amsterdam, the Netherlands.

Sommerdijk, J. L., A. Bril, and A. W. De Jager. 1974. Two photon luminescence with ultraviolet excitation of trivalent praseodymium. Journal of Luminescence 8: 341-343.

Strümpel, C., M. McCann, G. Beaucarne, et al. 2007. Modifying the solar spectrum to enhance silicon solar cell efficiency--An overview of available materials. Solar Energy Materials and Solar Cells 91: 238.

Strümpel, C., M. McCann, C. Del Cañizo, I. Tobias, and P. Fath. 2005. Erbium-doped up-converters on silicon solar cells: assesment of the potential. In Proceedings of Twentieth European Photovoltaic Solar Energy Conference, Eds. W. Hoffmann, J.-L. Bal, H. Ossenbrink, W. Palz, and P. Helm, 43-46. WIP, Munich, Germany.

Stupca, M., M. Alsalhi, T. Al Saud, A. Almuhanna, and M. H. Nayfeh. 2007. Enhancement of polycrystalline silicon solar cells using ultrathin films of silicon nanoparticle Applied Physics Letters 91: 063107-1 - 063107-3.

Suyver, J. F., A. Aebischer, D. Biner, et al. 2005a. Novel materials doped with trivalent lanthanides and transition metal ions showing near-infrared to visible photon upconversion Optical Materials 27: 1111-1130.

Suyver, J. F., J. Grimm, K. W. Krämer, and H. U. Güdel. 2005b. Highly efficient near-infrared to visible up-conversion process in NaYF4:Er3+, Yb3+. Journal of Luminescence 114: 53-59.

Svrcek, V., A. Slaoui, and J.-C. Muller. 2004. Silicon nanocrystals as light converter for solar cells. Thin Solid Films 451-452: 384-388.

Timmerman, D., I. Izeddin, P. Stallinga, I. N. Yassievich, and T. Gregorkiewicz. 2008. Space-separated quantum cutting with silicon nanocrystals for photovoltaic applications. Nature Photonics 2: 105-109.

Timmerman, D., J. Valenta, K. Dohnalova, W.D.A.M. de Boer and T. Gregorkiewicz. 2011. Step-like enhancement of luminescence quantum yield of silicon nanocrystals. Nature Nanotechnology 6.

Trupke, T., M. A. Green, and P. Würfel. 2002a. Improving solar cell efficiencies by down-conversion of high-energy photons. Journal of Applied Physics 92: 1668-1674.

Trupke, T., M. A. Green, and P. Würfel. 2002b. Improving solar cell efficiencies by up-conversion of sub-band-gap light. Journal of Applied Physics 92: 4117-4122.

Trupke, T., A. Shalav, B. S. Richards, P. Wurfel, and M. A. Green. 2006. Efficiency enhancement of solar cells by luminescent up-conversion of sunlight. Solar Energy Materials and Solar Cells 90: 3327.

Tsakalakos, L. 2008. Nanostructures for photovoltaics. Materials Science and Engineering: R: Reports 62: 175–189.

Van der Ende, B. M., L. Aarts, and A. Meijerink. 2009a. Near infrared quantum cutting for photovoltaics. Advanced Materials 21: 3073-3077.

Van der Ende, B.M., L. Aarts and A. Meijerink. 2009b. Lanthanide ions as spectral converters for solar cells, Phys. Chem. Chem. Phys. 11: 11081-11095.

Van Sark, W. G. J. H. M. 2002. Methods of Deposition of Hydrogenated Amorphous Silicon for Device Applications. in Thin Films and Nanostructures, Eds. M. H. Francombe, 1-215. San Diego: Academic Press.

Van Sark, W. G. J. H. M. 2005. Enhancement of solar cell performance by employing planar spectral converters. Applied Physics Letters 87: 151117.

Van Sark, W. G. J. H. M. 2006. Optimization of the performance of solar cells with spectral down converters. In Proceedings of Twentyfirst European Photovoltaic Solar Energy Conference, Eds. J. Poortmans, H. Ossenbrink, E. Dunlop, and P. Helm, 155-159. WIP, Munich, Germany.

Van Sark, W. G. J. H. M. 2007. Calculation of the performance of solar cells with spectral down shifters using realistic outdoor solar spectra. In Proceedings of Twenty

Second European Photovoltaic Solar Energy Conference,, Eds. G. Willeke, H. Ossenbrink, and P. Helm, 566-570. WIP, Munich, Germany.

Van Sark, W. G. J. H. M. 2008. Simulating performance of solar cells with spectral downshifting layers. Thin Solid Films 516: 6808–6812.

Van Sark, W. G. J. H. M., K. W. J. Barnham, L. H. Slooff, et al. 2008a. Luminescent Solar Concentrators - A review of recent results. Optics Express 16: 21773-21792.

Van Sark, W. G. J. H. M., P. L. T. M. Frederix, A. A. Bol, H. C. Gerritsen, and A. Meijerink. 2002. Blinking, Blueing, and Bleaching of single CdSe/ZnS Quantum Dots. ChemPhysChem 3: 871-879.

Van Sark, W. G. J. H. M., G. F. M. G. Hellenbrand, E. E. Bende, A. R. Burgers, and L. H. Slooff. 2008b. Annual energy yield of the fluorescent solar concentrator. In Proceedings of Twenty third European Photovoltaic Solar Energy Conference, Eds. G. Willeke, H. Ossenbrink, and P. Helm, 198-202. WIP, Munich, Germany.

Van Sark, W. G. J. H. M., A. Meijerink, R. E. I. Schropp, J. A. M. Van Roosmalen, and E. H. Lysen. 2005. Enhancing solar cell efficiency by using spectral converters. Solar Energy Materials and Solar Cells 87: 395-409.

Van Wijngaarden, J. T., Scheidelaar, S.; Vlugt, T. J. H.; Reid, M. F.; Meijerink. 2010. Energy transfer mechanism for downconversion in the (Pr(3+), Yb(3+)) couple Phys. Rev. B 81 (15): 155112.

Vergeer, P., T. J. H. Vlugt, M. H. F. Kox, et al. 2005. Quantum cutting by cooperative energy transfer in YbxY1−xPO4 : Tb3+ Physical Review B 71: 014119-1 - 014119-11.

Walker, G. W., V. C. Sundar, C. M. Rudzinski, et al. 2003. Quantum-dot optical temperature probes. Applied Physics Letters 83: 3555-3557.

Wang, T., J. Zhang, W. Ma, Y. Luo et al. 2011. Luminescent solar concentrator employing rare earth complex with zero self-absorption loss. Solar Energy 85: 2571-2579.

Wegh, R., H. Donker, A. Meijerink, R. J. Lamminmäki, and J. Hölsä. 1997. Vacuum-ultraviolet spectroscopy and quantum cutting for Gd3+ in LiYF4. Physical Review B 56: 13841-13848.

Wegh, R. T., H. Donker, K. D. Oskam, and A. Meijerink. 1999. Visible Quantum Cutting in LiGdF4:Eu3+ Through Downconversion. Science 283: 663-666.

Wegh, R. T., A. Meijerink, R. J. Lamminmäki, and J. Hölsä. 2000. Extending Dieke's diagram. 87-89: 1002-1004.

Wittwer, V., W. Stahl, and A. Goetzberger. 1984. Fluorescent planar concentrators Solar Energy Materials 11: 187-197.

Wolf, M. 1971. New look at silicon solar cell performance. Energy Conversion 11: 63-73.

Yablonovitch, E. 1980. Thermodynamics of the fluorescent planar concentrator. Journal of the Optical Society of America 70: 1362-1363.

Yuan, Z., G. Pucker, A. Marconi, F. Sgrignuoli, et al. 2011. Silicon nanocrystals as a photoluminescence down shifter for solar cells: 95, 1224-1227.

Zastrow, A. 1994. The physics and applications of fluorescent concentrators: a review. Proceedings of SPIE 2255: 534-547.

Zhang, Q. Y., G. F. Yang, and Y. X. Pan. 2007. Cooperative quantum cutting in one-dimensional (YbxGd1−x)Al3(BO3)4):Tb3+ nanorods Applied Physics Letters 90: 021107-1 - 021107-3.

2

Multi-Scale Modeling of Bulk Heterojunctions for Organic Photovoltaic Applications

Varuni Dantanarayana[1], David M. Huang[2], Jennifer A. Staton[3], Adam J. Moulé[3] and Roland Faller[3]
[1]*Department of Chemistry, University of California–Davis, One Shields Avenue, Davis, CA*
[2]*School of Chemistry & Physics, The University of Adelaide, Adelaide, SA*
[3]*Department of Chemical Engineering & Materials Science, University of California–Davis, One Shields Avenue, Davis, CA*
[1,3]*USA*
[2]*Australia*

1. Introduction

We discuss a variety of recent approaches to molecular modeling of bulk heterojunctions (BHJs) in organic photovoltaics (OPV). These include quantum chemical calculations of the electron donor and acceptor molecules (such as polythiophenes and fullerenes), molecular simulations of their interactions in atomistic detail, mapping between different levels of resolution, coarse–grained (CG) modeling of larger scale structure, as well as electrical modeling. These calculations give a holistic view of these systems from local interactions and structure up to morphology and phase behavior. We focus especially on the evolution of the mesoscale morphology that is characteristic of BHJs. The simulations are compared to experimental results, both structural and dynamic, wherever such results are available.

It turns out that CG models, which are clearly necessary to reach the length scales for morphology development, can accurately capture the large scale structure of atomistic systems in wide areas of the mixture phase diagram. They can also be used to study the dynamic evolution of the microstructure of a BHJ in a system approaching the device scale. On the other hand, atomistic simulations can lead to a geometrically based understanding of local structure which is crucial for charge separation and transport.

Donor/acceptor mixtures for use in OPV devices have been the subject of intense investigation for the past decade. Recent focus on polymer based solar cells has been accelerated by excitement about mixtures of polythiophenes, specifically poly(3–hexylthiophene) (P3HT) with fullerenes, specifically [6,6]–phenyl–C_{61}–butyric acid methyl ester (PCBM) (Padinger et al., 2003; Sariciftci et al., 1992). The most interesting aspect of the P3HT/PCBM mixture is that the power conversion efficiency (PCE) of OPVs resulting from this mixture can vary by almost an order of magnitude (Ma et al., 2005b). This range of efficiency values comes from variability in the hole mobility in the polymer (Mihailetchi et al., 2006), crystallinity of the polymer (Erb et al., 2005), domain sizes (Hoppe & Sariciftci, 2006; Yang et al., 2005), and absorbance of the polymer in the visible range (Al-Ibrahim et al., 2005; Zen et al., 2004) as a result of changes in processing conditions. In essence, the efficiency of the device is controlled by the nanoscale morphology of the blend.

Since these initial studies of P3HT/PCBM morphology, there have been numerous articles that show how morphology develops (Li et al., 2007) and how the morphology can be controlled (Moulé & Meerholz, 2009). A large variety of techniques for measuring aspects of morphology have been applied. X–ray crystallography techniques can observe crystal packing in semicrystalline samples, but amorphous samples are not measurable (Chabinyc, 2008). NMR can give local (nearest neighbor) information, but does not extend to large enough distances to resolve domain size (Nieuwendaal et al., 2010; Yang et al., 2006). Electron microscopy can image domain sizes, but suffers from poor signal–to–noise due to very little contrast between P3HT and PCBM (Andersson et al., 2009; Oosterhout et al., 2009; van Bavel et al., 2010; 2008; Yang et al., 2005). Scanning probe techniques mostly give information about the top surface, though some charge transport information can be derived from specialized scanning probe techniques (Groves et al., 2010; Pingree et al., 2009; Reid et al., 2008). The vertical segregation of materials can be determined non–destructively using refraction techniques (including ellipsometry (Campoy-Quiles et al., 2008; Madsen et al., 2011), X–ray (Andersen, 2009; Germack et al., 2010), and neutron scattering (Lee et al., 2010; Mitchell et al., 2004)). Alternatively, the composition of exposed surfaces can be determined using X–ray photoelectron spectroscopy (Xu et al., 2009). The interior composition of a mixed donor/acceptor layer has also been determined by etching away material using an ion beam and then measuring dynamic secondary ion mass spectrometry (SIMS) (Björstrom et al., 2005; 2007).

The measurement tools that materials science has to offer have been applied in abundance to the determination of the device morphology. However, each measurement technique has sample preparation requirements that can make it difficult to answer desired scientific questions. For example, a transmission electron microscopy (TEM) sample must be removed from the substrate, which requires that the morphology does not change during that sample removal process. One problem that cannot be solved even with a combination of measurement techniques is the determination of nano–scale information about the molecular arrangement in amorphous domains. In fact, the only techniques that can provide complete molecular scale information about amorphous organic domains are computational.

Computer simulations provide means to better understand the relationship between microscopic details and macroscopic properties in solar cells. Most importantly, simulations can be used to explain experimental observations and to elucidate mechanisms that cannot be captured experimentally. In this chapter, it is intended to provide an overview of relevant computational techniques along with advantages and limitations of these techniques with regard to OPV materials, and summarize the research that has been performed to provide insight into the molecular mechanisms in such materials. We also discuss the continuum drift–diffusion model and kinetic Monte–Carlo models used to describe the electric characteristics of OPV devices.

2. Modeling techniques

Computational modeling of soft matter in general and of OPV in particular is clearly not a "one size fits all" problem. Even though the computational expense of detailed modeling is continuously decreasing, it is still the limiting factor in addressing the size and time characteristics of device–scale systems. The most abundant molecular models – atomistic united–atom models – have one (classical) interaction site for every non–hydrogen atom. If one would like to model a seemingly small piece of OPV material with dimensions of (100 μm)^3, it would have about 10^{17} interaction sites in the system (1 nm^3 contains about 100 interaction sites). As the integration time–step in atomistic simulations is on the order of 1 fs for accuracy reasons, simulating for about a second would take at least 10^{33} floating point operations (flops), assuming only 10 floating point operations per particle per time–step. Even with the most powerful computer in the world (Fujitsu K at Riken in Japan, which, with just under 9,000 Tflops/s, tops the *TOP 500* in June 2011) this would take about 10^{17} s or 3×10^{10} years, which is roughly double the age of the universe. Obviously, this is completely impossible. Routinely one can perform simulations with 10^6 interaction sites for about 10^8 time–steps. Therefore one must increase the size of the interaction site and of the time–step to be able to model larger pieces of the system.

Presently, there is a range of modeling techniques that vary in terms of calculated system details and time scale: from electronic structure calculations to coarse–grained methods. These techniques are often used in tandem to obtain a broad understanding of the system.

Electronic structure methods provide the highest degree of accuracy, but are also the most computationally intensive calculations. All electronic structure methods, such as Hartree–Fock (HF) or Density–Functional Theory (DFT), mainly give an estimation to the ground state energy of the system by solving the Schrödinger equation involving some approximations. They also provide accurate structural information, various electronic properties and reaction mechanisms. When used to model organic semiconductors there are a number of limitations, which include the underestimation of the band gap and the inability to accurately describe weak van der Waals forces. Also, due to the high resource requirements, device scale modeling is not possible.

Molecular mechanics (MM) provides information on the structure of large molecules. It is based on empirical force fields that are optimized to reproduce molecular potential energy (PE) surfaces. These potentials are divided into two categories, bonded interactions (including bond, angle and dihedral interactions, $E_{bond} + E_{angle} + E_{dihedral}$) and non–bonded interactions ($E_{electrostatic} + E_{vdW}$). Bonds and angles are usually described by harmonic potentials or kept as a constant value as they just serve for keeping the local geometry correct. The dihedral interaction on the other hand is crucial for describing chain conformations as these interactions are typically comparable to the thermal energy. The non–bonded interactions describe interactions between atoms not stemming from any bonding. There is a vast effort being undertaken to constantly improve these potentials (so–called force–fields). We will not go into detail here. The energies from all interactions are summed up to obtain the intramolecular energy of the molecule: as a superposition of two–, three– and four–body interactions (Leclère et al., 2003).

Energy minimization (EM) techniques provide a simple way to calculate the minimum energy of a system. In a MM calculation, the PE of the system is essentially considered at 0 K. Based on a hypothetical structure based on crystallographic data or another initial condition, the total energy of the initial system is calculated as well as the gradient of the PE as a function of position of each particle (which corresponds to the force). The total energy of the system is minimized by allowing the particles to move in small steps to a lower energy configuration. This motion may either be directly along the force (e.g. steepest descent EM technique) or using other mathematical optimizers (e.g. conjugate gradient EM technique).

The local energy minimum of the structure is found by making small steps towards the local energy minimization. A global structural minimum can be obtained by varying the initial configuration and re–minimizing for each one until a statistically relevant number of starting points lead to a common deep structural minimum. The strength of MM calculations is that they can be used to obtain increased molecular detail for systems for which a partial crystal structure is available. For example, if the stacking distance between adjacent polymer chains is known, MM/EM calculations could provide details on the location of the side chains or the exact angle of backbones with respect to each other. EM calculations are meaningless without a starting configuration that is very close to the ideal because locally rather than globally minimized structures will be the result of the calculation. This means that non–crystaline structures cannot normally be determined using EM.

Molecular dynamics (MD) calculations take kinetic energy (KE) into account and can provide structural insight that is not predetermined from the initial configuration. The movement of each particle is determined from the sum of the bonded and non–bonded forces from surrounding atoms. In many cases (especially on the coarse–grained level) also partially random forces are used to thermostat the system, leading to Brownian motion. Since KE is determined by setting a temperature, the structures formed are not the lowest PE, but an ensemble of all possible structures at a given temperature weighted to minimize free energy. MD are often performed from an ordered starting point that results from MM/EM calculations. MD generates (in the case of constant volume) a set of configurations representing a canonical ensemble in correct time order, i.e. dynamic information is available. The disadvantage of MD calculations is that equilibration (when the system only moves within the correct ensemble) may require a long time, so due to limited computation speed and resources, most systems must be kept small and one can only afford short equilibrations ($\leq$ 100ns).

MD is a step–wise process. At each time step, the forces on each atom are calculated. Then the Hamiltonian equations of motions (which in the typical case of Cartesian co–ordinates are mathematically the same as Newton's equations) are integrated over each time–step. In most simulations temperature and/or pressure is fixed. Thus, an algorithm called a thermostat and/or barostat is needed to correct temperature/pressure and then the step is repeated. This generates a time series of conformations representing the chosen statistical ensemble.

Most molecular simulations are performed using atomistic or semi–atomistic models. The advantages are obvious: there is a unique and strong connection to the underlying chemistry and the models are in many cases easily available (can be downloaded from various websites). Atomistic models normally contain Lennard–Jones (LJ) interactions between non–bonded atoms in addition to a variety of bonded interactions. The most important bonded interaction is the dihedral as the dihedral energies are in the range of thermal energy whereas all other bonded interactions are much larger compared with kT. Quantum–chemistry is often used to obtain the dihedral interaction parameters. As proper dihedral interactions must be periodic with respect to 360°, a Fourier series is typically used to describe the dihedral potential. Also, quantum–chemistry is often used to calculate partial charges, which are needed for electrostatic interactions. Most other interactions are fit to empirical data as no simple and unique theoretical technique is available. A large number of simulation packages (e.g. Gromacs, Amber, LAMMPS, DL_POLY, Tinker etc) have been developed for this kind of modeling and are either freely available or for only a nominal charge. For details the reader is referred to a number of books on simulations (Allen & Tildesley, 1987; Frenkel & Smit, 1996).

The macroscopic properties of semiconductors used in OPV active layers depend on a number of factors. It is important to relate the chemical structure of the components of an active layer to the overall behavior of the system. For OPV applications, it is of interest to study the structural development of active layer mixtures on time and length scales that are relevant to device studies in order to accurately aid in interpretation of experimental findings. However, the computational limitations of atomistic simulations only allow for systems of the size of a few nanometers to be studied. The length scale of interest is on the order of 5–10 nm for the purposes of studying charge transport, and on the order of 100 nm for analyzing the phase separation behavior of the active layer components. Coarse–graining is the process of systematically grouping atoms from the atomistic model into "super–atoms." This reduces the number of interactions and degrees of freedom in the system allowing for simulations to be analyzed over longer length and time scales. For the simulation of P3HT/PCBM, our group achieved a 100× increase in computational speed by using a coarse–grained (CG) model, which could be applied to photovoltaic device–scale systems (Huang et al., 2010).
After this brief overview of the techniques we will now discuss a number of recent modeling approaches using atomistic, CG, and continuum models.

3. Atomistic scale modeling

Organic photovoltaics are characterized by charge carrier mobility, which in turn defines their device efficiency. In BHJ devices, the minimum requirement for high charge carrier mobility is the presence of percolating pathways which enable the holes and electrons to reach the electrodes producing a photocurrent. Thus, device performance is strongly dependent on system morphology, both at the molecular level and over larger length scales.
There are a number of factors that can affect the morphology and their description hinges not only on atomistic, but also on coarse–grained, descriptions. Morphology is critically dependent on the underlying substrates, percolation, thickness of the thin film, and on the processing and annealing methods and conditions. The latter is a significant challenge to simulate, as basically in simulations we always try to approach equilibrium as close as possible, but often in OPVs, equilibrium is neither attainable nor actually desired.
Atomistic level simulations play a critical role in elucidating detailed mechanisms at the nano–scale level. Such calculations eventually can lead to systematic improvement of performance of solar cells. Fullerene derivatives are important electron acceptors and have been studied with different models in varying degrees of detail (Andersson et al., 2008; Arif et al., 2007; Choudhury, 2006; Girifalco, 1992; Hagen et al., 1993; Kim & Tománek, 1994; Qiao et al., 2007; Wong-Ekkabut et al., 2008). Polythiophenes have only recently been studied by computer simulations (Akaike et al., 2010; Botiz & Darling, 2009; Cheung et al., 2009a; Curco & Aleman, 2007; Do et al., 2010; Gus'kova et al., 2009; Widge et al., 2007; 2008). We will discuss below simulations of different organic semiconductor materials where we first discuss fullerenes and thiophenes and then discuss a few other commonly used materials.

3.1 Modeling of fullerenes as electron acceptors

Buckminister fullerene and its derivatives are the most widely established and effective electron acceptors in OPV. Their ability to accept electrons from commonly used donors, such as photoexcited conjugated polymers, on a picosecond time scale and their high charge mobility (Singh et al., 2007) make them appealing as electron acceptors. Since the first

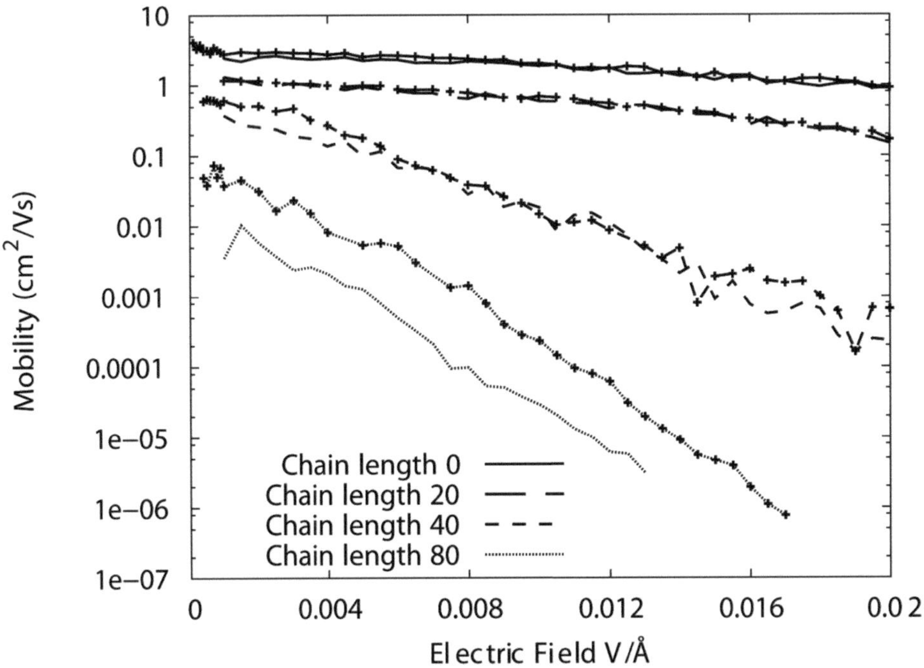

Fig. 1. Electron mobility for C_{60} with different chain lengths at 300 K. The lines without the points superimposed had the site energy differences calculated from a first–order approximation to the site energy differences between the two hopping sites. The lines with points superimposed, are the calculated mobilities when the site energy differences have been forced to 0 meV (MacKenzie et al., 2010). Reprinted with permission from J. Chem. Phys. 132, 064904, 2010. Copyright 2010 American Institute of Physics.

demonstration of polymer/fullerene solar cells (Sariciftci et al., 1993), there has been a significant effort to improve their processability from solutions and to improve performance by optimizing their morphology and energy levels (Troshin et al., 2009).

Molecular packing of the OPV active layer is greatly affected by the conformation of the side groups for both small molecules and polymers. MacKenzie *et al.* (MacKenzie et al., 2010) have studied the effect of charge mobility with different side chains on C_{60}. MD has been performed to generate a realistic material morphology for a series of four C_{60} derivatives. This series of four C_{60} derivatives are formed by attaching a saturated hydrocarbon chain of lengths 0, 20, 40, and 80 carbons to the C_{60} via a methano bridge. The addition of a side chain has been found to disrupt the optimal packing of C_{60}. Furthermore, Figure 1 shows how increasing the hydrocarbon chain length reduces charge mobility. This may stem from the increased size of the functional group, which pushes the C_{60} molecules away from each other and decreases the number of neighbors close enough to electronically interact. As a result, as the functional group size increases, the overall transfer rate decreases thereby reducing the charge mobility (MacKenzie et al., 2010).

The effect of temperature on a thin film of C_{60} has been studied by performing a Monte-Carlo calculation simulating the physical vapor deposition process (Kwiatkowski et al., 2009). The calculations performed at 298, 523, and 748 K reveal morphologies that seem quite different: crystalline regions are estimated to be around 2 nm, 4 nm, and 6 nm in length, respectively.

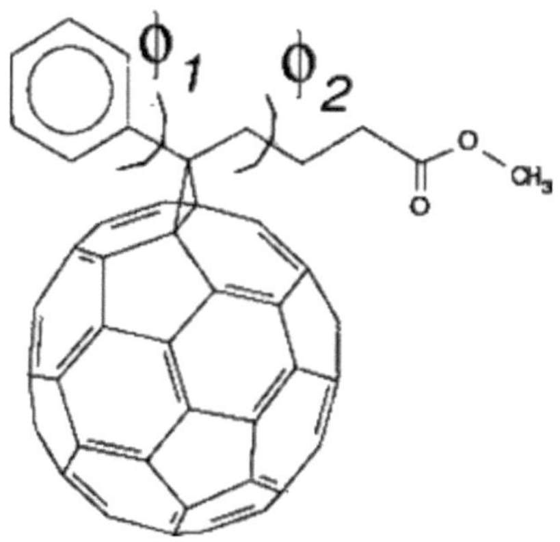

Fig. 2. Structure of PCBM, illustrating the two dihedral angles ϕ_1 and ϕ_2 (Cheung & Troisi, 2010). Reprinted with permission from J. Phys. Chem. C 114, 20479–20488, 2010. Copyright 2010 American Chemical Society.

This follows the same trend as the experimental measurements (Chen. et al., 2001; Cheng et al., 2003), even though the exact grain lengths are smaller than the measured values as it is computationally too expensive to achieve device scale equilibration. However, the calculated radial distribution function (RDF) for the three morphologies are quite similar to each other and to the single crystal RDF (Kwiatkowski et al., 2009). This can be attributed to the fact that C_{60} is a rotationally symmetric molecule with isotropic intermolecular interactions (Girifalco, 1992). This indicates that these spheres can pack efficiently, that even in partially disordered films the molecules are well connected to each other (Kwiatkowski et al., 2009), giving rise to the observed high electron mobility of 8×10^{-6} $m^2/(Vs)$ in thin films of evaporated C_{60} (Haddon et al., 1995) .

One of the main advantages of OPVs is the ease of fabrication, due to their solution processability. Recently, it has been shown that a highly soluble derivative of C_{60}, phenyl–C_{61}–butyric acid methyl ester (PCBM, Figure 2), performs better than C_{60} in solution processed OPV devices (Yu et al., 1995). The addition of the functional side group renders the acceptor more soluble and increased the acceptor strength, thus leading to a higher open–circuit voltage (V_{OC}) in the resulting device (Hummelen et al., 1995). PCBM has a high electron mobility of 2×10^{-3} $cm^2/(Vs)$ at room temperature (Mihailetchi et al., 2003) and since balanced hole and electron charge mobilities reduce space charge build–up and increase the filling factor (FF) of devices, donor polymers with higher hole mobilities are desired for use with PCBM. This functionalized fullerene is relatively inexpensive and it forms segregated phases with many common donors to form mixed layers with ideal morphology (i.e., a grain size separation on the order of the exciton diffusion length and a bicontinuous network formed from the two components) (Ballantyne et al., 2010; Ma et al., 2005b; Thompson & Frchet, 2008).

In order to properly understand the elementary processes in photovoltaics, a correct description of the electron acceptor morphology is important. MD calculations (with the all–atom OPLS force field parameters) performed at 300 K by Cheung *et al.* (Cheung & Troisi, 2010) correctly predict the experimental low–temperature crystal structure (Rispens et al., 2003) of PCBM (Figure 3), in which the electron hopping is facilitated by the molecular arrangement. Figure 4 shows the RDFs between the fullerene centers of mass, the phenyl

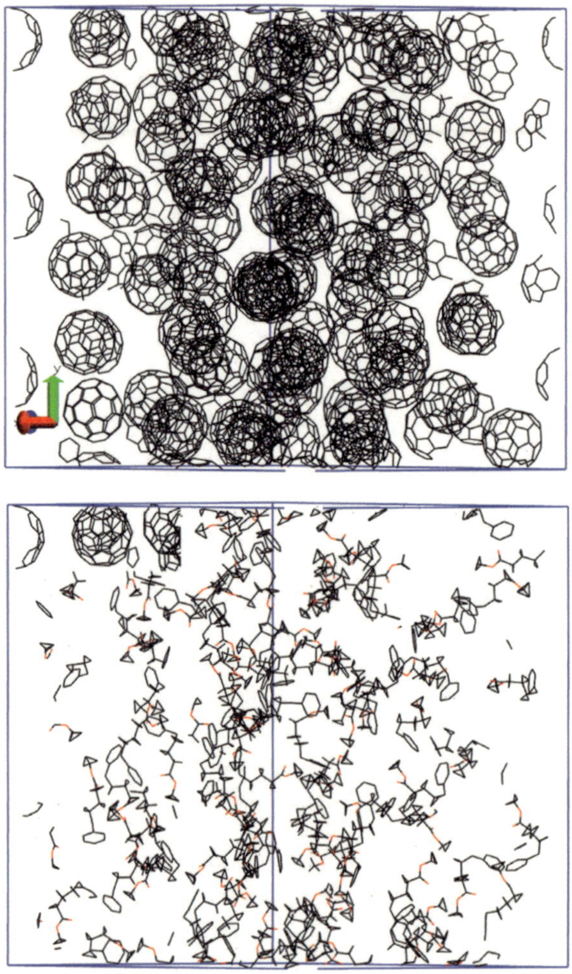

Fig. 3. PCBM system assembly at T = 300 K, showing only the fullerenes (top) and the side chains (bottom). Layering of PCBM molecules in a zigzag pattern is evident (Cheung & Troisi, 2010). Reprinted with permission from J. Phys. Chem. C 114, 20479–20488, 2010. Copyright 2010 American Chemical Society.

ring centers of mass, and between the fullerenes and phenyl rings (Cheung & Troisi, 2010). The fullerene–fullerene and fullerene–phenyl RDFs at 300 K, 400 K and 500 K suggest at best a weak influence of temperature due to the rigid structure of the fullerene cage. For the fullerene–fullerene RDF (Figure 4(a)) the peak is around 10.2 Å, in agreement with experiments (Xu et al., 1993) and previous simulation work carried out by MacKenzie *et al.* (MacKenzie et al., 2010). The first shell in the phenyl–phenyl RDF is around 5 Å (Figure 4(b)). This is significantly larger than the π–stacking distance in aromatic organic crystals (r = 3.8 Å), which suggests that the aromatic moiety in the PCBM side chain does not play a major role in the charge transfer process (Cheung & Troisi, 2010). The tail of PCBM helps to further stabilize the crystal, by the formation of weak hydrogen bonds between aromatic rings and oxygen in addition to the van der Waals interactions (Ecija et al., 2007; Napoles-Duarte

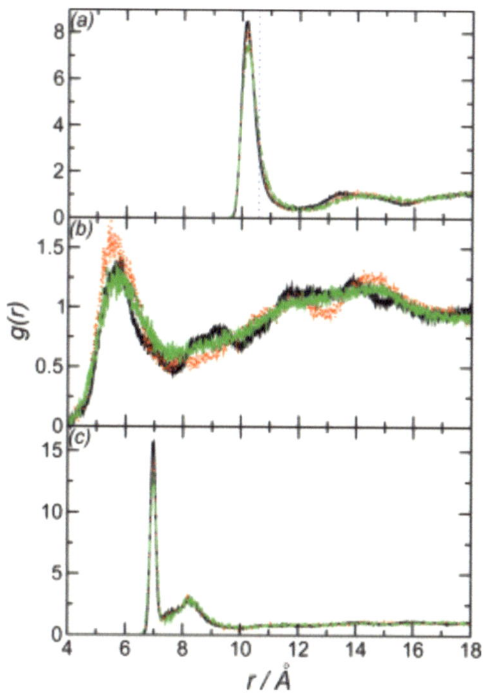

Fig. 4. RDFs for PCBM between, (a) Fullerene–fullerene (b) phenyl–phenyl (c) fullerene–phenyl, at T = 300 K (solid line, black), T = 400 K (dotted line, red), and T = 500 K (dashed line, green) (Cheung & Troisi, 2010). Reprinted with permission from J. Phys. Chem. C 114, 20479–20488, 2010. Copyright 2010 American Chemical Society.

et al., 2009). Since only the fullerene cage in PCBM is likely to be involved in charge transport, morphology related calculations may be used to describe the charge separation, transfer, and recombination processes in OPV BHJ.

MD calculations have also been performed to examine the structural variation between PCBM molecules. Figure 5 shows the dihedral angle distribution for the phenyl group and ester group of the PCBM with respect to the fullerene. The probability distribution of ϕ_1, the angle between the alkyl chain and phenyl ring, has two equally populated peaks at 90° and 270° (Figure 5(a)), indicating that the ring orients parallel to the fullerene surface to reduce steric hinderance. The side group alkyl chain predominately adopts a trans conformation (ϕ_2 = 180°). While at 300 K the population of the gauche conformations is around 15 %, this increases to around 20 % percent at 500 K (Cheung & Troisi, 2010). This is in agreement with crystallographic studies (Rispens et al., 2003). That is, this structure provides a good representation of the solid–state PCBM system.

It is well-known that the PCE of a solar cell depends on V_{OC}, which in turn is a function of the energy gap between the highest occupied orbital (HOMO) in the donor and the lowest unoccupied orbital (LUMO) in the acceptor. That is, increasing the LUMO in the acceptor improves the device PCE (Kooistra et al., 2007). An effective approach to increase the donor LUMO is to alter the fullerene cage by directly adding multiple substituents (Lenes et al., 2008). On the other hand, this increases the structural disorder of the fullerenes. A DFT study has been conducted (Frost et al., 2010) to study the balance between these two effects

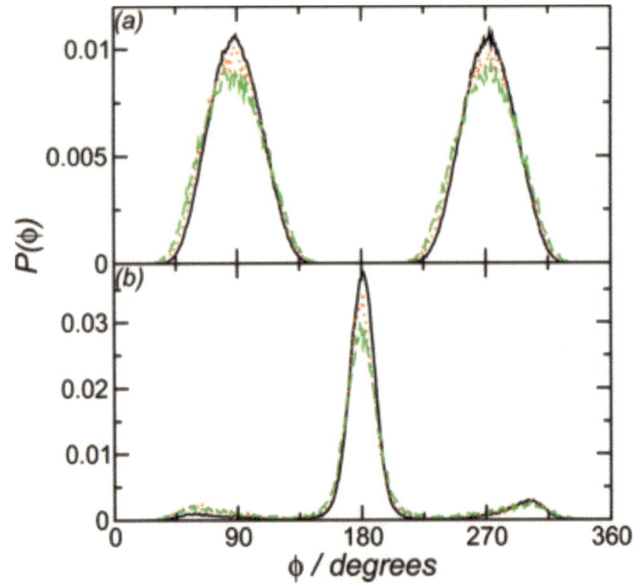

Fig. 5. Side chain dihedral angle distributions for (a) ϕ_1 and (b) ϕ_2. Solid line (black) denotes T = 300 K, dotted line (red) denotes T = 400 K, and dashed line (green) denotes T = 500 K (Cheung & Troisi, 2010). Reprinted with permission from J. Phys. Chem. C 114, 20479–20488, 2010. Copyright 2010 American Chemical Society.

on device performance (Lenes et al., 2009). The authors of that work calculated the energy levels of all eight unique isomers of bis adduct of PCBM and ten representative isomers of tris–PCBM (Djojo et al., 1999). Indeed, a generally lower range of LUMO energies was calculated as a function of the isomers compared to mono-PCBM (Frost et al., 2010). Yet this large range of energetic disorder, in accordance with cyclic voltammetry measurements of the adducts and device current-voltage characteristics (Lenes et al., 2009), decreases the device FF imposing an adverse effect on performance (Frost et al., 2010). It is suggested that designing higher adduct fullerenes with shorter sidechains or hetero–adducts with distinct roles for each of the sidegroups or manipulating energy levels, will lead to fullerene derivatives with better performance as acceptors in solar cells without the need for expensive isolation of isomers (Frost et al., 2010). In that sense, atomistic level calculations can provide insight into the design and optimization of novel electron acceptors.

3.2 Atomistic structure and dynamics of polythiophene semiconductors

The electrooptical properties of semiconducting conjugated polymer films depend critically on molecular structure, doping, and morphology. Whereas the former two can be relatively well controlled during synthesis, the latter is far from well controlled or understood in experiments. Thus, accurate computer simulation models play an important role in elucidating the phase behavior of polymers. They can determine the most likely structures that a polymer will form and calculate the energetic price for assuming another structure. They aid the interpretation of experimental data, because particle positions can be tracked exactly during the course of a simulation.

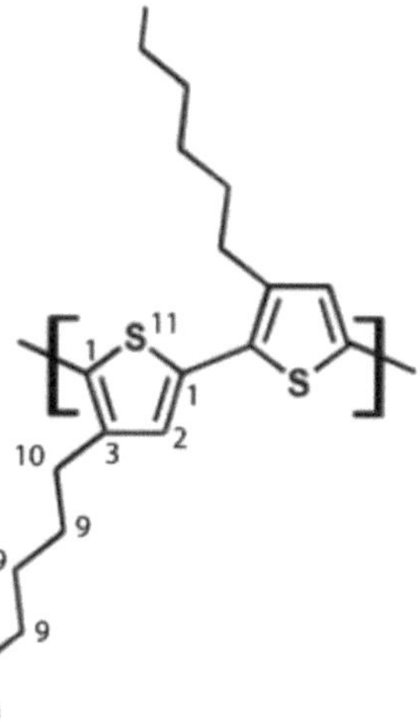

Fig. 6. Structure of P3HT, illustrating the atom type numbers used in a typical force field (Cheung et al., 2009b). Reprinted with permission from J. Phys. Chem. B 113, 9393–9401, 2009. Copyright 2009 American Chemical Society.

Poly(3–alkylthiophene)s [P3ATs] are an abundant class of polymer semiconductors. In particular, the hexyl derivative poly(3–hexylthiophene), P3HT (Figure 6), has become one of the most widely used electron donor materials in OPV. P3HT exhibits high charge mobilities (Sandberg et al., 2002). The charge mobility in polymer semiconductors is strongly dependent on the packing of polymer chains in P3HT films (Kline & McGehee, 2006; Li et al., 2005). Low molecular weight (LMW) (Meille et al., 1997) P3HT exhibits a preferred interdigitated structure, with the distance between the chains 3.8 Å perpendicular to the rings and 16 Å parallel to the rings (Kline et al., 2007; Prosa et al., 1996; Yamamoto et al., 1998). Radial distribution functions (RDFs), $g(r)$, are a standard tool of quantitative structural characterization. RDFs describe how the atomic density varies as a function of distance from a reference atom. For a molecule containing several different atom types, partial RDFs can be defined between any two atom types.

Cheung et al (Cheung et al., 2009b) have performed classical MD simulations at several different temperatures to study the microstructure of P3HT, using a force field previously employed in tetrathiophene calculations (Marcon & Raos, 2006; Marcon et al., 2006). This study on LMW P3HT (20 monomers per chain, molecular weight around 6700 Da) with an initially interdigitated configuration of three polymer layers depicts some important characteristics of P3HT. RDFs between several atoms in the same layer of polymer are illustrated in Figure 7. A common feature among all the RDFs is the peak at $r = 4$ Å, which corresponds to the separation between monomers on the same chain. As temperature increases, it is expected that the interlayer lattice spacing will increase with the expansion of the alkyl chain. This is clearly seen in Figure 7 (c), where the position of the first peak grows with temperature. Such an increase in the interchain spacing with temperature is bound to decrease the P3HT charge mobility, because the probability for charge transfer between adjacent sites in the polymer matrix has an exponential dependence on distance. The sharp peaks at short distances in Figure 7 (a) and (d) are due to the bonded neighbors of atom types C_1 and C_9. Additionally, the widening of peaks with temperature (Figure 7 (c) and (d))

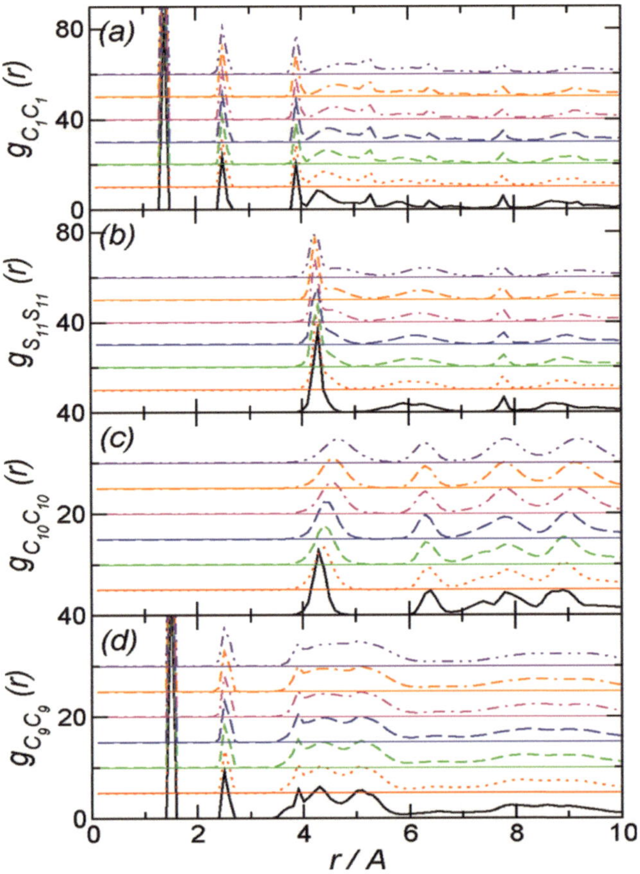

Fig. 7. Atomic radial distribution functions (a) C_1–C_1, (b) S_{11}–S_{11}, (c) C_{10}–C_{10}, and (d) C_9–C_9. In all cases, $T = 100$ K denoted by solid line (black), $T = 200$ K dotted line (red), $T = 225$ K dashed line (green), $T = 250$ K long dashed line (blue), $T = 275$ K dot–dashed line (magenta), $T = 300$ K dot-double-dashed line (orange), and $T = 350$ K double-dot-dashed line (violet). Successive curves are offset by 10 (a and b) or 5 (c and d) units along the y–axis for better representation. Reprinted with permission from J. Phys. Chem. B 113, 9393–9401, 2009. Copyright 2009 American Chemical Society.

represents a less well–defined molecular structure (increased disorder) at high temperatures for LMW P3HT (Cheung et al., 2009b).

Furthermore, at low temperatures and high temperatures, the system density shows a roughly linear decrease with temperature, with a distortion from linearity around 250 K between two linear regions. This is accompanied by a peak in the molar heat capacity, C_p, and isothermal compressibility, k_T, (Figure 8) around the same region. These features signify a conformational transition in the side chains, in accordance with X–ray measurements (Prosa et al., 1992): at low temperatures the ring–tail dihedral angle is around 100°, but above 250 K, this distribution changes with two separate peaks around 100° and 260° (Cheung et al., 2009b). The side–chain disorder directly affects the molecular packing, which influences the morphology and charge mobility.

Atomistic molecular dynamics simulations of P3HT and poly(2,5–bis(3–tetradecyllthiophen–2–yl)thieno[3,2–b]thiophene) (PBTTT) at finite temperatures have been compared to investigate the nanoscale structural properties that lead to the higher measured hole mobility in PBTTT than in P3HT field–effect transistors (FET). Simulations of the two polymer melts show that the structural properties in PBTTT facilitate both intra– and inter–chain charge transport compared with P3HT, due to greater degree of planarity, closer and more parallel stacking of the thiophene and thienothiophene rings, and possible interdigitation of the dodecyl side chains. X–ray diffraction studies have shown that PBTTT indeed forms interdigitated alkyl side chains (Brocorens et al., 2009; Kline et al., 2007). Thus, the crucial role played by the bulky dodecyl side chain and thienothiophene ring, respectively, in determining intra–chain and inter–chain structural order is clarified through these simulations (Do et al., 2010).

In addition to determination of the interaction of a polymer with itself or with the fullerene, MD modeling can address anisotropic interactions such as the interaction of the polymer with a surface. For the most part, MD calculations of interactions between oligomers of polymers with a surface have been performed starting from a crystalline polymer structure that is then lowered onto the surface. MD simulations of thiophenes on surfaces have been recently reviewed (Gus'kova et al., 2009). Modelling of P3HT on a ZnO with the specific application of OPV devices was modelled by Saba et al. (2011) This group allowed a slab of P3HT 12-mers interact with a ZnO ($10\bar{1}0$) surface at 2 K and determined that the structure formed by the P3HT is face on to the ZnO and also that the P3HT crystal spacing increases to lattice match to the ZnO. This article shows that surface properties can significantly influence the structure of polymers. The results, however, would be more believable if MD modeling had been performed at higher temperatures where the polymer had the opportunity to freely rearrange.

It is clear that atomistic level calculations can accurately elucidate mechanistic details of the polymer matrix morphology. Due to the direct access to the microscopic scale that atomistic simulation affords, it can provide a vital link between molecular structure and charge mobility. Atomistic simulations run much more quickly for small molecules than for polymers due to the lower particle number and faster dynamics of the molecules near room temperature. There is a large body of literature involving atomistic modeling of other organic and organometalic compounds for evaporated photovoltaic devices. Discussion of this literature is beyond the scope of this book chapter.

3.3 Some other materials used in OPVs

Scanning force microscope (SFM) measurements on asymmetrically substituted poly[2-methoxy–5-(3′,7′–dimethyloctyloxy)–1,4–phenylene vinylene] (OC_1C_{10}–PPV), also known as MDMO–PPV, shows spiraling chains and no aggregation. Both atomistic scale MD and Monte-Carlo calculations consistently portray this connected ring–forming structure formation of MDMO–PPV (Figure 9) (Kemerink et al., 2003). The bending force present in MDMO–PPV can be explained by an interaction, either attractive or repulsive, between the aliphatic side chains of successive monomer units. These seemingly connected rings could be due to the presence of a conformational or configurational defect. At such a defect, the orientation of the side chain is flipped from one side of the polymer chain to the other, and the bending direction is reversed (Kemerink et al., 2003). This molecular conformation and

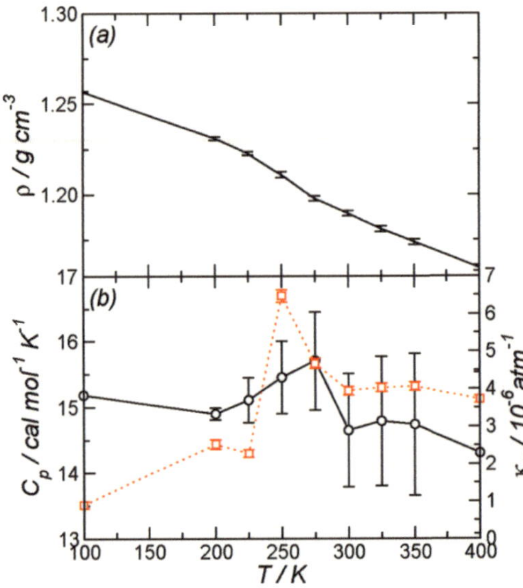

Fig. 8. (a) Density against temperature for P3HT (b) Molar heat capacity C_p (black circles) and isothermal compressibility κ_T (red squares) against temperature (Cheung et al., 2009b). Reprinted with permission from J. Phys. Chem. B 113, 9393–9401, 2009. Copyright 2009 American Chemical Society.

the weakly ordered stacking of the molecules, which result in a high disorder energy and poor $\pi - \pi$ interaction, might be the origin of the poor transport properties in MDMO–PPV. The hole mobility of neat MDMO–PPV is around 5×10^{-7} $cm^2/(Vs)$ (Blom et al., 1997; Martens et al., 1999; Vissenberg & Blom, 1999), whereas neat PCBM has an electron mobility that is 4000 times larger than that (Mihailetchi et al., 2003). Thus, charge transport in a BHJ OPV with an active layer of a MDMO–PPV and PCBM blend is expected to be strongly unbalanced. As the electrons do not neutralize the holes in the device, this results in an augmentation of space-charges and a limited photocurrent. However, hole mobility measurements in a blend of MDMO–PPV with an excess PCBM shows that hole mobility is increased. When mixed with PCBM, the ring formation in MDMO–PPV is hindered due to interactions between MDMO–PPV and PCBM (Figure 10). The change in morphology with an enhanced number of percolating pathways results in improved charge-transfer properties in the blend (Melzer et al., 2004).

Hexabezocoronene (HBC) is a well–known discotic liquid crystal which self–organizes to form columns with strong π-orbital interactions between molecules within columns. Such materials, often referred to as an alternative to thin films of organic semiconductors, possess many characteristics that are important in solar cells (Funahashi, 2009): improved vertical mobility, self-annealing and self-assembling characteristics that give rise to larger domains thus reducing charge-trapping sites. Time–resolved microwave conductivity (TRMC) measurements have shown that hole mobility in HBC is well above 0.1 $cm^2V^{-1}s^{-1}$ and depends on side chains (van de Craats et al., 1999). Atomistic simulations (Andrienko et al., 2006; Kirkpatrick et al., 2008; Marcon et al., 2008) performed on several HBC derivatives with different types of side chains (Figure 11) illustrate the dependence of morphology on semiconductor side chains. Alkyl side chains are generally added to polymers to improve

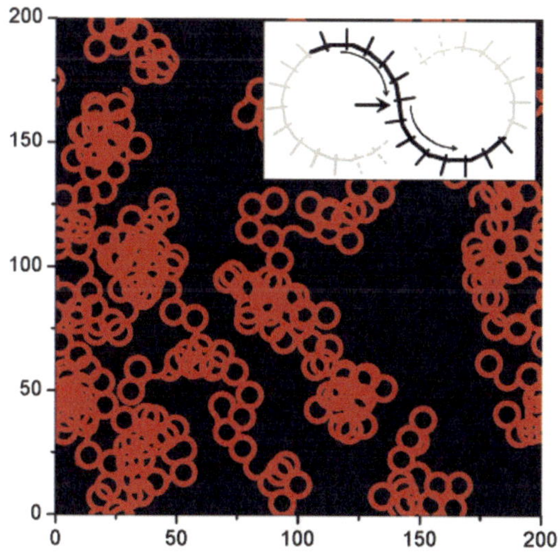

Fig. 9. Simulation of the surface morphology of MDMO–PPV using a Monte Carlo model. The inset illustrates the proposed side chain orientation. At a conformational defect (thick arrow), the sense of spiralling (thin arrows) reverses, leading to a morphology of connected rings (Kemerink et al., 2003). Reprinted with permission from Nano Lett. 3, 1191–1196, 2003. Copyright 2003 American Chemical Society.

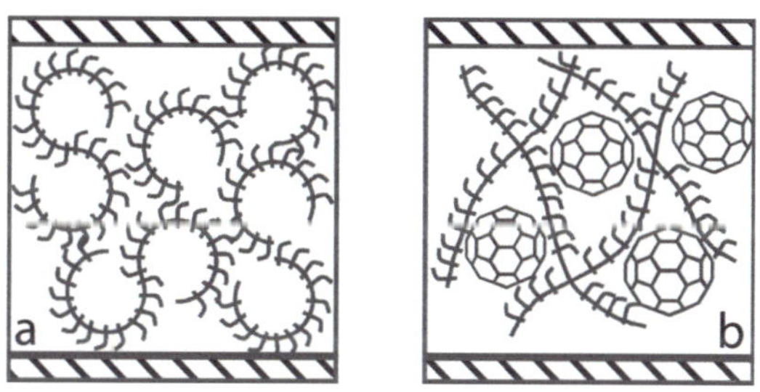

Fig. 10. (a) The molecular conformation of neat MDMO–PPV (b) The molecular conformation of a mixture of MDMO–PPV and PCBM (Melzer et al., 2004). Reprinted with permission from Adv. Func. Mater. 14, 865–870, 2004. Copyright 2004 Wiley.

solubility and reduce the melting point. However, this increases the distance between separate polymer chains, resulting in lower charge mobility. As shown by both experiments and simulations, linear side chains show maximum packing order, while packing is significantly reduced for branched side chains. Consequently, long linear side chains show higher charge mobility than HBC with bulky side chains. Thus, whether it be a crystalline material or smectic liquid crystalline phase, the higher the degree of order, the higher the charge mobility of the material.

4. Coarse–grained modeling

A wide variety of different techniques for coarse–graining, especially for polymeric systems, have been developed (Baschnagel et al., 2000; Faller, 2004). The common main theme is

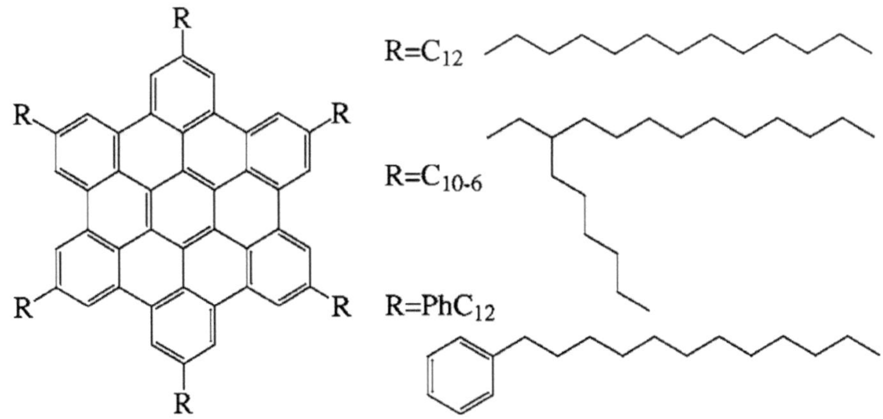

Fig. 11. Structure of the HBC derivatives: R=C_{10}, C_{12}, C_{14}, C_{16}, C_{10-6}, PhC_{12}. Only R=C_{12}, C_{10-6} and PhC_{12} are shown in this figure. Reprinted with permission from J. Chem. Phys. 125, 124902, 2006. Copyright 2006 American Institute of Physics

to circumvent the problems associated with the small time and length scales in atomistic simulations by coarsening the model and decreasing the number of degrees of freedom. These mesoscale or CG models must conserve the chemical nature of the atomistic model. The interactions that describe the polymer properties can be determined for the coarse–grained system from a microscale reference simulation using, e.g., the Iterative Boltzmann Inversion (IBI) method (Reith et al., 2003). One groups atoms appropriately into "super–atoms" or CG beads using chemical intuition. The set of effective potentials to describe the structural distributions of the CG polymer chain can then be generated from a set of correlation functions obtained from a corresponding atomistic simulation. The IBI method is best described with an example involving a non–bonded potential, $V_0(r)$ (Faller, 2007; Reith et al., 2003). If we restrict ourselves to pair potentials non–bonded interactions can be fully described by a radial distribution function (RDF) which, describes how on average the particles in a system radially pack around each other. This radial packing illustrates the correlation between the packing of particles and the forces the particles exert on each other. Mathematically RDFs can be calculated by choosing a reference molecule in a system and a series of concentric spheres around it. RDFs are a measure of the number of sites in a sphere at distance r from the reference center divided by the number in an ideal gas at the same density:

$$g(r) = \frac{1}{d}\frac{n(r)}{4\pi r^2 \Delta r} \tag{1}$$

Here, $g(r)$ is the RDF, $n(r)$ is the average number of molecules in the shell which are counted based on the position of the center of mass of the molecule, d is the overall density, and $4r^2\Delta r$ is the volume of a spherical shell. The IBI aims to match the RDF from the coarse grained model onto the atomistic RDF by iteratively altering the CG potentials. In order to obtain an accurate reproduction of structural details in the CG model, it is important to determine the proper set of potentials.

$$V_{i+1}(r) = V_i(r) + k_B T \ln\left(\frac{g_i^{\mathrm{CG}}(r)}{g_i^{\mathrm{A}}(r)}\right) \tag{2}$$

This equation defines the iterations, where, $V_{i+1}(r)$ and $V_i(r)$ are the potential energies at the $i+1^{th}$ and i^{th} iteration steps, respectively. The RDF of the coarse grained model at the ith step is described by $g_i^{CG}(r)$ and is calibrated against the target atomistic RDF, $g^A(r)$. A reasonable initial guess for $V_0(r)$ is first required. The potential of mean force, $F(r)$, based on the atomistic simulation will usually suffice:

$$F(r) = -k_B T \ln g^{A}(r) \tag{3}$$

The iteration continues until the difference in the RDFs between the CG and the target is below a pre–defined tolerance. This method can be applied to any set of interactions by replacing the radial distribution function by the appropriate probability distribution and the potential by the correct interaction correlation. In dense systems, the interaction distributions are interdependent; thus, one cannot determine each potential separately. Instead, one normally performs the iteration on one of the potentials constant while keeping the rest constant. There must, in the end, be a readjustment of potentials after all are individually iterated. The speed at which this iterative process converges relies on the order in which one optimizes the set of potentials; it is better to start with potentials that are least affected by changes to the rest of the set (Huang et al., 2010; Sun & Faller, 2006; Sun et al., 2008).

4.1 Polymer/fullerene coarse-graining

Our group recently published the first systematic CG–MD model of OPV relevant compounds (Huang et al., 2010; 2011), namely P3HT and the simplest fullerene, C_{60}. CG simulations allow an increase in system size that could reasonably be simulated using a computer cluster to about $25 \times 25 \times 25\,\text{nm}^3$ (compared to about $6 \times 6 \times 6\,\text{nm}^3$ atomistically). This larger volume allows the formation of morphological features approaching what is expected in a device. The larger volume also allows simulation of polymers with a molecular weight approaching typical experimental systems (9–18 kDa) (Brinkmann & Rannou, 2007; Schilinsky et al., 2005; Zen et al., 2004). In a follow up study, we simulated random mixtures of P3HT and C_{60} that were equilibrated at high temperatures and cooled down to temperatures at which C_{60} formed clusters. The coarse–grained model replicated many thermodynamic features that were physically expected. C_{60} did not form large clusters in low MW P3HT but did so with higher MW P3HT. Also, C_{60} did not form clusters until its concentration in P3HT reached a certain threshold (Fig. 12)(Huang et al., 2011).

For the first time a system approaching the size needed for domain formation was studied. The simulations clearly were able to demonstrate that fullerene forms disordered clusters as expected. But now the shape and size distribution of such clusters can be measured. It also turned out that the polymer conformations were much more heterogeneous than expected. In particular, the heterogeneity (e.g. in gyration radius and anisotropy) was found to vary with chain length.

5. Electrical modeling

Given a model for the BHJ morphology and its dependence on material properties and processing conditions, a theoretical understanding of the relationship between the morphology and charge–carrier and exciton transport is needed in order to predict the morphology dependence of device characteristics. A truly accurate theoretical description

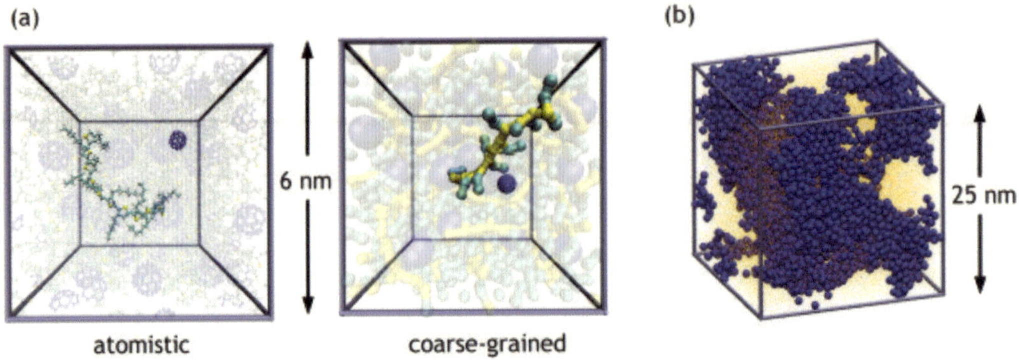

Fig. 12. (a) Snapshots of configurations from atomistic and coarse–grained simulations of small systems with P3HT:C_{60} = 1.85:1 w/w with $N_{\mathrm{mono}} = 12$ at 550 K and 1 atm (Huang et al., 2010). A single molecule of each type is highlighted. (b) C_{60} aggregation in a simulation containing P3HT:C_{60} = 1.27:1 w/w with $N_{\mathrm{mono}} = 48$. The largest C_{60} cluster is highlighted in blue and all other particles in the system are shown as dots (Huang et al., 2011). Reprinted with permission from J. Chem. Theor. Comp. 6, 527–537, 2010 and Fluid Phase Equilib. 302, 21–25, 2011. Copyright 2010 American Chemical Society and 2011 Elsevier, respectively.

of charge–carrier and exciton transport must inherently be multi–scale in nature, because it must account for the anisotropic transport in conjugated polymers on the molecular scale (e.g. higher intra- versus inter-chain charge mobility), exciton dissociation and diffusion between electron donor and acceptor domains on the 10 nm scale, and charge–carrier transport within donor and acceptor domains on the 100 nm scale.

5.1 Continuum drift–diffusion models

The modeling of the electrical characteristics of OPVs has been heavily influenced by the more developed field of inorganic semiconductor physics. One common approach to device–scale electrical modeling of organic solar cells that has been translated almost directly from inorganic semiconductor field is the continuum drift–diffusion model (Sze & Ng, 2006). This approach involves solving the continuity equations for the electron and hole densities, n and p:

$$\frac{\partial n}{\partial t} = \frac{1}{e}\nabla \cdot \boldsymbol{J}_n + D - R, \tag{4}$$

$$\frac{\partial p}{\partial t} = -\frac{1}{e}\nabla \cdot \boldsymbol{J}_p + D - R, \tag{5}$$

assuming that the electron and hole current densities, $\boldsymbol{J}_n$ and $\boldsymbol{J}_p$, consist of a drift term that is proportional to the gradient of the electrical potential ψ and a diffusion term proportional to the gradient of the carrier densities,

$$J_n = -en\boldsymbol{\mu}_n \cdot \nabla\psi + e\mathbf{D}_n \cdot \nabla n, \tag{6}$$

$$J_p = -ep\boldsymbol{\mu}_p \cdot \nabla\psi - e\mathbf{D}_p \cdot \nabla p. \tag{7}$$

Here e is the elementary charge, D is the charge–carrier generation rate (which is equal to the exciton dissociation rate for organic solar cells), R is the recombination rate, $\boldsymbol{\mu}_n$ and $\boldsymbol{\mu}_p$

are the electron and hole mobilities respectively, and $\mathbf{D}_n$ and $\mathbf{D}_p$ are the electron and hole diffusion coefficients. Typically, the mobilities and diffusion coefficients are assumed to obey the Einstein relation, $\mathbf{D}_{n,p} = k_B T \boldsymbol{\mu}_{n,p}/e$. In general, the mobilities and diffusion coefficients in anisotropic materials such as conjugated polymers are tensors, but they have been assumed to be scalars in most theoretical treatments. The drift–diffusion model is completed by the Poisson equation relating the electrostatic potential to the charge–carrier densities,

$$-\nabla \cdot (\epsilon \nabla \psi) = e\,(p - n)\,, \tag{8}$$

where ϵ is the dielectric permittivity of the medium.

In contrast to inorganic semiconductor devices, in which light absorption produces free charge carriers directly, electrical modeling of organic photovoltaics must take into account the dynamics of (singlet) excitons and their finite probability of dissociating into free carriers. Some calculations have avoided explicit consideration of exciton dynamics by taking the charge–carrier generation rate to be a (constant) parameter that is chosen to fit device electrical characteristics (Maturová et al., 2009a;b). In the framework of the drift–diffusion model, the time evolution of the exciton density x is given by

$$\frac{\partial x}{\partial t} = \frac{1}{e}\nabla\left(k_B T \boldsymbol{\mu}_x \cdot \nabla x\right) + G - R_d - D + \frac{1}{4}R\,, \tag{9}$$

where $\boldsymbol{\mu}_x$ is the exciton mobility, G the exciton generation rate, and R_d the exciton decay rate, which is usually assumed to have the form $R_d = x/\tau_x$, where τ_x is the exciton lifetime. The factor of $1/4$ in front of the charge–carrier recombination rate in Equation 9 accounts for the fact that only a fraction of charge–carrier pairs recombine to form singlet excitons (Buxton & Clarke, 2006; Shah & Ganesan, 2009). The recombination rate of free charge carriers is generally assumed to be of the bimolecular Langevin form, $R = \gamma np$, where $\gamma = e\left(\mu_p + \mu_n\right)/\epsilon$, while the exciton dissociation rate D is often given by the Onsager theory of electrolytic dissociation (Onsager, 1934), generalized to account for electric field dependence by (Braun, 1984). More sophisticated approaches to the drift–diffusion model have also been employed, accounting for phenomena such as trapped charges (Hwang et al., 2009) and various forms of the density of states (MacKenzie et al., 2011).

The exciton generation rate G is equal to the dissipation rate of optical power and is thus proportional to the absorbed light intensity. Some drift–diffusion calculations of BHJ solar cells have assumed an exponential attenuation of the light intensity in the active layer with attenuation rate proportional to the absorption coefficient (Buxton & Clarke, 2006; Shah & Ganesan, 2009), while others have used a more sophisticated optical transfer matrix approach (Moulé & Meerholz, 2007; Pettersson et al., 1999) to account for interference effects between ingoing and outcoming light waves (Kotlarski et al., 2008; Nam et al., 2010). Interference results in significant variations in the light intensity in thin-film devices for layer thicknesses on the order of the wavelength of light. Some calculations have simply assumed a constant generation rate within the active layer (Koster et al., 2005; Maturová et al., 2009a;b). While this may seem a drastic approximation, assuming a constant generation rate or using the optical transfer matrix method with the same total generation rate has been shown to make little difference to the the electrical characteristics of MDMO–PPV:PCBM BHJ solar cells (Kotlarski et al., 2008).

To solve the drift–diffusion equations, boundary conditions at the electrode interfaces for the electrical potential and charge–carrier densities (or current densities) must be specified, which in general depend on the applied voltage, electrode work functions, and electron donor and acceptor HOMO and LUMO levels (Koster et al., 2005; Shah & Ganesan, 2009). For charge transfer at internal interfaces between donor and acceptor domains, a transfer rate that depends exponentially on the energy difference between the donor and acceptor HOMO levels (for hole transport) or LUMO levels (for electron transport) is often assumed (Ruhstaller et al., 2001). Device characteristics are generally calculated under steady-state conditions, in which the time derivatives on the left-hand sides of Equations 4, 5, and 9 are set to zero.

The parameters used in the drift–diffusion model, such as the charge–carrier mobilities, exciton lifetime, dielectric permittivity, electrode work functions, and donor and acceptor HOMO and LUMO levels, are generally taken from experimental measurements of the pure donor, acceptor, and electrode materials. These properties are therefore assumed to be isotropic and spatially invariant, except that differences between the values in donor and acceptor domains are sometimes taken into account. In some cases spatial variation of the carrier mobility is introduced in the form of a Poole–Frenkel electric field dependence, $\mu_{n,p} = \mu_{0n,p} \exp\left(\sqrt{E/E_{0n,p}}\right)$, where $E = |\nabla\psi|$ is the magnitude of the electric field and $\mu_{0n,p}$ and $E_{0n,p}$ are constants (Buxton & Clarke, 2006).

In the simplest examples of drift–diffusion modeling of BHJ solar cells, the structure of the BHJ has been ignored entirely and the active layer has been assumed to be completely uniform (Hwang et al., 2009; Koster et al., 2005; Sievers et al., 2006). In this case, the model simplifies considerably, with the electric field and charge carrier densities depending only on one spatial dimension. Such models can be useful for understanding the dependence of device characteristics on parameters such as the active layer thickness, injection barriers, and carrier mobilities. However, they completely ignore the sensitivity of device characteristics to the BHJ morphology, which is demonstrated by the strong dependence of solar cell efficiency on processing conditions for the same donor/acceptor blend ratios (Saunders & Turner, 2008).

The BHJ morphology has been accounted for in some drift–diffusion calculations (Figure 13) by assuming a regular two-dimensional array of interdigitated donor and acceptor domains characterized by a domain width or widths (Maturová et al., 2009a;b; Nam et al., 2010). The material properties within the domains in these calculations have been taken to be uniform. These models allow the dependence of the device characteristics on the level of phase separation in the BHJ to be determined but they assume a BHJ morphology rather than determining it from some structural model.

Within the framework of drift–diffusion models, some calculations have accounted for the effect of material properties and thermodynamic conditions on the heterojunction morphology formed and the consequences for electrical characteristics. Buxton et. al. (Buxton & Clarke, 2006) used a Flory–Huggins Cahn–Hilliard model to determine the morphology of an active layer consisting of a donor–acceptor block copolymer under various conditions and a two-dimensional drift–diffusion model as described above to calculate device properties. Adding another level of complexity (Shah & Ganesan, 2009), used self–consistent field theory (SCFT) to calculate the morphology of a donor–acceptor rod–coil block copolymer, accounting for the orientational anisotropy of the rods and the resulting charge mobility anisotropy. With this model, they found a significant interplay of domain size and anisotropy in optimizing device characteristics.

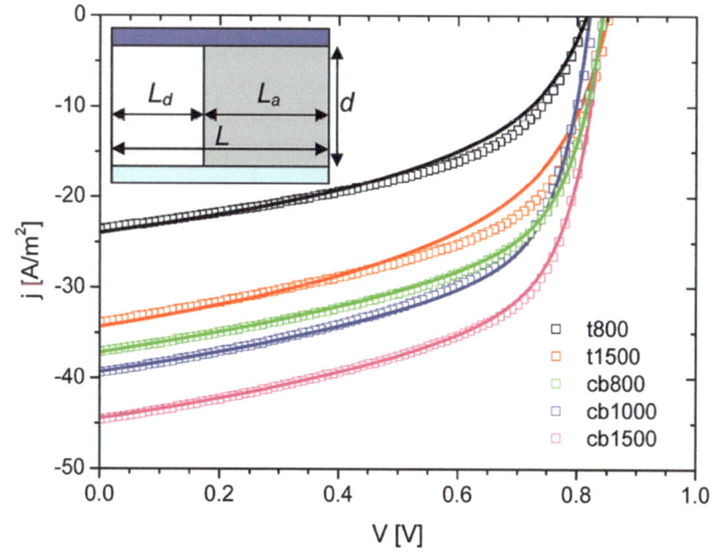

Fig. 13. Low field current–voltage characteristics of BHJ solar cells with different length scales of phase separation. Letters and numbers in the legend refer to solvent [toluene (t) or chlorobenzene (cb)] and spin–casting speed (in rpm), respectively. The symbols are the experimental data, and the lines were calculated using the parameters employed in the numerical modeling. The inset shows the sample layout for the numerical model. Each slab has a thickness d and repeats with a dimension L where L_d and L_a represent the width of the donor and acceptor slabs, respectively (Maturová et al., 2009b). Reprinted with permission from Nano Lett. 9, 3032–3037, 2009. Copyright 2009 American Chemical Society.

The structural models of Buxton & Clarke (2006) and Shah & Ganesan (2009) assumed generic material properties rather than considering specific realistic systems. They were also restricted to BHJs consisting of block copolymers, for which the equilibrium morphologies, which are readily obtained by methods such as SCFT, consist of phase-separated donor and acceptor domains. For typical BHJs consisting of donor and acceptor materials that are not chemically bonded and for which the equilibrium morphology is two completely phase–separated donor and acceptor domains (i.e. a bilayer), such approaches are less suitable. In these cases, an accurate means for calculating non-equilibrium morphologies is required. Non-equilibrium continuum field theories of polymer mixtures or polymer/fullerene mixtures are not well developed, although progress has been made on this front (Ceniceros et al., 2009).
Molecular dynamics simulations, such as those described above for atomistic and coarse–grained models, provide a well–established means for obtaining non–equilibrium morphologies. In particular, coarse–grained simulations like those described in the previous sections provide a means of obtaining non–equilibrium BHJ morphologies on length scales close to the device scale. Continuum drift–diffusion models of charge–carrier and exciton cannot, however, be directly combined with BHJ morphologies from molecular simulations to study the morphology dependence of device characteristics. The next section describes how molecular dynamics models can be used to describe charge–carrier and exciton transport.

5.2 Kinetic Monte Carlo models

At the other extreme to continuum drift–diffusion modeling, charge–carrier transport has been studied on the atomic scale using quantum chemical calculations and time–dependent perturbation theory (Brédas et al., 2004; Cheung & Troisi, 2008; Coropceanu et al., 2007). Such calculations are computationally intensive and are generally restricted to systems on the nanometer scale and therefore cannot be used to study transport between donor–acceptor domains on the 10 nm scale, let alone device–scale transport. Vukmirovic et. al. (Vukmirović & Wang, 2008; 2009) have developed a multi–scale ab initio method of simulating charge–carrier transport that provides a systematic means of combining density functional calculations of small-scale local structural motifs to obtain the electronic structure and charge–transport properties of organic semiconductors up to the 100 nm length scale. But this method has only been applied to studying pure materials so far.

The equations of time–dependent perturbation theory can be simplified in the case of weak electronic coupling between donor and acceptor sites. In the limit of weak electron–phonon coupling and low temperature, the charge–transfer rate can be described by the Miller–Abrahams formalism (Coropceanu et al., 2007), in which the hopping rate from site i to j is

$$k_{ij} = \nu \exp\left(-\frac{2r_{ij}}{a_0}\right) \times \begin{cases} \exp\left(-\frac{E_j - E_i}{k_B T}\right), & E_j > E_i \\ 1, & E_j < E_i \end{cases}, \tag{10}$$

where ν is the attempt-to-escape frequency, a_0 the localization radius, r_{ij} is the hopping distance, and E_i and E_j are the energies of sites i and j respectively. In the limit of strong electron–phonon coupling and high temperature, the charge-transfer rate can be described by Marcus theory (Coropceanu et al., 2007; Nelson et al., 2009),

$$k_{ij} = \frac{|J|^2}{\hbar} \sqrt{\frac{\pi}{\lambda k_B T}} \exp\left[-\frac{(\Delta G + \lambda)^2}{4\lambda k_B T}\right], \tag{11}$$

where J is the electronic coupling between initial and final states (which depends, among other things, on the hopping distance r_{ij}), λ is the reorganisation energy, and ΔG is the difference in free energy between the initial and final states, which is often assumed to be equal to the difference in energy $(E_j - E_i)$. In the weak–coupling limit, exciton transport can be described by Förster resonant energy transfer (FRET) between sites, which, in its simplest form, gives an exciton transfer rate that is inversely proportional to the sixth power of the hopping distance r_{ij} and decays exponentially with the energy difference $(E_j - E_i)$ between the sites (Watkins et al., 2005). The parameters in the transport equations can be estimated from quantum chemical calculations or experimental measurements (Coropceanu et al., 2007; Nelson et al., 2009; Westenhoff et al., 2006; 2005).

Such models of exciton and charge–carrier hopping between discrete sites have been used to study exciton and charge–carrier dynamics in BHJs, which can be simulated with a kinetic Monte Carlo algorithm (Frost et al., 2006; Groves et al., 2009; Meng et al., 2010; 2011; Watkins et al., 2005). These models can account for the molecular nature of the active layer, the morphology of the BHJ, and the anisotropy of exciton and charge transport in a straightforward fashion. However, all implementations of such models so far (Frost et al., 2006; Groves et al., 2009; Meng et al., 2010; 2011; Watkins et al., 2005) have treated the dependence of the BHJ morphology on material properties and thermodynamic conditions in an approximate, albeit physically motivated, fashion: electron donor and acceptor sites were assumed to occupy sites on a cubic lattice and their interactions tuned to produce varying levels of phase separation after donor and acceptor sites were moved according to a Monte Carlo algorithm. Groves et. al used a kinetic Monte Carlo model in a cubic lattice to coarsen a random mixture of donor and acceptor sites into extended domains by allowing high energy interfacial sites to "exchange" position with lower energy sites. Once the coarsened morphology was generated, a drift–diffusion model was used to electrically model the device. When energetic disorder of isolated or smaller domains/sites was used to reduce charge mobility locally, the effects of local morphology on electrical properties could be modelled (Groves et al., 2009). Charge mobility was shown to be the most important determinant of efficiency using this model.

Recently, the Marcus charge–hopping model has been coupled with morphologies obtained from molecular dynamics or Monte Carlo simulations of more realistic molecular models to study charge transport in various physical pure materials, including hexabezocoronene (Kirkpatrick et al., 2007) and fullerenes (Kwiatkowski et al., 2009; MacKenzie et al., 2010; Nelson et al., 2009). But so far, such hopping models have not been combined with realistic structural models of electron donor/acceptor mixtures to study exciton and charge–carrier transport in BHJs. The atomistic and coarse–grained structural models described in the previous sections are amenable to such a treatment. By back–mapping the coarse–grained simulation configurations on to an atomistic description (Baschnagel et al., 2000) and estimating exciton and charge transport parameters from quantum chemical calculations of small atomistic systems as functions of relevant structural parameters, molecular–scale exciton and charge–carrier transport over an entire coarse–grained system can be studied. Such a strategy could be used to develop a multi-scale model of exciton and charge–carrier transport that accurately accounts for the BHJ morphology and its relationship to electrical characteristics from the molecular level up to the device scale.

6. Conclusions

It is obvious that the realistic description of OPVs is a true multi–scale problem. Therefore, the question of which technique is the best to describe the system is meaningless. We need all of them and much work needs to be done to combine them in a meaningful manner. We will always need to identify the right time and length scale for any given problem and then select the model to describe it, never the other way round. It may be tempting to take a model that has been validated in other contexts and just apply it. But we always have to make sure that it contains the right physics and chemistry and that it can realistically describe the relevant aspects of the system. This chapter has clearly been constrained to a small number of problems. But it should have given the reader a flavor of what is possible and how to approach modeling organic photovoltaics in a meaningful manner.

7. References

Akaike, K., Kanai, K., Ouchi, Y. & Seki, K. (2010). Impact of ground–state charge transfer and polarization energy change on energy band offsets at donor/acceptor interface in organic photovoltaics, *Advanced Functional Materials* 20(5): 715–721.

Al-Ibrahim, M., Ambacher, O., Sensfuss, S. & Gobsch, G. (2005). Effects of solvent and annealing on the improved performance of solar cells based on poly(3-hexylthiophene): Fullerene, *Applied Physics Letters* 86(20): 201120.

Allen, M. P. & Tildesley, D. J. (1987). *Computer Simulation of Liquids*, Clarendon Press, Oxford.

Andersen, P. D.; Skarhøj, J. C.; Andreassen J. W, Krebs F. C. (2009). Investigation of optical spacer layers from solution based precursors for polymer solar cells using x-ray reflectometry., *Optical Materials* 31: 1007–1012.

Andersson, B. V., Herland, A., Masich, S. & Inganäs, O. (2009). Imaging of the 3d nanostructure of a polymer solar cell by electron tomography, *Nano Letters* 9(2): 853–855.

Andersson, B. V., Persson, N. K. & Inganas, O. (2008). Comparative study of organic thin film tandem solar cells in alternative geometries, *Journal of Applied Physics* 104(12): 124508.

Andrienko, D., Marcon, V. & Kremer, K. (2006). Atomistic simulation of structure and dynamics of columnar phases of hexabenzocoronene derivatives, *Journal of Chemical Physics* 125(12): 124902.

Arif, M., Yun, M., Gangopadhyay, S., Ghosh, K., Fadiga, L., Galbrecht, F., Scherf, U. & Guha, S. (2007). Polyfluorene as a model system for space-charge-limited conduction, *Physical Review B* 75(19) 195202.

Ballantyne, A. M., Ferenczi, T. A. M., Campoy-Quiles, M., Clarke, T. M., Maurano, A., Wong, K. H., Zhang, W., Stingelin-Stutzmann, N., Kim, J.-S., Bradley, D. D. C., Durrant, J. R., McCulloch, I., Heeney, M. & Nelson, J. (2010). Understanding the influence of morphology on poly(3-hexylselenothiophene):PCBM solar cells, *Macromolecules* 43: 1169–1174.

Baschnagel, J., Binder, K., Doruker, P., Gusev, A. A., Hahn, O., Kremer, K., Mattice, W. L., Müller-Plathe, F., Murat, M., Paul, W., Santos, S., Suter, U. W. & Tries, V. (2000). Bridging the gap between atomistic and coarse–grained models of polymers: Status and perspectives, *Advances in Polymer Science: Viscoelasticity, Atomistic Models, Statistical Chemistry*, Vol. 152 of *Advances in Polymer Science*, Springer, pp. 41–156.

Björstrom, C. M., Bernasik, A., Rysz, J., Budkowski, A., Nilsson, S., Svensson, M., Andersson, M. R., Magnusson, K. O. & Moons, E. (2005). Multilayer formation in spin–coated thin films of low–bandgap polyfluorene : PCBM blends, *Journal of Physics – Condensed Matter* 17(50): L529–L534.

Björstrom, C. M., Nilsson, S., Bernasik, A., Budkowski, A., Andersson, M., Magnusson, K. O. & Moons, E. (2007). Vertical phase separation in spin-coated films of a low bandgap polyfluorene/pcbm blend - effects of specific substrate interaction, *Applied Surface Science* 253(8): 3906–3912.

Blom, P. W. M., de Jong, M. J. M. & van Munster, M. G. (1997). Electric–field and temperature dependence of the hole mobility in poly(p–phenylene) vinylene, *Physical Review B* 55(2): R656.

Botiz, I. & Darling, S. B. (2009). Self–assembly of poly(3-hexylthiophene)-block-polylactide block copolymer and subsequent incorporation of electron acceptor material, *Macromolecules* 42(21): 8211–8217.

Braun, C. L. (1984). Electric–field assisted dissociation of charge–transfer states as a mechanism of photocarrier production, *Journal of Chemical Physics* 80(9): 4157–4161.

Brédas, J. L., Beljonne, D., Coropceanu, V. & Cornil, J. (2004). Charge-transfer and energy-transfer processes in pi-conjugated oligomers and polymers: A molecular picture, *Chem. Rev.* 104(11): 4971–5003.

Brinkmann, M. & Rannou, P. (2007). Effect of molecular weight on the structure and morphology of oriented thin films of regioregular poly(3-hexylthiophene) grown by directional epitaxial solidification, *Advanced Functional Materials* 17(1): 101–108.

Brocorens, P., Vooren, A. V., Chabinyc, M. L., Toney, M. F., Shkunov, M., Heeney, M., McCulloch, I., Cornil, J. & Lazzoroni, R. (2009). *Advanced Materials* 21(1193-1198).

Buxton, G. A. & Clarke, N. (2006). Predicting structure and property relations in polymeric photovoltaic devices, *Physical Review B* 74(8): 085207.

Campoy–Quiles, M., Ferenczi, T., Agostinelli, T., Etchegoin, P. G., Kim, Y., Anthopoulos, T. D., Stavrinou, P. N., Bradley, D. D. C. & Nelson, J. (2008). Morphology evolution via self–organization and lateral and vertical diffusion in polymer:fullerene solar cell blends, *Nature Materials* 7(2): 158–164.

Ceniceros, H. D., Fredrickson, G. H. & Mohler, G. O. (2009). Coupled flow–polymer dynamics via statistical field theory: Modeling and computation, *Journal of Computational Physics* 228(5): 1624–1638.

Chabinyc, M. L. (2008). X-ray scattering from films of semiconducting polymers, *Polymer Reviews* 48(3): 463–492.

Chen, R. S., Lin, Y. J., Su, Y. C. & Chiu, K. C. (2001). Surface morphology of C-60 polycrystalline films from physical vapor deposition *Thin Solid Films* 396: 103–108.

Cheng, W. R., Tang, S. J., Su, Y. C., Lin, Y. J. & Chiu, K. C. (2003). Effects of substrate temperature on the growth of C_60 polycrystalline films by physical vapor deposition, *Journal of Crystal Growth* 247: 401–407.

Cheung, D. L., McMahon, D. P. & Troisi, A. (2009a). Computational study of the structure and charge–transfer parameters in low–molecular–mass P3HT, *Journal of Physical Chemistry B* 113(28): 9393–9401.

Cheung, D. L., McMahon, D. P. & Troisi, A. (2009b). Computational study of the structure and charge-transfer parameters in low-molecular-mass P3HT, *Journal of Physical Chemistry B* 113: 9393–9401.

Cheung, D. L. & Troisi, A. (2008). Modelling charge transport in organic semiconductors: from quantum dynamics to soft matter, *Physical Chemistry Chemical Physics* 10(39): 5941–5952.

Cheung, D. L. & Troisi, A. (2010). Theoretical study of the organic photovoltaic electron acceptor PCBM: Morphology, electronic structure, and charge localization, *Journal of Physical Chemistry C* 114: 20479–20488.

Choudhury, N. (2006). A molecular dynamics simulation study of buckyballs in water: Atomistic versus coarse–grained models of C-60, *Journal of Chemical Physics* 125(3): 034502.

Coropceanu, V., Cornil, J., da Silva, D. A., Olivier, Y., Silbey, R. & Brédas, J. L. (2007). Charge transport in organic semiconductors, *Chemical Reviews* 107(4): 926–952.

Curco, D. & Aleman, C. (2007). Computational tool to model the packing of polycyclic chains: Structural analysis of amorphous polythiophene, *Journal of Computational Chemistry* 28(10): 1743–1749.

Djojo, F., Hirsch, A. & Grimme, S. (1999). *European Journal of Organic Chemistry* 1999: 113027.

Do, K., Huang, D. M., Faller, R. & Moulé, A. J. (2010). A comparative MD study of the local structure of polymer semiconductors P3HT and PBTTT, *Physical Chemistry Chemical Physics* 12: 14735–14739.

Ecija, D., Otero, R., Sánchez, L., Gallego, J. M., Wang, Y., Alcamí, M., Martín, F., Martín, N. & Miranda, R. (2007). Crossover site-selectivity in the adsorption of the fullerene derivative PCBM on Au(111), *Angewandte Chemie International Edition* 46: 7874.

Erb, T., Zhokhavets, U., Gobsch, G., Raleva, S., Stuhn, B., Schilinsky, P., Waldauf, C. & Brabec, C. J. (2005). Correlation between structural and optical properties of composite polymer/fullerene films for organic solar cells, *Advanced Functional Materials* 15(7): 1193–1196.

Faller, R. (2004). Automatic coarse graining of polymers, *Polymer* 45(11): 3869–3876.

Faller, R. (2007). *Reviews in Computational Chemistry*, Vol. 23, Wilwy-VCH, chapter 4: Coarse-Grain Modeling of Polymers, pp. 233–262.

Frenkel, D. & Smit, B. (1996). *Understanding Molecular Simulation: From Basic Algorithms to Applications*, Academic Press, San Diego, CA.

Frost, J. M., Cheynis, F., Tuladhar, S. M. & Nelson, J. (2006). Influence of polymer–blend morphology on charge transport and photocurrent generation in donor–acceptor polymer blends, *Nano Letters* 6(8): 1674–1681.

Frost, J. M., Faist, M. A. & Nelson, J. (2010). Energetic disorder in higher fullerene adducts: A quantum chemical and voltammetric study, *Advanced Materials* 22: 4881.

Funahashi, M. (2009). Development of Liquid–Crystalline Semiconductors with High Carrier Mobilities and Their Application to Thin–film Transistors, *Polymer Journal* 41(6): 459–469.

Germack, D. S., Chan, C. K., Kline, R. J., Fischer, D. A., Gundlach, D. J., Toney, M. F., Richter, L. J. & DeLongchamp, D. M. (2010). Interfacial segregation in polymer/fullerene blend films for photovoltaic devices, *Macromolecules* 43(8): 3828–3836.

Girifalco, L. A. (1992). Molecular properties of C–60 in the gas and solid–phases, *Journal of Physical Chemistry* 96: 858–861.

Groves, C., Koster, L. J. A. & Greenham, N. C. (2009). The effect of morphology upon mobility: Implications for bulk heterojunction solar cells with nonuniform blend morphology, *Journal of Applied Physics* 105(9): 094510.

Groves, C., Reid, O. G. & Ginger, D. S. (2010). Heterogeneity in polymer solar cells: Local morphology and performance in organic photovoltaics studied with scanning probe microscopy, *Accounts of Chemical Research* 43(5): 612–620.

Gus'kova, O. A., Khalatur, P. G. & Khokhlov, A. R. (2009). Self–assembled polythiophene–based nanostructures: Numerical studies, *Macromolecular Theory and Simulation* 18: 219–246.

Haddon, R. C., Perel, A. S., Morris, R. C., Palstra, T. T. M., Hebard, A. F. & Fleming, R. M. (1995). C_{60} thin film transistors, *Applied Physics Letters* 67(1): 121–123.

Hagen, M. H. J., Meijer, E. J., Mooij, G. C. A. M., Frenkel, D. & Lekkerkerker, H. N. W. (1993). Does C-60 have a liquid-phase, *Nature* 365: 425–426.

Hoppe, H. & Sariciftci, N. S. (2006). Morphology of polymer/fullerene bulk heterojunction solar cells, *Journal of Materials Chemistry* 16(1): 45–61.

Huang, D. M., Faller, R., Do, K. & Moulé, A. J. (2010). Coarse–grained computer simulations of polymer/fullerene bulk heterojunctions for organic photovoltaic applications, *Journal of Chemical Theory and Computation* 6(2): 526–537.

Huang, D. M., Moulé, A. J. & Faller, R. (2011). Characterization of polymer–fullerene mixtures for organic photovoltaics by systematically coarse–grained molecular simulations, *Fluid Phase Equilibria* 302(21–25).

Hummelen, J. C., Knight, B. W., LePeq, F. & Wudl, F. (1995). Preparation and characterization of fulleroid and methanofullerene derivatives, *Journal of Organic Chemistry* 60(3): 532.

Hwang, I., McNeill, C. R. & Greenham, N. C. (2009). Drift–diffusion modeling of photocurrent transients in bulk heterojunction solar cells, *Journal of Applied Physics* 106(9): 094506.

Kemerink, M., van Duren, J. K. J., Jonkheijm, P., Pasveer, W. F., Koenraad, P. M., Janssen, R. A. J., Salemink, H. W. M. & Wolter, J. H. (2003). Relating substitution to single–chain conformation and aggregation in poly(p-phenylene vinylene) films, *Nano Letters* 3(9): 1191–1196.

Kim, S. G. & Tománek, D. (1994). Melting the fullerenes: A molecular dynamics study, *Physical Review Letters* 72(15): 2418.

Kirkpatrick, J., Marcon, V., Kremer, K., Nelson, J. & Andrienko, D. (2008). Columnar mesophases of hexabenzocoronene derivatives. II. charge carrier mobility, *Journal Of Chemical Physics* 129: 094506.

Kirkpatrick, J., Marcon, V., Nelson, J., Kremer, K. & Andrienko, D. (2007). Charge mobility of discotic mesophases: A multiscale quantum and classical study, *Physical Review Letters* 98(22): 227402.

Kline, R. J., DeLongchamp, D. M., Fischer, D. A., Lin, E. K., Richter, L. J., Chabinyc, M. K., Toney, M. F., Heeney, M. & McCulloch, I. (2007). *Macromolecules* 40: 7960–7965.

Kline, R. J. & McGehee, M. D. (2006). Morphology and charge transport in conjugated polymers, *Journal of Macromolecular Science, Part C: Polymer Reviews* 46: 27.

Kooistra, F. B., Knol, J., Kastenberg, F., andW. J. H. Verhees, L. M. P., Kroon, J. M., & Hummelen, J. C. (2007). Increasing the open circuit voltage of bulk-heterojunction solar cells by raising the LUMO level of the acceptor, *Org. Lett.* 9(4): 551–554.

Koster, L. J. A., Smits, E. C. P., Mihailetchi, V. D. & Blom, P. W. M. (2005). Device model for the operation of polymer/fullerene bulk heterojunction solar cells, *Phys. Rev. B* 72(8): 085205.

Kotlarski, J. D., Blom, P. W. M., Koster, L. J. A., Lenes, M. & Slooff, L. H. (2008). Combined optical and electrical modeling of polymer : fullerene bulk heterojunction solar cells, *J. Appl. Phys.* 103(8): 084502.

Kwiatkowski, J. J., Frost, J. M. & Nelson, J. (2009). The effect of morphology on electron field-effect mobility in disordered C60 thin films, *Nano Letters* 9(3): 1085–1090.

Leclère, P., Hennebicq, E., Calderone, A., Brocorens, P., Grimsdale, A. C., Müllen, K., Brédas, J. L. & Lazzaroni, R. (2003). Supramolecular organization in block copolymers containing a conjugated segment: a joint AFM/molecular modeling study, *Progress in Polymer Science* 28: 55–81.

Lee, K. H., Schwenn, P. E., Smith, A. R. G., Cavaye, H., Shaw, P. E., James, M., Krueger, K. B., Gentle, I. R., Meredith, P. & Burn, P. L. (2010). Morphology of all–solution-processed "bilayer" organic solar cells, *Advanced Materials* 23: 766–770.

Lenes, M., Shelton, S. W., Sieval, A. B., Kronholm, D. F., Hummelen, J. C. & Blom., P. W. (2009). Electron Trapping in Higher Adduct Fullerene–Based Solar Cells *Advanced Functional Materials* 19(18): 3002–3007.

Lenes, M., Wetzelaer, G.-J. A. H., Kooistra, F. B., Veenstra, S. C., Hummelen, J. C. & Blom, P. W. M. (2008). Fullerene bisadducts for enhanced open-circuit voltages and efficiencies in polymer solar cells, *Advanced Materials* 20: 2116–2119.

Li, G., Shrotriya, V., Huang, J., Yao, Y., Moriarty, T., Emery, K. & Yang, Y. (2005). High-efficiency solution processable polymer photovoltaic cells by self-organization of polymer blends, *Nature Materials* 4: 864.

Li, G., Shrotriya, V., Yao, Y., Huang, J. S. & Yang, Y. (2007). Manipulating regioregular poly(3-hexylthiophene): [6,6]-phenyl-c-61-butyric acid methyl ester blends - route towards high efficiency polymer solar cells, *Journal of Materials Chemistry* 17(30): 3126–3140.

Ma, W., Yang, C., Gong, X., Lee, K. & Heeger, A. J. (2005b). Thermally stable, efficient polymer solar cells with nanoscale controle of the interpenetrating network morphology, *Advanced Functional Materials* 15: 1617–1622.

MacKenzie, R. C. I., Frost, J. M. & Nelson, J. (2010). A numerical study of mobility in thin films of fullerene derivatives, *Journal Of Chemical Physics* 132(6): 064904.

MacKenzie, R. C. I., Kirchartz, T., Dibb, G. F. A. & Nelson, J. (2011). Modeling nongeminate recombination in P3HT:PCBM solar cells, *Journal of Physical Chemistry C* 115(19): 9806–9813.

Madsen, M. V., Sylvester-Hvid, K. O., Dastmalchi, B., Hingerl, K., Norrman, K., Tromholt, T., Manceau, M., Angmo, D. & Krebs, F. C. (2011). Ellipsometry as a nondestructive depth profiling tool for roll-to-roll manufactured flexible solar cells, *Journal of Physical Chemistry C* 115: 10817–10822.

Marcon, V. & Raos, G. (2006). Free energies of molecular crystal surfaces by computer simulation: Application to tetrathiophene, *Journal of the American Chemical Society* 128: 1408–1409.

Marcon, V., Raos, G., Campione, M. & Sassella, A. (2006). Incommensurate epitaxy of tetrathiophene on potassium hydrogen phthalate: Insights from molecular simulation, *Crystal Growth & Design* 6: 1826–1832.

Marcon, V., Vehoff, T., Kirkpatrick, J., Jeong, C., Yoon, D. Y., Kremer, K. & Andrienko, D. (2008). Columnar mesophases of hexabenzocoronene derivatives. I. phase transitions, *Journal of Chemical Physics* 129(9): 094505.

Martens, H. C. F., Brom, H. B. & Blom, P. W. M. (1999). Frequency–dependent electrical response of holes in poly(p-phenylene vinylene), *Physical Review B* 60(12): R8489.

Maturová, K., Kemerink, M., Wienk, M. M., Charrier, D. S. H. & Janssen, R. A. J. (2009). Scanning Kelvin probe microscopy on bulk heterojunction polymer blendsmicroscopy on bulk heterojunction polymer blends, *Advanced Functional Materials* 19(9): 1379–1386.

Maturová, K., van Bavel, S. S., Wienk, M. M., Janssen, R. A. J. & Kemerink, M. (2009). Morphological device model for organic bulk heterojunction solar cells, *Nano Letters* 9(8): 3032–3037.

Meille, S. V., Romita, V., Caronna, T., Lovinger, A. J., Catellani, M. & Belobrzeckaja, L. (1997). *Macromolecules* 30: 7898–7905.

Melzer, C., Koop, E. J., Mahailetchi, V. D. & Blom, P. W. M. (2004). Hole transport in poly(phenylene vinylene)/methanofullerene bulk-heterojunction solar cells, *Advanced Functional Materials* 14(9): 865–870.

Meng, L., Shang, Y., Li, Q., Li, Y., Zhan, X., Shuai, Z., Kimber, R. G. E. & Walker, A. B. (2010). Dynamic Monte Carlo simulation for highly efficient polymer blend photovoltaics, *Journal of Physical Chemistry B* 114(1): 36–41.

Meng, L., Wang, D., Li, Q., Yi, Y., Brédas, J.-L. & Shuai, Z. (2011). An improved dynamic monte carlo model coupled with poisson equation to simulate the performance of organic photovoltaic devices, *Journal of Chemical Physics* 134(12): 124102.

Mihailetchi, V. D., can Duren, J. K. J., Blom, P. W. M., Hummelen, J. C., Janssen, R. A. J., Kroon, J. M., Rispens, M. T., Verhees, W. J. H. & Wienk, M. M. (2003). Electron transport in a methanofullerene, *Advanced Functional Materials* 13(1): 43–46.

Mihailetchi, V. D., Xie, H. X., de Boer, B., Koster, L. J. A. & Blom, P. W. M. (2006). Charge transport and photocurrent generation in poly (3-hexylthiophene): Methanofullerene bulk-heterojunction solar cells, *Advanced Functional Materials* 16(5): 699–708.

Mitchell, W. J., Burn, P. L., Thomas, R. K., Fragneto, G., Markham, J. P. J. & Samuel, I. D. W. (2004). Relating the physical structure and optical properties of conjugated polymers using neutron reflectivity in combination with photoluminescence spectroscopy, *Journal of Applied Physics* 95(5): 2391–2396.

Moulé, A. J. & Meerholz, K. (2009). Morphology control in solution–processed bulk-heterojunction solar cell mixtures, *Advanced Functional Materials* 19(9999): 3028–3036.

Moulé, A. J. & Meerholz, K. (2007). Minimizing optical losses in bulk heterojunction polymer solar cells, *Applied Physics B* 86(4): 771–777.

Nam, Y. M., Huh, J. & Ho Jo, W. (2010). Optimization of thickness and morphology of active layer for high performance of bulk-heterojunction organic solar cells, *Solar Energy Materials & Solar Cells* 94(6): 1118–1124.

Napoles–Duarte, J. M., Lopez-Sandoval, R., Gorbatchev, A. Y., Reyes–Reyes, M. & Carroll, D. L. (2009). Encapsulation of the fullerene derivative [6,6]-phenyl-c61-butyric acid methyl ester inside micellar structures, *J. Phys. Chem. C* 113: 13677.

Nelson, J., Kwiatkowski, J. J., Kirkpatrick, J. & Frost, J. M. (2009). Modeling charge transport in organic photovoltaic materials, *Acc. Chem. Res.* 42(11): 1768–1778.

Nieuwendaal, R. C., Snyder, C. R., Kline, R. J., Lin, E. K., VanderHart, D. L. & DeLongchamp, D. M. (2010). Measuring the extent of phase separation in poly-3-hexylthiophene/phenyl-c-61-butyric acid methyl ester photovoltaic blends with h-1 spin diffusion nmr spectroscopy, *Chemistry of Materials* 22(9): 2930–2936.

Onsager, L. (1934). Initial recombination of ions, *Physical Review* 54: 554–557.

Oosterhout, S. D., Wienk, M. M., van Bavel, S. S., Thiedmann, R., Koster, L. J. A., Gilot, J., Loos, J., Schmidt, V. & Janssen, R. A. J. (2009). The effect of three-dimensional morphology on the efficiency of hybrid polymer solar cells, *Nature Materials* 8(10): 818–824.

Padinger, F., Rittberger, R. S. & Sariciftci, N. S. (2003). Effects of postproduction treatment on plastic solar cells, *Advanced Functional Materials* 13(1): 85–88.

Pettersson, L. A. A., Roman, L. S. & Inganäs, O. (1999). Modeling photocurrent action spectra of photovoltaic devices based on organic thin films, *Journal of Applied Physics* 86(1): 487–496.

Pingree, L. S. C., Reid, O. G. & Ginger, D. S. (2009). Imaging the evolution of nanoscale photocurrent collection and transport networks during annealing of polythiophene/fullerene solar cells, *Nano Letters* 9(8): 2946–2952.

Prosa, T. J., Winokur, M. J. & McCullough, R. D. (1996). Evidence of a novel Side chain structure in regioregular poly(J-alkyl thiophenes), *Macromolecules* 29: 3654–3656.

Prosa, T. J., Winokur, M. J., Moulton, J., Smith, P. & Heeger, A. J. (1992). X-Ray structural studies of poly (3-alkylthiophenes) – An example of an inverse comb, *Macromolecules* 25: 4364–4372.

Qiao, R., Roberts, A. P., Mount, A. S., Klaine, S. J. & Ke, P. C. (2007). Translocation of C–60 and its derivatives across a lipid bilayer, *Nano Letters* 7: 614–619.

Reid, O. G., Munechika, K. & Ginger, D. S. (2008). Space charge limited current measurements on conjugated polymer films using conductive atomic force microscopy, *Nano Letters* 8(6): 1602–1609.

Reith, D., Pütz, M. & Müller-Plathe, F. (2003). Deriving effective meso–scale coarse graining potentials from atomistic simulations, *Journal of Computational Chemistry* 24(13): 1624–1636.

Rispens, M. T., Meetsma, A., Rittberger, R., Brabec, C. J., Sariciftci, N. S. & Hummelen, J. C. (2003). Influence of the solvent on the crystal structure of PCBM and the efficiency of MDMO-PPV:PCBM 'plastic' solar cells, *Chemical Communications* 17: 2116–2118.

Ruhstaller, B., Carter, S. A., Barth, S., Riel, H., Riess, W. & Scott, J. C. (2001). Transient and steady-state behavior of space charges in multilayer organic light-emitting diodes, *Journal of Applied Physics* 89(8): 4575–4586.

Saba, M. I., Melis, C., Colombo, L., Malloci, G. & Mattoni, A. (2011). Polymer crystallinity and transport properties at the poly(3-hexylthiophene)/zinc oxide interface, *J Phys Chem C* 115(19): 9651–9655.

Sandberg, H. G. O., Frey, G. L., Shkunov, M. N., Sirringhaus, H., Friend, R. H., Nielsen, M. M. & Kumpf, C. (2002). Ultrathin regioregular poly(3-hexyl thiophene) field-effect transistors, *Langmuir* 18: 10176–10182.

Sariciftci, N. S., Braun, D., Zhang, C., Srdanov, V. I., Heeger, A. J., Stucky, G. & Wudl, F. (1993). Semiconducting polymer-buckminsterfullerene heterojunctions: Diodes, hotodiodes, and photovoltaic cells, *Applied Physics Letters* 62(6): 585.

Sariciftci, N. S., Smilowitz, L., Heeger, A. J. & Wudl, F. (1992). Photoinduced electron-transfer from a conducting polymer to buckminsterfullerene, *Science* 258(5087): 1474–1476.

Saunders, B. R. & Turner, M. L. (2008). Nanoparticle-polymer photovoltaic cells, *Advances in Colloid and Interface Science* 138(1): 1–23.

Schilinsky, P., Asawapirom, U., Scherf, U., Biele, M. & Brabec, C. J. (2005). Influence of the molecular weight of poly(3-hexylthiophene) on the performance of bulk heterojunction solar cells, *Chemistry of Materials* 17(8): 2175–2180.

Shah, M. & Ganesan, V. (2009). Correlations between morphologies and photovoltaic properties of rodâĹŠcoil block copolymers, *Macromolecules* 43(1): 543–552.

Sievers, D. W., Shrotriya, V. & Yang, Y. (2006). Modeling optical effects and thickness dependent current in polymer bulk-heterojunction solar cells, *J. Appl. Phys.* 100(11): 114509.

Singh, T. B., Sariciftci, N. S., Yang, H., Yang, L., Plochberger, B. & Sitter, H. (2007). Correlation of crystalline and structural properties of c60 thin films grown at various temperature with charge carrier mobility, *Applied Physics Letters* 90: 213512.

Sun, Q. & Faller, R. (2006). Systematic coarse–graining of a polymer blend: Polyisoprene and polystyrene, *Journal of Chemical Theory and Computation* 2(3): 607–615.

Sun, Q., Ghosh, J. & Faller, R. (2008). *Coarse–Graining of Condensed Phase and Biomolecular Systems*, Chapman and Hall/CRC Press, Taylor and Francis Group, Chapter 6. State Point Dependence and Transferability of Potentials in Systematic Structural Coarse–Graining, pp. 69–82.

Sze, S. M. & Ng, K. K. (2006). *Physics of semiconductor devices*, 3rd edn, Wiley–Interscience, Hoboken.

Thompson, B. C. & Frchet, J. M. J. (2008). Polymer–fullerene composite solar cells, *Angewandte Chemie International Edition* 47: 58–77.

Troshin, P. A., Hoppe, H., Renz, J., Egginger, M., Mayorova, J. Y., Goryachev, A. E., Peregudov, A. S., Lyubovskaya, R. N., Gobsch, G., Sariciftci, N. S. & Razumov, V. F. (2009). Material solubility–photovoltaic performance relationship in the design of novel fullerene derivatives for bulk heterojunction solar cells, *Advanced Functional Materials* 19: 779–788.

van Bavel, S. S., Barenklau, M., de With, G., Hoppe, H. & Loos, J. (2010). P3HT/PCBM bulk heterojunction solar cells: Impact of blend composition and 3D morphology on device performance, *Advanced Functional Materials* 20(9): 1458–1463.

van Bavel, S. S., Sourty, E., With, G. d. & Loos, J. (2008). Three–dimensional nanoscale organization of bulk heterojunction polymer solar cells, *Nano Letters* 9(2): 507–513.

van de Craats, A. M., Warman, J. M., Fechtenkötter, A., Brand, J. D., Harbison, M. A. & Müllen, K. (1999). Record charge carrier mobility in a room–temperature discotic liquid–crystalline derivative of hexabenzocoronene, *Advanced Materials* 11(17): 1469–1472.

Vissenberg, M. & Blom, P. W. (1999). Transient hole transport in poly(-p-phenylene vinylene) leds, *Synthetic Metals* 102: 1053–1054.

Vukmirović, N. & Wang, L.-W. (2008). Charge patching method for electronic structure of organic systems, *Journal of Chemical Physics* 128(12): 121102.

Vukmirović, N. & Wang, L.-W. (2009). Charge carrier motion in disordered conjugated polymers: A multiscale ab initio study, *Nano Letters* 9(12): 3996–4000.

Watkins, P. K., Walker, A. B. & Verschoor, G. L. B. (2005). Dynamical Monte Carlo modelling of organic solar cells: The dependence of internal quantum efficiency on morphology, *Nano Lett.* 5(9): 1814–1818.

Westenhoff, S., Beenken, W. J. D., Yartsev, A. & Greenham, N. C. (2006). Conformational disorder of conjugated polymers, *Journal of Chemical Physics* 125(15): 154903.

Westenhoff, S., Daniel, C., Friend, R. H., Silva, C., Sundstrom, V. & Yartsev, A. (2005). Exciton migration in a polythiophene: Probing the spatial and energy domain by line–dipole Forster–type energy transfer, *Journal of Chemical Physics* 122(9): 094903.

Widge, A. S., Matsuoka, Y. & Kurnikova, M. (2007). Computational modeling of poly(alkylthiophene) conductive polymer insertion into phospholipid bilayers, *Langmuir* 23: 10672–10681.

Widge, A. S., Matsuoka, Y. & Kurnikova, M. (2008). Development and initial testing of an empirical forcefield for simulation of poly(alkylthiophenes), *Journal of Molecular Graphics and Modelling* 27: 34–44.

Wong-Ekkabut, J., Baoukina, S., Triampo, W., Tang, I. M., Tieleman, D. P. & Monticelli, L. (2008). Computer simulation study of fullerene translocation through lipid membranes, *Nature Nanotechnology* 3: 363–368.

Xu, H., Chen, D. M. & Creager, W. N. (1993). Double domain solid C_{60} on Si (111) 7 x 7, *Physical Review Letters* 70(12): 1850.

Xu, Z., Chen, L. M., Yang, G. W., Huang, C. H., Hou, J. H., Wu, Y., Li, G., Hsu, C. S. & Yang, Y. (2009). Vertical phase separation in poly(3-hexylthiophene): Fullerene derivative blends and its advantage for inverted structure solar cells, *Advanced Functional Materials* 19(8): 1227–1234.

Yamamoto, T., Komarudin, D., Arai, M., Lee, B.-L., Suganuma, H., Asakawa, N., Inoue, Y., Kubota, K., Sasaki, S., Fukuda, T. & Matsuda, H. (1998). *J. Am. Chem. Soc.* 120: 2047–2058.

Yang, C. Y., Hu, J. G. & Heeger, A. J. (2006). Molecular structure and dynamics at the interfaces within bulk heterojunction materials for solar cells, *Journal of the American Chemical Society* 128(36): 12007–12013.

Yang, X. N., Loos, J., Veenstra, S. C., Verhees, W. J. H., Wienk, M. M., Kroon, J. M., Michels, M. A. J. & Janssen, R. A. J. (2005). Nanoscale morphology of high–performance polymer solar cells, *Nano Letters* 5(4): 579–583.

Yu, G., Gao, J., Hummelen, J. C., Wudl, F. & Heeger, A. J. (1995). Polymer photovoltaic cells: Enhanced efficiencies via a network of internal donor-acceptor heterojunctions, *Science* 270(5243): 1789–1791.

Zen, A., Pflaum, J., Hirschmann, S., Zhuang, W., Jaiser, F., Asawapirom, U., Rabe, J. P., Scherf, U. & Neher, D. (2004). Effect of molecular weight and annealing of poly (3-hexylthiophene)s on the performance of organic field–effect transistors, *Advanced Functional Materials* 14(8): 757–764.

3

Organic Photovoltaics: Technologies and Manufacturing

Yulia Galagan and Ronn Andriessen
Holst Centre / TNO
Netherlands

1. Introduction

It is assumed, that the organic electronics industries and organic solar cells in particular, are in the transition stage towards commercialization. The companies and R&D institutes in this area are moving now from research and development stage to manufacturing. The biggest challenges are how to scale from laboratory to full production, how to select the right tools and processes, and how to use tests and measurements to improve yield and quality. The step from lab-scale to volume production requires adjusting and optimizing of many system aspects, such as: a) deposition techniques and drying conditions, b) substrates, c) ink and solvent systems, d) electrodes and e) dimension of individual cells and modules. Each parameter has its influence on the performance of the final product. In this chapter we discuss a knowledge base concerning the influence of different parameters and process conditions on the performance, cost and lifetime of polymer solar cells.

Deposition techniques

The first step towards mass production is the choice of deposition techniques which will provide high yield and low cost of production. The key property which makes organic photovoltaics so attractive is the potential of roll-to-roll processing on low cost substrates with standard coating and printing processes. Printing or coating techniques like screen, inkjet, offset, gravure, slot die, spray and others are being established and demonstrated for organic photovoltaic (OPV) devices on lab scale. The next step is to transfer the lab scale know-how to industrial roll-to-roll production. Each technique has certain advantages and disadvantages, which makes a given technique more or less attractive in comparison with others. We will provide a short overview of different deposition techniques and evaluate the applicability of them for industrial manufacturing of different functional layers in organic photovoltaic devices. Different process aspects as line speed, stability of the process, capability, robustness and simplicity of it will be taken into account. However, in addition to process aspects, for successful benchmarking of the different deposition techniques, production aspects are equally important. Uniformity of the deposited layer, layer thickness and overall accuracy as well as excluding possible contamination and destroying the underlayers during printing, are also very important parameters which will determine the final choice. Apart from the process and production aspects, the properties of the ink can also have an influence on the final choice. The inks for each functional layer can be formulated taking into account the requirements of the deposition techniques. At the same time the properties of inks such as viscosity, surface tension, solubility, etc. can limit sometimes the usage of certain deposition techniques. The relation between the deposition techniques and the inks will be discussed.

Inks and solvents

One of the main requirements for the industrial production is a safe and environmentally friendly process. It is well known, that currently the best solvents for organic photoactive materials are chlorobenzene or dichlorobenzene. However, these present barriers to mass-production because of health and environmental issues. That is why, one of the major topics for the technology development is searching for alternative solvents which will provide the appropriate blend morphology and hence, high efficiency of the solar cell devices. The nature of the solvent used for the ink formulation of the OPV blend can have a dramatic effect on the ultimate performance of the solar cell as the solvent system is responsible for good wettability of the photoactive blend on the previous deposited layer, drying and phase separation in the photoactive layer. To provide effective donor-acceptor charge transfer and transport in bulk heterojunction solar cell, the photoactive layer needs to have the right morphology, which means the appropriate domain size, crystallinity and vertical distribution of both components. The choice of solvent or solvent mixture, the conditions of the drying process and the annealing temperature are the most critical factors determining the morphology for a given OPV blend system. The influence of different solvents and solvent mixtures on the performance of the photovoltaic devices will be analyzed. Also, the influence of the solvent on drying and morphology on both lab and roll-to-roll scale will be discussed.

Substrates

Substitution of glass substrates by flexible foils introduces several changes that will affect the ultimate performance of the OPV devices. The two most important are: (1) the transparency of the substrate and (2) the conductivity of the thin conductive oxide (TCO), which is deposited on the substrate and used as a transparent electrode. The influence of transparency and flexibility on the device performance will be evaluated. Moreover, the conductivity of TCO on flexible substrates will be discussed.

Electrodes

The strongest motivation for the development of organic photovoltaic (OPV) cell technology is the low cost potential, based on the use of low-cost materials and substrates, the use of non-vacuum and relative low temperature processes (< 120°C) as well as the very high production speeds that can be reached by using roll-to-roll printing and coating techniques. Indium tin oxide (ITO) is one of the main cost consuming elements in present OPV devices due to high cost of indium, the necessity of vacuum deposition and annealing with relative low yields (typically sputtering is used with an average target consumption of about 50 - 60%) and its multi-step patterning. Omitting ITO from the device layout will significantly contribute towards lower production cost. A second argument to omit ITO from OPV devices is mechanical flexibility. The brittle ITO layer can be easily cracked, leading to a decrease in conductivity and, as a result, degradation of the device performance. This fact of course conflicts with the idea of having highly flexible OPV cells. A third argument to omit ITO in OPV applications is its limited conductivity when deposited on plastic foils. With typical sheet resistances of 40-80 Ohm/square, it only allows maximum individual cell widths around 6-10 mm. Larger cell widths will have an immediate negative effect on efficiency. A forth argument to omit ITO in flexible OPV devices relates to the limited annealing temperature that can be used on plastic substrates. This causes the higher sheet conductivities as mentioned before compared to 10–20 Ohm/square on glass. But moreover, these results also in a rougher ITO surface compared to ITO on glass. These unwanted roughness and occurring surface peaks will in turn mortgage the reliable deposition of the thin and homogeneous electro-active layers on top of the ITO. Alternative ITO-free electrodes will be presented and evaluated for use in large-area OPV devices. Moreover, methods of producing these ITO-free electrodes by printing and coating methods will be shown.

Size (scaling up)

The step from lab-scale processing towards industrial manufacturing implies upscaling of single cell sizes and producing modules. What happens when the size of the cell increases; what is the influence of the size on the performance of the OPV devices; what is the maximum size of a cell without substantial efficiency losses; how to minimize efficiency losses during upscaling, and what is the optimum dimension of a module? Answers on these questions will be given in this chapter.
Roll-to-roll production of large area **modules:** For roll-to-roll production of large area solar cell modules, the correlation between all parameters and the final device performance has to be determined. It is important to find a balance and transfer this knowledge to a roll-to-roll process. First of all, inks with the required properties should be formulated, and the deposition method appropriate for this ink should be chosen. The influence of the deposition method and parameters on the layer properties has to be determined. The influence of drying conditions on the layer properties, such as morphology and uniformity, has to be analyzed. Depending on the deposition method, it is important to determine how the patterning of the layers will be applied. The design and the cell architecture should be optimized. The balance between all parameters can guarantee the successful manufacturing of large area OPV modules.

2. Roll-to-roll deposition

The main objectives in the field of Organic Photovoltaics (OPV) are achieving high efficiency, long term stability and low cost. Low cost can be achieved by combining low cost materials with fast roll-to-roll (R2R) manufacturing techniques. Indeed compared with Si-based solar cells, organic solar cells should be less expensive and easier to manufacture, due to the non-vacuum processing, the relatively low temperatures uses during the processing and the possibility of direct patterning during coating or printing. However, in comparison to evaporation processes, solution based processes show quite more challenges in terms of ink formulation, wetting/dewetting, controlling the layer thickness and variations thereof, avoiding intermixing of layers, controlling the self assembly processes, controlling contamination and layer defects and ultimately developing a reliable and robust 24/7 production process.
Benchmarking of deposition techniques for roll-to-roll manufacturing is a complicated process. For the wet chemical deposition of organic semiconductors a large number of deposition techniques can be used. The choice is mainly based on the following factors:

- stability of printing/coating step for the desired materials (viscosity of the ink),
- the capacity of the technique to print/coat the desired feature sizes (lateral resolution, thickness and uniformity),
- throughput of the process.

Deposition technique	Roll-to-roll compatibility	Materials waste	Layer thickness accuracy (nm scale)	Viscosity requirements (mP·s)
Spin coating	-	very high	very good	1-40
Doctor blade coating	-	moderate	good	1-1000
Knife Over Roll Coating	+	moderate	moderate	50-1000
Metering Rod Coating	+	moderate	moderate	50-1000
Slot Die Coating	+	low	very good	1-10,000
Curtain Coating	+	low	moderate	10-500
Pad printing	+/-	low	moderate	10-200
Flexographic printing	+	low	moderate	15-200
Gravure Printing	+	low	good	15-500
Screen Printing	+	low	moderate	50-50,000
Offset printing	+	low	moderate	100-10,000
Ink jet Printing	+	low	good	1-40
Spray coating	+	high	low	1-40
Air Knife Coating	+	moderate	low	10-200
Immersion (Dip) Coating	+/-	low	low	1-200
Brush coating	-	moderate	low	1-50

Table 1. Comparison of different deposition techniques (Brabec et al., 2008; Gamota et al., 2004; Krebs 2009b).

Each layer in an OPV device stack has its own requirements (layer thickness variations, annealing conditions, etc.) but also its own limitations (solubility, solvents, viscosity, surface tension, etc.). This can have an effect on the choice of deposition technique. The choice of the technique should be done individually for the each layer. Although most of the deposition techniques mentioned above have been tested for the deposition of the photoactive layer in organic photovoltaic devices, not all of them are suitable for roll-to-roll mass production. The factors which limit the usage of some deposition techniques are bad compatibility with roll-to-roll processing, high materials waste, low speed, high contamination during the processing, low stability and robustness of the process, bad uniformity of the printed layer and minimum thickness of the wet layer. Sometimes the viscosity of the ink limits the usage of given technique for the deposition of a function layer. Typically the viscosity of the ink for a photoactive layer is very low due to the limited solubility of the photo-active compounds. For P3HT/PCBM (Poly3-hexylthiophene (P3HT) and [6,6]-phenyl-C61-butyric acid methyl ester (PCBM)) solutions it is in the range of 1-5 mP·s. The viscosity of PEDOT:PSS (poly(3,4-ethylenedioxythiophene):poly(4-styrene sulphonate) dispersions is typically in the range of 10-30 mP·s. However, also high viscous pastes of PEDOT/PSS are also commercially available with viscosities higher then 50 mP·s. ZnO solutions or dispersions typically also have very low viscosities down to 1-2 mP·s. These low viscosities already exclude the proper use of techniques like screen printing or offset printing. Silver inks are available both as low viscous inks (for example, ink jet printing) and as pastes, (for example, screen printing). An overview of typical ink viscosity requirements for the different deposition techniques is given in Table 1. The viscosity of the ink is one of the critical factors for the choice of the deposition technology. Viscosity affects the inks flow and how the ink deposits on the substrate. Some techniques require relatively low viscosity, so that each individual printed pixel or cell may merge together to form close layer. However, sometimes, too low viscosity can create problems, because it does not allow forming very sharp edges and to print very small features.

The most commonly used deposition technique for the manufacturing of lab scale devices is spin coating, which is not roll-to-roll compatible. However, spin coating is still widely used for studying and understanding the fundamental principles of OPV. Roll-to-roll compatible methods, especially blade and wire bar coatings, are employed as intermediate steps towards real roll-to-roll processing (Schilinsky et al., 2006), (Chang et al., 2009). Roll-to-roll coating and printing of organic semiconductors become a focus of many companies and research groups. Applications based on organic semiconductors, such as organic light emitting diodes and organic photovoltaic devices, have strict requirements to the layer properties, (e.g., thickness, uniformity and overlay accuracy). Many different printing and coating techniques are applied and developed for the deposition of thin layers of organic semiconductors (Krebs, 2009b). For example, jet printing is a well-studied method for the deposition of PEDOT:PSS (Steirer et al., 2009), (Eom et al., 2009), and polymer-fullerene blends (Aernouts et al., 2008; Hoth et al., 2007; Hoth et al., 2008). The advantage of this method is the possibility to deposit patterned layers in one printing step. The challenge is to find appropriate solvent systems for polymer-fullerene blends, which will provide appropriate spreading and wetting of the ink on the substrate and homogeneous drying with the required morphology of the active layer.

Gravure printing, which is widely used for the printing of organic transistors (Kaihovirta et al., 2008; Voigt et al., 2010), has also been applied for the deposition of OPV layers (Ding et al., 2009; Kopola et al., 2011; Kopola et al., 2010; Voigt et al., 2011). The main difficulty in gravure printing is the required viscosity of the ink, which is for most of the OPV blend systems difficult to reach due to the limited solubility of the components. The somewhat higher needed ink viscosity will also partially hamper the leveling process which is required to achieve high homogeneity of the layers after printing and drying.

There are a number of publications on spray coating of photovoltaic inks (Green et al., 2008; Hoth et al., 2009; Ishikawa et al., 2004; Steirer et al., 2009; Vak et al., 2007; Park et al., 2011; Susanna et al., 2011; Girotto et al., 2011). This deposition method is very efficient especially for low viscosity solutions and is less demanding in terms of ink formulation. However, control on layer homogeneity and the current lack of easy-for-use patterning strategies makes it currently somewhat less attractive for roll-to-roll processes.

Pad printing is a rather unconventional method which has been employed in a roll-to-roll process for the fabrication of OPV (Krebs, 2009c). Screen printing has been applied for the deposition of photoactive layers based on MDMO-PPV:C60-PCBM (Shaheen et al., 2001) and MEH-PPV:C60-PCBM (Krebs et al., 2004; Krebs et al., 2007). As well as a complete process for production of flexible large area polymer solar cells entirely using screen printing has been demonstrated (Krebs et al., 2009b). There are many publications where slot die coating was chosen for the deposition of several layers, including active layer, in polymer based solar cells (Blankenburg et al., 2009; Krebs & Norrman, 2010; Krebs, 2009e; Krebs, 2009a; Krebs, 2009d; Krebs et al., 2009a; Zimmermann et al., 2011). As one of the coating techniques, slot die deposition can provide very thin, uniform, non-patterned layers. One-dimensional patterning is possible by coating stripes which can be used for making OPV modules.

Each deposition technique has advantages and disadvantages. The biggest disadvantage of coating techniques is a requirement of post patterning, which is not a problem for printing methods. Coating techniques can provide very thin uniform layers, but patterning in most of the cases should be done in a separate process step which will add costs to the production process. The typical post-patterned methods are laser ablation, photolithography (Lim et al.,

2009), plasma etching (Colsmann et al., 2009) or solvent etching. Patterning can also be applied in one process step together with coating by self assembled coating based on wetting and dewetting process. However, printing, which typically can provide any feature or structure, can not compete in layer uniformity with coating techniques.

The choice of the deposition technique depends very much on the characteristics of the methods and the criteria which made these methods attractive. The main criteria of selection are based on:

- materials aspects such as viscosity, surface tension of the ink and surface energy of the substrate;
- products aspects, which include uniformity of the layer, layer thickness, possibility of patterning;
- process aspects, such as, roll-to-roll compatibility, speed of the process, stability of the process, capability, robustness and simplicity of the process.

However, selection of a technique at this point of technology development very much depends on how this technique fit with the know-how of organization, their experience and availability of equipment. Collaborations and partnerships very often make some of the deposition techniques more attractive. The market attractiveness also has a lot of influence on the selection of the deposition method. Moreover, the possibility to generate sustainable IP position, publications and innovation sometimes is a leading factor for the selection. The cost of the equipment and total cost of the process are the final criteria of the selection. The satisfaction of the aforementioned is a first step for successful selection, so that technological processes developed with the selected technique can be readily commercialized.

2.1 Roll-to-roll coated PEDOT:PSS and photoactive layers (am example)

The example of successful roll-to-roll coating of the hole transport layer (PEDOT:PSS) and the photoactive layer (P3HT/PCBM) by slot die is illustrated by (Galagan et al., 2011b). The PEDOT:PSS (OrgaconTM) formulation for slot die coating was delivered by Agfa-Gevaert. A thin layer of PEDOT:PSS was deposited with a speed of 10 m/min and dried with the same speed at 110°C. The thickness of the dry layers was about 110 nm and illustrates high uniformity. The layer thickness variation was within ±2%. The thickness and uniformity of the layer was checked by ellipsometry. The results of these measurements are shown in Fig. 1.

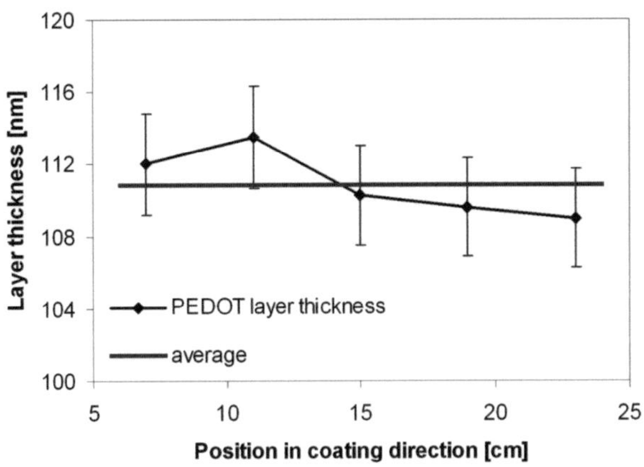

Fig. 1. Layer thickness of roll-to-roll slot die coated PEDOR:PSS layer, measured every 5 cm in coating direction (reproduced with permission from (Galagan et al., 2011b), Copyright 2010, Elsevier B.V.).

The photoactive layer was coated from o-xylene solution. The concentration of the inks was 10 mg/ml. The layer was coated and dried with a speed of 10 m/min. The drying has been performed at 90°C during 30 seconds. Fig. 2 illustrates the roll of the PET foil with slot die coated PEDOT:PSS and OPV layers.

Fig. 2. Roll of the PET foil with slot die coated PEDOT and photoactive layers (reproduced with permission from (Galagan et al., 2011b), Copyright 2010, Elsevier B.V.).

3. Substrates

The substrates for OPV devices must satisfy numerous requirements, such as: optical quality of transparency to let light reach the photoactive layer; substrate smoothness in the nanometer range to provide a surface that will promote high-quality deposition of subsequent layers and prevent the penetration of potential substrate spikes or irregularities into the device layers; the ability to support processing at high enough temperatures; good dimensional stability; good resistance to any chemicals used during processing; low water absorption.

Optical properties

Transparency is one of the main characteristics of substrates. Both transparent and non-transparent substrates are suitable for the manufacturing of organic solar cells. However, non transparent substrates constrain some device architectures. The usage of non-transparent substrates requires transparent top electrodes with transparent top barrier layers. Devices on transparent substrates can have their transparent electrode either on top or on the bottom side of the devices or even on both sides, in case of semi-transparent solar cells. However, in most cases, the transparent substrate is combined with a transparent electrode. Such device architecture requires high optical transparency of the electrode as well as the substrate. Also the barrier (directly attached to the substrate or laminated afterwards) should have high optical transparency. In addition, substrates for OPV devices should have low birefringence.

Surface roughness

Thin film devices are very sensitive to surface roughness. High roughness structures over short distances must be avoided, as it can create shorts in the devices. However, an intermediate roughness over long distance is acceptable. Standard metal substrates usually are rough on both scales, while plastic substrates may be rough only over long distances.

Thermal and thermo-mechanical properties

Some processes of solar cells fabrication such as drying annealing and thermal sintering require applying of high temperatures. It is important that maximum fabrication process temperature must be lower then the glass transition temperature of the substrate polymer.

Dimensional stability

Glass substrates have very high dimensional stability. But conventional processing on glass substrates can not be directly transferred on plastic. The thermal contraction mismatch between the substrates and the deposited device film and built-in stresses in these films lead to curving and changing in the in-plane dimension of the substrates. This change causes misalignment between the device layers. Plastic substrates will also change size on exposure to moisture. The first defense against these changes is the selection of substrates with low water absorption and low coefficients of hydroscopic expansion. Appropriate substrate selection and heat stabilization techniques can significantly reduce the size of changes, but they can not eliminate them completely.

Mechanical properties

A high elastic modulus makes the substrate stiff, and a hard surface support the device layer under impact.

Chemical properties

The substrates should not release contaminants and should be inert against process chemicals.

Barrier properties

Barrier properties of the substrates against permeation of atmospheric gasses such as water and oxygen can be the biggest advantage of the substrates.

Electrical properties

Some substrates, such as metal foil or ITO coated plastic foils are conductive. These conductive substrates may serve as an electrode in solar cell architecture.
The main substrates used for the manufacturing of organic photovoltaic devices are glass, plastic and metal foil. The properties of typical substrate materials are given in Table 2.

Property	Glass	PEN	Stainless steel
Weight, mg/m^2 (for 100-µm-thik film)	220	120	800
Transmission in the visible range (%)	92	90	0
Maximum process temperature (°C)	600	180	>600
TCE (ppm/°C)	5	16	10
Elastic modulus (Gpa)	70	5	200
Permeability for oxygen and moisture	No	Yes	No
Coefficient of hydrolytic expansion (ppm/%RH)	0	11	0
Planarization necessary	No	No	Yes
Electrical conductivity	None	None	High
Thermal conductivity	1	0.1	16

Table 2. Comparison of PEN, stainless steel and glass substrates (Lawrence et al., 2004).

3.1 Metal foil substrates

Metal foil substrates offer the advantages of higher process temperature capability, dimensional stability and excellent barrier properties for oxygen and water. The disadvantages and limitations of the metal foil substrates include non-transparency and surface roughness. However, non-transparent metal foils can be successfully used in some of the device architectures, where illumination of the OPV devices occurs from the top side. The smoothness of the metal substrates can be increased by polishing or by applying additional planarization layers. The planarization layer can be organic, inorganic, or a combination. With high content of organic components, thicker layer of planarization material can be applied without forming cracks and a very smooth surface can be achieved. It is very important for OPV devices, because the electro-active layers are very thin and the chance to have shorts in the devices is much higher with rougher surfaces.

The conductivity of metal substrates can be considered both as an advantage and disadvantage. Sometimes the metal substrate can be served as back contact, but very often the conductive property of the metal substrate is not used. In such case conducting substrates are completely insulated from OPV device by an additional planarizaton layer.

Stainless steel has been most commonly used in research because of its high resistance to corrosion and process chemicals, and its long record of application in amorphous silicon solar cells. Stainless steel substrates can tolerate process temperature as high as 1000°C. These substrates are dimensionally stable and have excellent barrier properties against water and oxygen. Generally, stainless steel substrates are more durable then plastic substrates.

3.2 Flexible plastic substrates

Polymer foil substrates are highly flexible, can be inexpensive and are roll-to-roll compatible. Transparent plastic substrates have the advantage of being compatible with any organic solar cell architecture. The most common plastic substrates for fabrication OPV devices are polyethylene terephthalate (PET), polyethylene naphthalate (PEN), polycarbonate (PC), polyethersulfone (PES) and polyimide (PI). While polyimide (e.g., Kapton® from DuPont) absorbs in the visible region (has yellow colour), which makes it not suitable for all device architectures. However, the advantage of this plastic substrate is its high process temperature capability (>350°C), while other plastic substrates have limited process temperature capabilities, as shown in Table 3.

Property	PET Melinex	PEN Teonex	PC Lexan	PES Sumilite	PI Kapton
Tg, °C	78	121	150	223	410
CTE* (-55 to 85 °C)	15	13	60-70	54	30-60
Transmission (400-700 nm), %	89	87	90	90	yellow
Moisture absorption, %	0.14	0.14	0.4	1.4	1.8
Young's modulus, Gpa	5.3	6.1	1.7	2.2	2.5
Tensile strength, Mpa	225	275	-	83	231
Density, gcm^{-3}	1.4	1.36	1.2	1.37	1.43
Refractive index	1.66	1.5-1.75	1.58	1.66	-
Birefringence, nm	46	-	14	13	-

Table 3. Properties of plastic substrates (Lawrence et al., 2004) (* - coefficient of thermal expansion).

Another disadvantage of plastic substrates is the lack of dimensional stability during processing at elevated temperatures. There are no polymeric substrates which meet the extremely demanding requirements for low moister and water permeability for OPV application. The typical water and oxygen permeation rates of flexible plastic substrates are 1-10 $g/m^2/day$ and 1-10 $cm^3/m^2/day$, respectively, instead of typically required for organic electronic devices 10^{-3}-10^{-6} $g/m^2/day$ and 10^{-3}-10^{-5} $cm^3/m^2/day$. Ideal plastic substrates required additional barrier layer coatings. Barrier coatings can reduce the absorption and permeability of the atmospheric gasses, can raise the resistance to process chemicals, can strengthen the adhesion of the device layers, and can reduce the surface roughness. Despite some of these disadvantages, plastic substrates have a lower cost potential compared to the metal foil substrates. Table 3 summarizes the properties of the most common substrates which can be used in organic solar cells manufacturing.

3.3 Glass substrates

Plastic substrates offer the property of flexibility that is very important for roll-to-roll manufacturing. But at the same time, there are a lot of applications for solar cells where rigid substrate can be used - first of all, in solar windows. In this case a glass substrate makes a logical choice. Glass has also many advantages over plastic substrates. Glass is highly transparent over the total visible spectrum, it has a high homogeneity of the refractive index and it shows a high UV resistance. Compared with plastic substrates, glass has more advanced thermal properties like high temperature stability, high dimensional stability and a low thermal expansion coefficient. Additionally, glass has a high chemical resistance and excellent barrier properties against water and oxygen. Good barrier properties of the glass substrate can probably significantly decrease the cost of the solar cells, as in such case an extra barrier layer will not required. Moreover, the good barrier properties of the glass can improve the lifetime of the solar cells. Due to the high mechanical stability, glass substrates have also advantage in high scratch resistance.

4. Inks and solvents

The most widely studied materials for the photoactive layer for OPV devices are P3HT and PCBM, which are typically blended in a solvent or solvent mixture. The optimal solvents to reach good solubility and stability and ultimately high device efficiencies for this mixture are o-dichlorobenzene and chlorobenzene. However, chlorinated aromatic solvents can not be used in mass-production due to health and environmental reasons. That is why, the first topic for the technology development is searching for alternative lower-toxicity solvents which will provide appropriate morphology and hence, high efficiencies of the solar cell devices. The choice of solvent for OPV blend is one of the major factors which have influence on the performance of the solar cell. Indeed, the solvent is responsible for appropriate wettability of the photoactive blend on the previously deposited layer, drying behaviour and phase separation in the photoactive layer. To provide effective donor-acceptor charge transfer and transport in bulk heterojunction solar cells, the photoactive layer has to demonstrate the right morphology, which means appropriate domain size, crystallinity and vertical distribution of both components. The choice of solvent, drying conditions and annealing temperature and time are the most critical factors determining the final morphology. The effect of morphology can be described by the formation of an effective network between donor and acceptor which creates effective routes for separated charge transport. The phase separation in P3HT/PCBM systems has been well studied in several solvent systems (Baek et al., 2009; Berson et al., 2007; Jang et al., 2009; Janssen et al., 2007; Kawano et al., 2009; Kim et al., 2011; He et al., 2011). The properties of the solvents, such as boiling point, vapor pressure, solubility and surface tension have a considerable impact on the final morphology of the photoactive layer. There are several publications where alternative more environmentally friendly solvents have been used for the deposition of the photo-active layer (Hoth et al., 2007; Hoth et al., 2009; Zhao et al., 2009). The use of solely toluene as a solvent demonstrated to yield a non-preferable morphology for P3HT/PCBM blend (Hoppe & Sariciftci, 2006). Xylene is also a well studied alternative solvent for this system and the solubility of both P3HT and PCBM in o-xylene is better in comparison with toluene. Moreover, the boiling point and viscosity of o-xylene are very close to the values of chlorobenzene (Table 4). For all these reasons we choose o-xylene as a solvent for this study.

Solvent	Vapor pressure [mm Hg]	Boiling point [°C]	Surface tension [dynes cm^{-1}]	Viscosity [mPa s]	Health hazards TLV-TWA [ppm]
Chlorobenzene	11.80 (25°C)	132	33.0	0.80 (20°C)	10
o-Xylene	5.10 (20°C)	144	28.7	0.76 (25°C)	100

Table 4. Solvent properties (Galagan et al. 2011b).

Bulk heterojunction solar cells have been prepared by spin coating of PEDOT and OPV layers on ITO coated glass substrate. A LiF/Al top electrode was vacuum evaporated. The OPV blend consists of a mixture of P3HT (purchased from Merck, Mw = 20050 g/mol) and PCBM (purchased from Solenne) in a ratio of 1:1. The photoactive layer was spin coated from chlorobenzene as comparison and o-xylene solutions. The concentration of tP3HT was 15 mg/ml.

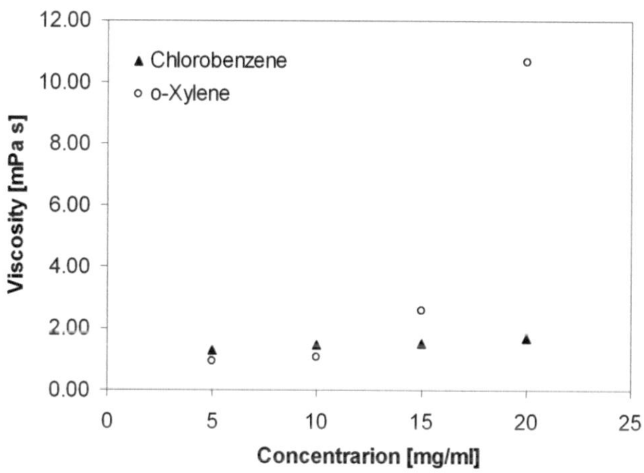

Fig. 3. Viscosity of the P3HT:[C60]PCBM (1:1) mixture in chlorobenzene and o-xylene as function of P3HT concentration (reproduced with permission from (Galagan et al., 2011b), Copyright 2010, Elsevier B.V.).

The OPV ink was processed in ambient conditions. The solutions were stirred for 14 hours at 70°C and subsequently cooled down to room temperature and processed within 30 min.

Upon cooling aggregation and gelation is often observed due to limited solubility of P3HT at room temperature. The time of gelation strongly depends on the molecular weight of the polymer (Koppe et al., 2009) and the solubility of the polymer in a certain solvent. The gelation of the P3HT:[C60]PCBM mixture, indicated by an increase in the viscosity of the mixture is shown in Fig. 3. The viscosity was measured after 30 min when the solution was cooled down. o-Xylene solution of P3HT/PCBM mixture shows up a high tendency for gelation when the concentration of the solution increases. P3HT in o-xylene solution shows a rather fast gel formation at room temperature, causing particles, film defects and the reduction of solution's shelf lifetime (Koppe et al., 2009; Malik et al., 2001).

AFM topographic images of a P3HT/PCBM layer, deposited from o-xylene on top of Pedot coated glass substrates, demonstrate higher roughness in comparison with chlorobenzene as-casted layer, as shown in Fig. 4 (a and c). The observed higher roughness of o-xylene deposited layer corresponds probably with the polymer aggregates which were formed in the solution. UV-Vis absorption spectra of as-casted P3HT/PSBM layers deposited from o-xylene show a blue-shift in comparison with the same layer deposited from chlorobenzene (Fig. 5, a). The shift of the π-π* transition absorption peaks to higher energy indicates an increasing density of conformational defects, equivalent to non-planarity, and causes loss in conjugation (Hotta et al., 1987; Inganäs et al., 1988). The observed red-shift of the absorption maximum in chlorobenzene deposited layers indicates a better crystallinity of P3HT.

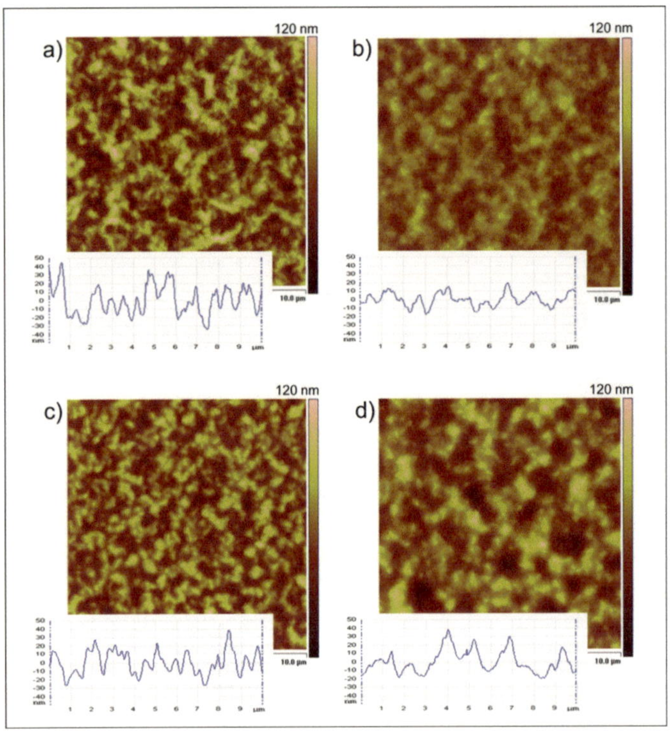

Fig. 4. AFM images of OPV layer deposited from a) o-xylene without annealing, b) o-xylene annealed during 10 min at 110°C, c) chlorobenzene without annealing and d) chlorobenzene annealed during 10 min at 110°C (reproduced with permission from (Galagan et al., 2011b), Copyright 2010, Elsevier B.V.). The measurements were performed with a VEECO Dimension 3100 scanning probe microscope, using Si probes in tapping mode.

Thermal annealing at 110°C during 10 min increases the crystallinity of the deposited photoactive layers both deposited from chlorobenzene or o-xylene. This is indicated in the UV-Vis spectra (Fig. 5, a). The topology of the chlorobenzene deposited layer after annealing shows an increase of roughness that typically also can be attributed to an increase in crystallinity. However, the roughness of the o-xylene deposited layer decreases after annealing. This might be caused by a (partial) destruction of amorphous aggregates.
The solar cells, made from both chlorobenzene and o-xylene, illustrate a clear increase of the performance after thermal annealing (Table 5). The thickness of the photoactive layers has been optimized to 200 nm. J-V curves of the devices after thermal annealing are given in (Fig. 5, b). Current-voltage curves were measured using simulated AM 1.5 global solar irradiation (100 mW/cm²), using a xenon-lamp-based solar simulator Oriel (LS0104) 150W. The lower current in the device produced from o-xylene compared to chlorobenzene after annealing can probably be explained by a difference in the morphology of photoactive layer. Morphology difference is mainly caused due to differences in solvent properties such as boiling point, vapor pressure, viscosity, surface tension and polarity. Another important parameter is solubility which can lead to the formation of aggregates already in the solution. Aggregation and gelation can have a big influence on the final morphology and device performance.

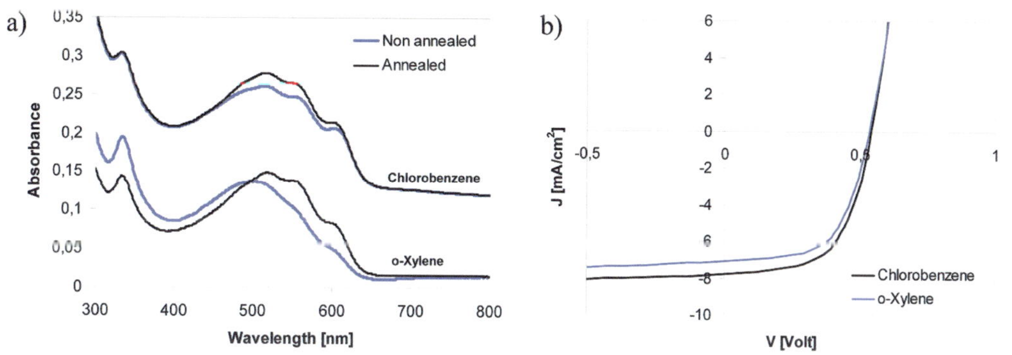

Fig. 5. a - UV-Vis spectra of P3HT/PCBM layers deposited from chlorobenzene and o-xylene before and after thermal annealing. The thicknesses of the layers are 200 nm (reproduced from (Galagan et al., 2011b) with permission); b- J-V curves of OPV devices, deposited from chlorobenzene and o-xylene solutions. The active area of the devices is 0.09 cm² (reproduced with permission from (Galagan et al., 2011b), Copyright 2010, Elsevier B.V.).

Device (solvent, annealing)	Jsc [mA/cm²]	Voc [Volt]	FF	η [%]
Chlorobenzene, non annealed	3.64	0.572	0.296	0.62
Chlorobenzene, annealed	7.75	0.546	0.572	2.42
o-Xylene, non annealed	4.71	0.507	0.332	0.79
o-Xylene, annealed	7.05	0.535	0.581	2.19

Table 5. The characteristics of photovoltaic devices, produced from different solvents, with and without thermal annealing (110°C during 10 minutes) (Galagan et al., 2011b).

5. Scaling up

Scaling up of the active area of the solar cell devices typically leads to efficiency losses (Al-Ibrahim et al., 2004; Gupta et al., 2008; Lungenschmied et al., 2007; Manor et al., 2011). The trend of the losses strongly depends on the sheet resistance of the electrode. Thus, scaling up of the square shaped active areas from 0.09 cm² to 6 cm², results in about 20% efficiency loss, for ITO-based devices prepared on glass substrates (Table 6). The JV curves of the devices are shown on Fig. 6.

Active area, cm² (cm x cm)	Jsc [mA/cm²]	Voc [Volt]	FF	η [%]
0.09 (0.3 x 0.3)	7.75	0.546	0.572	2.42
6 (2 x 3)	7.46	0.530	0.478	1.89

Table 6. The characteristics of photovoltaic devices with different active areas (Galagan et al., 2011b). Devices produced from chlorobenzene solution. Solar cell parameters are corresponding to Fig. 6.

The efficiency of the devices shows a rapid decay upon increasing the width of the solar cells (Lungenschmied et al., 2007). By adding on one side of a cell a charge collector or busbar for the anode and on the opposite side a charge collector or busbar for the cathode only the width of the electrode is relevant for the resulting device efficiency. To keep the resistivity losses in the electrode as small as possible, the width has to be narrow with the contacts taken on the long side. Based on the experimental findings (Al-Ibrahim et al., 2004; Gupta et al., 2008; Lungenschmied et al., 2007) and theoretical modeling (Lungenschmied et al., 2007), it was illustrated that significant loses in the efficiency of the solar cells are expected when the width of the cell is higher than 1 cm. The effective coverage of a large area is only possible by interconnecting several devices.

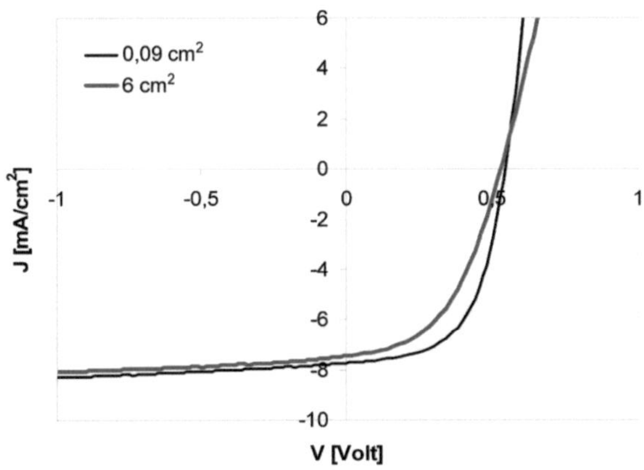

Fig. 6. J-V curves of OPV devices, deposited from chlorobenzene solution with different sizes of active area: 0.09 cm² and 6 cm² (reproduced with permission from (Galagan et al., 2011b), Copyright 2010, Elsevier B.V.).

6. Electrodes

The main attractiveness of organic solar cells is the low cost potential due to roll-to-roll solution based fabrication. However, currently fabrication of the (transparent) electrodes is usually done by vacuum deposition. Combination of printed organic layers and vacuum deposited electrodes will increase the cost of such OPV devices. Thus, to reach the goal of low-cost manufacturing, all layers in OPV devices should be solution processed. There are a lot of publications where all-solution processed OPV devices with inverted structures have been reported (Krebs et al., 2011; Ming-Yi et al., 2010; Hagemann et al., 2008; Hsiao et al., 2009). But in these publications only the top electrode is printed. The bottom electrode is still ITO, which is deposited in vacuum. ITO-free OPV devices with solution processed top electrode has been illustrated in (Manceau et al., 2011; Krebs, 2009e), but in these cases, the bottom electrode is a metal layer deposited in vacuum. All-solution processed polymer solar cells free from ITO and vacuum coating steps have been reported (Krebs, 2009a), where the bottom electrode is a solid layer of printed Ag and ZnO. Printing of a top electrode in a conventional device configuration is less studied, however, successful preparation of solar cells with nano-scale ZnO as a buffer layer and inkjet printed silver cathode have been described (Eom et al., 2008). Roll-to-roll processing of solution processable electrodes will significantly contribute towards low-cost manufacturing of OPV devices.

6.1 ITO electrode

The efficiency of a solar cell depends on the electrode dimensions and its sheet resistance. Typically, the sheet resistance of ITO/glass substrate is 10-15 Ω/□, while the sheet resistance of ITO on PET substrate is around 60 Ω/□. A rapid decay of the efficiency was shown upon increasing the width of the solar cell (Fig. 7, a). Moreover, a much faster decay was observed in case of a foil substrate with a higher sheet resistance. Thus, the efficiency of solar cells on plastic substrate drops down almost two times in comparison with the cell on glass substrate, for the devices with an active area of 2x2 cm². The series resistance induced by the sheet resistance of the electrode can dramatically reduce the charge collection in the device. The J-V curves of the devices on glass and foil substrate are shown in Fig. 7, b.

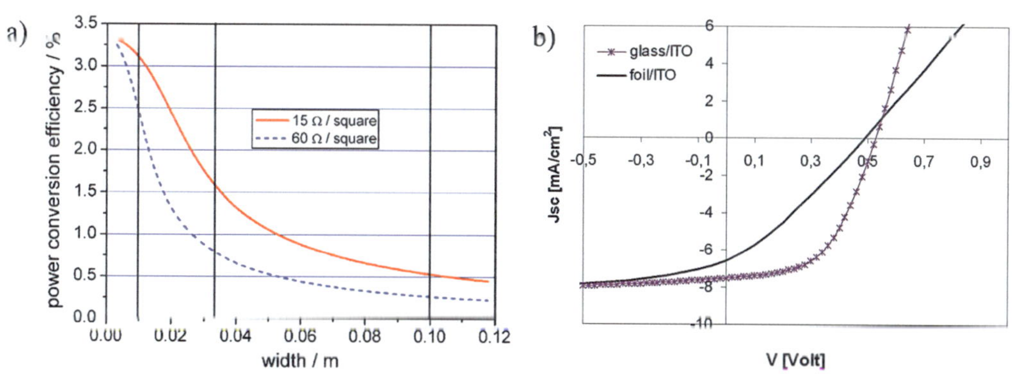

Fig. 7. a - Calculated theoretical power conversion efficiency of a P3HT:PCBM-based single rectangular organic solar cell as a function of the width of the electrode. Two cases are shown, an ITO sheet resistance of 15 Ohm/square (solid line) and 60 Ohm/square (dashed line) (reproduced with permission from (Lungenschmied et al., 2007), Copyright 2006, Elsevier B.V.), b - J-V curves of OPV devices with active area of 4 cm², prepared on glass and foil substrate. The sheet resistance of ITO on glass is 13 Ohm/square, and on foil is 60 Ohm/square (reproduced with permission from (Galagan et al., 2011b), Copyright 2010, Elsevier B.V.).

Within realistic ranges of substrate conductivities (15-60 Ω/□), the overall device efficiency heavily depends on the width of the electrode. To keep the resistivity losses in the electrode as small as possible, the width has to be narrow with the contacts taken on the long side. This minimizes the distance where the charges, extracted from the active layer, have to travel in the resistive ITO electrode.

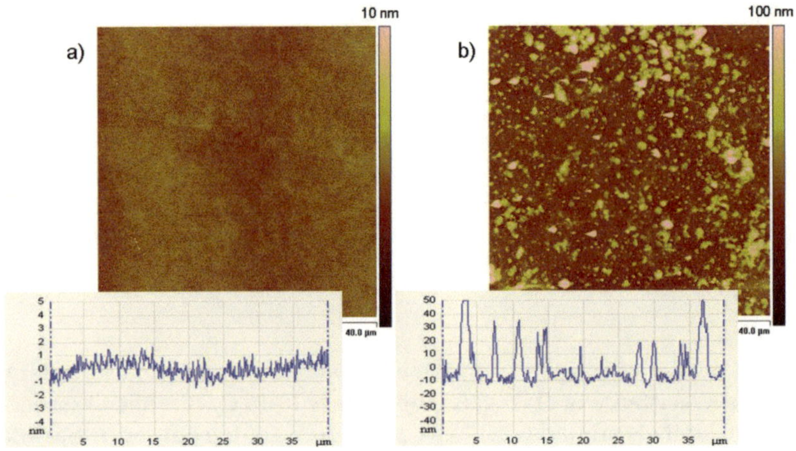

Fig. 8. AFM topology image of (a) ITO coated glass and (b) ITO coated PET foil (reproduced with permission from (Galagan et al., 2011b), Copyright 2010, Elsevier B.V.).

Processing on plastic foils such as PEN and PET typically is limited to temperatures generally below 100°C. Thus, the quality of ITO layer is not optimal. First of all, this results in a higher sheet resistance of ITO on plastic substrates, as is shown above. Additionally, low temperature processing leads to an increased surface roughness compared to ITO annealed at higher temperatures as for ITO on glass. The ITO coated PET foil is typically characterized by the presence of a huge amount of spikes, which create unwanted shorts in the OPV devices. Fig. 8 depicts the AFM topology images of ITO on glass and ITO on PET foil. The current leakage due to shorts is indicated by the lower fill factor and decreased value of short circuit current.

Another disadvantage of TCO layers and ITO in particular, is their brittleness. The ITO anode has a critical bending radius of about 8 mm, up to which the film can stand the load without increasing in resistance (Paetzold et al., 2003). However, the number of bending cycles, even with low bending radius, will affect the resistance of ITO. ITO breaks down rather quickly and shows an increase in resistance, which will negatively affect the performance of the organic solar cells. Furthermore, the limited availability of materials, such as indium, results in a high cost of ITO coatings. In (Krebs et al., 2010) it was estimated that the cost of typical commercial available PET/ITO substrates, used for roll-to-roll manufacturing of organic solar cells, takes more then half of the total material cost required for the manufacturing of OPV modules.

6.2 ITO-free transparent electrode

The biggest motivation for the development of organic solar cell technology is the low cost potential, based on the use of low-cost materials and substrates and the very high potential production speeds that can be reached by roll-to-roll printing and coating techniques

(Medford et al., 2010; Krebs, 2009d; Krebs, 2009e; Krebs, 2009b; Galagan et al., 2011b; Blankenburg et al., 2009). However, indium-tin oxide (ITO), which is commonly used as a transparent electrode, is one of the main cost consuming elements in present photovoltaic devices (Krebs, 2009e; Krebs et al., 2010; Espinosa et al., 2011). The second argument to omit ITO from OPV devices is mechanical flexibility. The brittle ITO layer can be easily cracked, leading to a decrease in conductivity and as a result degradation of the device performance. A third argument is the multi-step patterning of the ITO layer which involves a lot of chemicals. A lot of efforts have been directed on the development of highly conductive polymeric materials such as poly(3,4-ethylenedioxythiophene):poly(4-styrene sulphonate) (PEDOT:PSS). Replacement of ITO by highly conductive PEDOT:PSS has been intensively investigated and reported (Zhou et al., 2009; Winther-Jensen & Krebs, 2006; Hau et al., 2009; Chang et al., 2008; Ahlswede et al., 2008; Jun-Seok, 2011). However, organic photovoltaic devices with only a PEDOT:PSS electrode do not provide high efficiency for large area devices due to the limited conductivity of the PEDOT:PSS, which is typically up to 500 S/m. Increasing of the layer thickness can reduce the sheet resistance, but at the same time this decreases the transparency. Decreasing of the transparency has a negative influence on the short circuit current and results in efficiency losses (Hau et al., 2009). The JV curve of the 2 cm^2 device with the high conductive PEDOT:PSS (HC-Pedot) as electrode is shown in Fig. 9. Such type of device is characterized by a low fill factor due to the limited conductivity of the PEDOT:PSS layer. The fill factor depends on the series resistance (Rs) of solar cells. Here, the high sheet resistance of the electrode may, therefore, reduce the fill factor. At the same time, the Voc does not change. The decreased Jsc can be partly explained by the very high value of the series resistance. Normally, Jsc does not change much with increasing of Rs, and only starts to change with very large values of Rs (Servaites et al., 2010), as in this case.

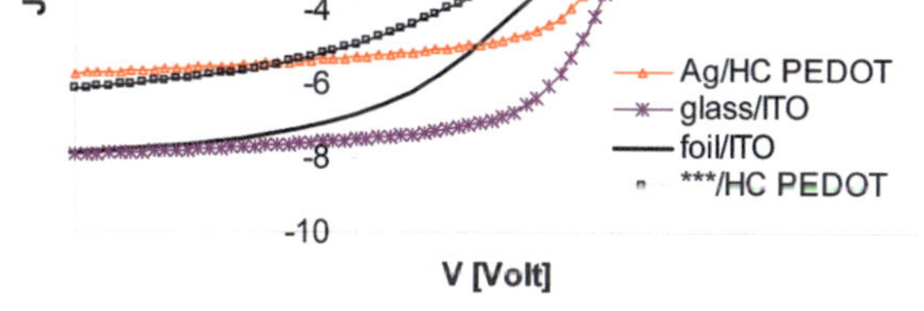

Fig. 9. JV-curves of OPV devices with an active area of 4 cm^2 with different anodes: a) ITO on glass substrate, b) ITO on PEN/barrier substrate, c) high conductive Pedot on glass substrate and d) high conductive Pedot with metal grid on glass substrate.

Improving the conductivity of such polymeric electrode is possible by combining it with a metal grid, which is either thermally evaporated through micro structured shadow masks (Zimmermann et al., 2007; Glatthaar et al., 2005) or patterned by a lithographic method (Zou et al., 2010; Tvingstedt & Inganäs, 2007). However, photolithography is a complex and time consuming multiple step process, which requires expensive facilities. Moreover, the photolithographic process generates large volume of hazardous waste. Vacuum deposition techniques for the fabrication of current collecting grids are not as costly as photolithography, but are energy intensive and give patterns with low resolution. Metal deposition by electroless plating is simpler than photolithography, but it is rather slow, environmentally undesirable as it creates a lot of waste. The deposition of an Ag grid by diffusion transfer reversal, which is a specific photographic technique, has been reported (Aernouts et al., 2004). This technique is fully compatible with large area roll-to-roll processing, yielding metallic conducting Ag-patterns on a plastic support. Another way of making grids is using laser ablation. However, this technique is not the best choice for the preparation of the grids due to low speed of the process, where more than 90 % of the surface has to be removed. Additionally, debris formation will complicate preparation of thin organic solar cells. Printing of the metal grid, however, is a good alterative and a prerequisite for fully printed OPV devices, enabling low-cost manufacturing. Screen printed silver grids (Krebs, 2009d; Krebs et al., 2011) were demonstrated in a roll-to-roll processed inverted OPV device, where the grid is the last printed layer in the devices. Inkjet printed current collecting grids as a part of a composite anode in a conventional OPV device significantly improved efficiency of ITO-free devices with PEDOT based transparent electrodes (Fig. 9). In this case the grid is the first printed layer in the device (Galagan et al., 2010). Integration of a conductive grid significantly decreases the resistance of the polymer anode. Despite the fact that single pass inkjet printed current collection grids demonstrate a sheet resistance of 15Ω/□ for an grid area coverage of 7.2%, which is still a high value for large area devices, but anyhow, it is better than what can be reached with ITO on a flexible substrate and comparable with ITO on glass. The conductivity of the grid can be improved by increasing the line height, but the increase in topology of the grid might make it impossible to overcoat the grid with the subsequent thin electro-active device layers.

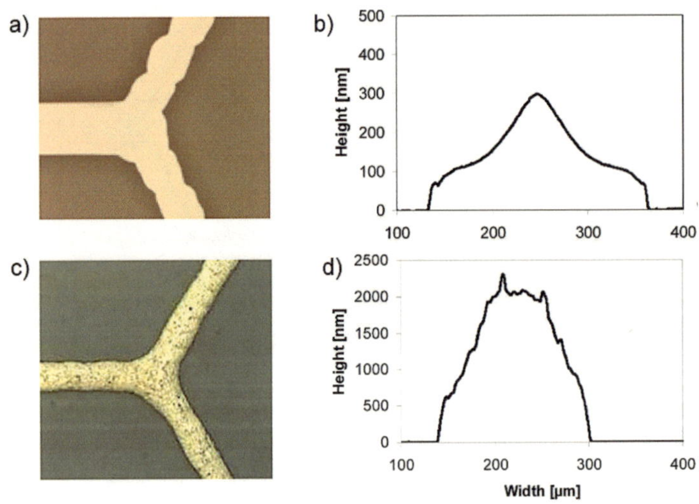

Fig. 10. a) 50x optical microscope image and b) typical line profile of inkjet printed silver lines (grid) on barrier coated PEN film. (c) 50x microscope image and (d) typical line profile of screen printed Inktec TEC-PA-010 ink (c and d reproduced with permission from (Galagan et al., 2011a), Copyright 2010, Elsevier B.V.).

Integration of silver grids into the structure of organic solar cell, however, has some restrictions. Such silver grid covers a part of the surface, making this part non-transparent for light. That is why the surface coverage should be limited; another reason is the high price of Ag. Ink jet printed honeycombs grids with a pitch size of 5 mm and the set value of the line width of 100 μm, should provide a surface coverage of approximately 4%. However, there is observed a mismatch in the theoretical and experimental values. Inkjet printed grids reported by (Galagan et al., 2010) have a mismatch geometry factor of 1.48. First of all, due to the bad pinning of the Ag ink on the receiving substrate (in this case the top layer of the barrier, Si_xN_y) the observed line width is 180 μm instead of 100 μm. This increases the expected surface coverage from 4 % to 7.2%. Also, the measured sheet resistance of such grids is roughly a factor 2.4 higher than predicted. This can be partly explained by the amount of deposited ink which was factor 1.5 lower than expected and partly by the dependency of the line width on the resistivity (the thinner the lines, the higher the resistivity of printed conductive lines sintered at a given condition (time and temperature). The optical microscopic image and typical line profile of inkjet printed silver grid lines on barrier coated PEN film is shown in Fig. 10, a, b.
Further improvement of the device performance is possible with improving the conductivity of the grid, which is possible by applying a larger amount of silver without increasing the area coverage of the grid. With inkjet printing this is possible by using multi-pass printing. Another alternative is screen printing of the grids (Galagan et al., 2011a). Screen printing inks are more viscous, have higher content of solid material and as a result have higher lines profile (Fig. 10, c, d) and better conductivity. The effective line width of the screen printed grid lines is 160 μm (although set at 100 μm) resulting in 6.4% area coverage instead of 4%. The average printed line height was higher than expected (1.5 μm instead of 1 μm. After sintering, the grid showed a sheet resistance of 1 Ω/□, which is in a very agreement with the theoretically expected value. The low value of Rs should be very efficient for larger area OPV devices. However, over-coating of the grid with PEDOT:PSS using spin coating was not successful due to the height of the lines. This problem was solved by embedding the grid in the barrier layer as shown in Fig. 11. In this way, a smooth surface, containing the current collecting grid, was obtained (See Fig. 11 c).

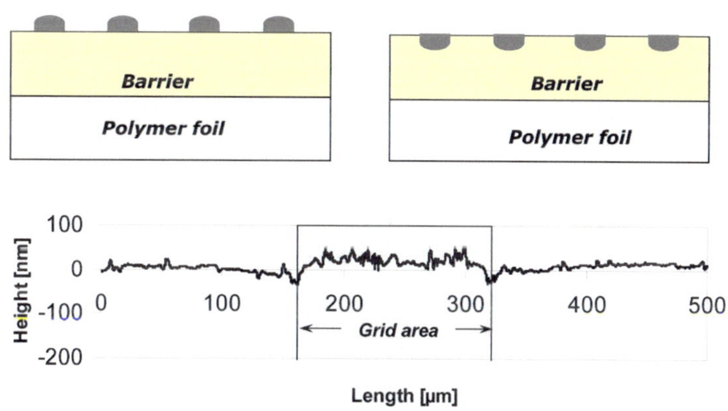

Fig. 11. Schematic illustration of current collection grids: (a) printed on top and (b) embedded into the barrier. (c) Dektak profile of the surface with embedded grid (reproduced with permission from (Galagan et al., 2011a), Copyright 2010, Elsevier B.V.).

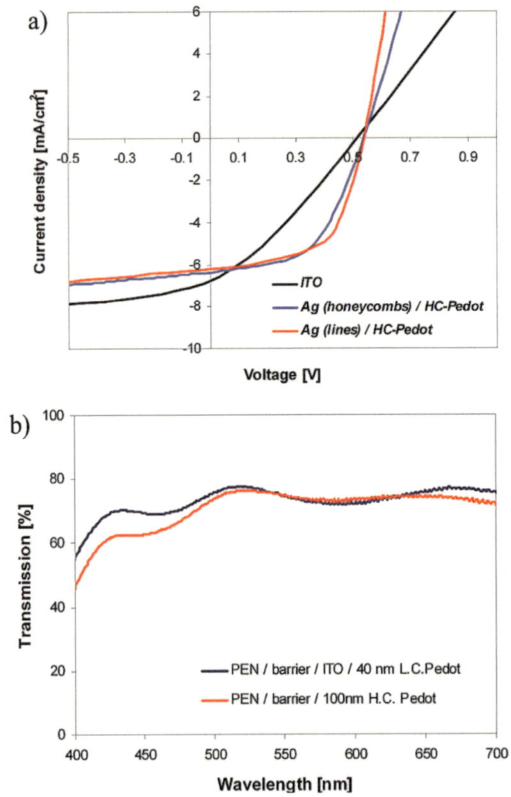

Fig. 12. a - JV-curves of flexible OPV devices with an active area of 4 cm2 with different anodes: (a) ITO on glass substrate, (b) and (c) high conductive PEDOT:PSS with, respectively, a honeycomb and a line pattern current collecting grid (reproduced with permission from (Galagan et al., 2011a), Copyright 2010, Elsevier B.V.); b - Transmittance of the device stacks: PEN/barrier/ITOwith40nmLC-PEDOT:PSS versus PEN/barrier/with 100 nm HC-PEDOT:PSS (reproduced with permission from (Galagan et al., 2011a), Copyright 2010, Elsevier B.V.).

The JV curves of the 2x2 cm2 devices with embedded screen printed current collecting grids are shown in Fig. 12, a. The difference in current density for devices with such grids versus ITO-based devices can be partly explained by shadow losses due to the grids. With the honeycomb pattern, 6.4% of the surface is covered by the grid, while the line pattern covers 8% of the surface. This explains the difference in the J_{sc} between two grid-based devices.

Apart from shadow losses by the grid, the transparency difference between the double layer ITO/low conductive PEDOT and the highly conductive PEDOT needs to be taken into account. The 120 nm layer of ITO on PEN provides an average transmittance of 90% over the visible spectrum (400-700 nm). The subsequent 40 nm layer of low conductive PEDOT also absorbs a part of the solar spectrum. The question now is: will this double layer yield lower or higher transparency compared to the 100 nm thick layer of the highly conductive PEDOT combined with a grid with a given area coverage? The transmission spectra of the stacks PEN/barrier/ITO/40 nm LC PEDOT and PEN/barrier/100 nm HC PEDOT are shown in Fig. 12, b.

The transmission spectra indicate that the transmittance of the light into the photoactive layer is higher in case of the ITO based device. The 100 nm thick layer of high conductive Pedot has a relative high absorption in the visible region. Combined with a grid makes it only worse. This fact can partly explain the difference in the measured current in both types of devices. Also interference effects should be taking into account. The thickness of the active layer (~220 nm), according to optical modeling (Moule et al., 2006; Moulé & Meerholz, 2007) was optimal for the ITO based devices with 40 nm thick PEDOT layer. Such ratio in the layer thicknesses provides a maximum value of Jsc. Omitting of ITO and replacement of low conductive Pedot by high conductive with a different layer thickness could cause the interference induced maximum Jsc to be shifted. Only additional modeling on the optimal thicknesses of the layers can help explaining these results further. Non-optimized layer thicknesses can also contribute to the slightly lower values of Jsc in ITO-free devices. Additionally, as already mentioned above, the shadow effect of the grids results in a lower current density in the ITO-free devices. The sum of all these factors explains why the current density in ITO-based and ITO-free devices is different. Further improvements of the current density in ITO-free devices are possible by decreasing the shadow effect (by minimizing the line width in the grids), by increasing the transparency of the high conductive PEDOT and by further optimization of layer thicknesses.

Solar cells containing the composite anode versus ITO-based devices show almost the same open-circuit voltage (V_{oc}) values, but the fill factors significantly differ. Introduction of a conductive grid with a sheet resistance of 1 Ω/□ into the photovoltaic devices substantially improves the fill factor for 2 x 2 cm^2 devices. As the HC-PEDOT is still responsible for the current collection in the area between the grid lines, the high conductivity of this PEDOT is very important. Moreover, the distance between the grid lines is an important parameter for successful current collection. In (Zimmermann et al., 2007) the optimum pitch size has been calculated in wrap-through OPV devices for PEDOT formulations with different conductivities. In (Galagan et al., 2011a) two different pitch sizes have been tested. The results illustrate that the device with the line pattern, with a pitch size of 2 mm, has a higher fill factor that the honeycomb pattern, which has a pitch size of 5 mm.

As the data show, replacing the ITO by a composite anode consisting of combination of a metal grid and HC-PEDOT, results in a significant increase in efficiency for devices of 2 x 2 cm^2. Future work will concentrate in maximizing the cell area without substantial efficiency losses by using optimized grid structures. This will enable a substantial increase of the active area of OPV modules which in turn will increase the final Wp/m^2.

An additional and very significant argument to support the ITO-free approach with current collecting grids and high conducting PEDOT is the observed improved stability of such devices. In (Galagan et al., 2010) lifetimes of ITO based and ITO-free devices have been compared. A study of the intrinsic stability of the ITO-free devices and a comparison with the "standard" ITO-based devices has been performed on devices with the following layouts: "glass/Ag-grid/HC-Pedot/P3HT:PCBM/LiF/Al/metal encapsulation" and glass/ITO/Pedot/P3HT:PCBM/LiF/Al/metal encapsulation", respectively. Both types of devices were prepared on glass substrates, as glass provides good barrier properties from one side of the devices. The other side of the devices was protected by a stainless steel lid containng a getter material which was sealed by using epoxy glue. Identical sets of devices have been tested under three different conditions: in dark at room temperature, in dark at a temperature of 45°C and illuminated (1.5AM condition) at 45°C. The cells, which were kept in the dark at room temperature, remained stable over time. At an elevated temperature in the dark, the efficiency of ITO-based devices drops. The efficiency of the ITO-free devices did not change substantially. The combination of light and elevated temperature shows a very rapid degradation of the ITO-based devices. At the same time the efficiency of the ITO-free devices remains almost unchanged. The relative changes of the V_{oc}, J_{sc}, FF and efficiency of the devices stored at 45°C under the light, are given in Fig. 13.

Light and temperature are the two main factors which speed up the degradation of the solar cells. In the described experiment encapsulated devices were used. Due to the presence of a getter material and the quality of the sealant, leakage of water and oxygen from the environment does not play a significant role during the test period. All observed changes in the device performance are related to the intrinsic stability of the devices. It is clear that devices containing ITO degrade much faster than ITO-free devices. This fast degradation of ITO-based devices is explained by the indium diffusion through all layers in such devices (Jørgensen et al., 2008; Krebs & Norrman, 2007). Indium migration results in conductivity losses in the ITO anode. These conductivity losses lead to an increase of the series resistances in the devices, and as a result the fill factor decreases. But it should be stated that for some ITO/PEDOT:PSS combinations, this degradation is not observed.

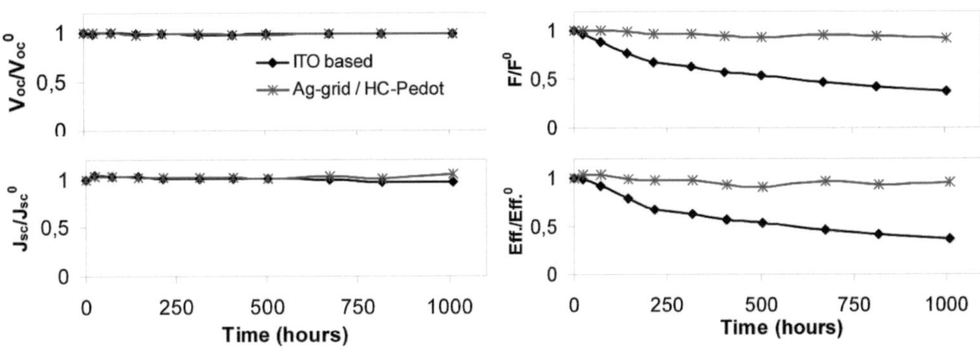

Fig. 13. Relative changes of the V_{oc}, J_{sc}, FF and efficiency of device over time, stored at 45°C with light exposure.

This work demonstrates the possibility of excluding the expensive and brittle ITO electrode from organic photovoltaic devices and replace it by a composite anode containing a combination of a printed metal grid and a printed high conducting PEDOT:PSS layer. This should enable complete roll-to-roll printing of solar cells on large flexible plastic substrates. Devices with a composite electrode illustrate a reasonable efficiency. Integration of conducting grids significantly decreases the resistance of the anode. Moreover, the observed intrinsic stability of studied ITO-free devices is much higher than the stability of studied devices based on ITO. After 1000 hours at elevated temperature and light exposure, the encapsulated ITO-free devices maintained more than 90% of their initial efficiency, whereas the efficiency of devices with an ITO electrode was reduced to 40% of their initial efficiency. In the end, taking into account the cost, compatibility and simplicity of roll-to-roll production of the patterned electrode, the devices with silver grids and highly conductive PEDOT:PSS are evidently preferred.

7. Module design and size

A standard single organic solar cell has an open circuit voltage below 1 V. In solar cells based on P3HT/PCBM blend this value is typical around 0.5 V. The voltage at maximum power point is lower, even at high illumination level. The value of current strongly depends on the light intensity and the size of active area of the device. Scaling up of the active area can increase the actual output current, but the voltage delivered by photovoltaic device will remain unchanged. For electrical powering of electronic tools and devices, very often a much higher voltage is required. This can be achieved by serial interconnection of several single cells for delivering a higher voltage. The interconnection can be performed by external wiring, which is commonly used in wafer-based Si solar cells. However, application coating and printing techniques for manufacturing of solar cells creates the possibility of direct patterning of the layers. Patterned printing opens new possibilities of manufacturing solar sell modules with internal interconnection (Fig. 14). Printing of complete modules provides a significant potential to decrease the manufacturing cost and to increase the stability of the solar cell modules.

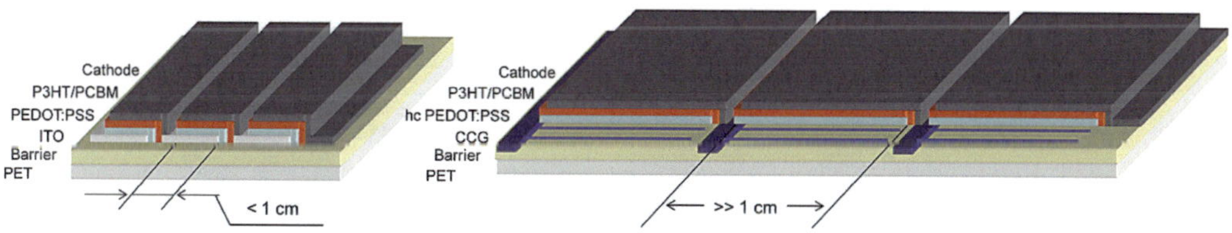

Fig. 14. a - Schematic illustration of the organic solar cell modules with ITO electrode and b - current collecting grids/high conducting PEDOT electrode.

As explained before, due to the relatively high sheet resistivity of ITO, the width of individual efficient cells typically is limited to 0.5-1 cm. By this way the Ohmic losses are reduced. By serial interconnection of individual cells the total active area coverage of the module will be rather low as, certainly with printing techniques, the non-active interconnections will be in the same range, i.e. between 0.5–1 cm. The larger the ratio between active cell area and interconnection area, the lower the losses are. Hence, for reducing the losses, minimizing of the interconnection area and increasing the active area of single cell are the focus points. The concept with the current collecting grids can provide cell widths much larger then 1 cm without a substantial decrease in efficiency. In such a way, the active area of the module can be significantly improved (Fig. 14, b). This is one more advantage of using a composite electrode with current collecting grids in stead of ITO or other TCO's.

8. Conclusions

Commercialization of organic photovoltaics is complicated as many parameters change when moving from lab scale manufacturing of devices towards low-cost, roll-to-roll manufacturing. . The main factors which have influence on the efficiency, stability, and cost of the final product of the devices are: solvent, layer thickness, deposition technique, drying time, thermal annealing, substrate, electrode roughness and sheet resistance, size, design and architecture of the solar cells. Omitting expensive materials, such as ITO, and the development of a technology that allows for solution processed (composite) transparent electrodes seems to be a large step towards low cost manufacturing. ITO-free electrodes not only significantly decrease the cost but also provide better scalability and higher stability of the organic photovoltaic devices. Selection of the right deposition technique for each electro-active layer can enable successful low-cost manufacturing of complete OPV modules. We built a knowledge base concerning the influence of different parameters and process conditions on the performance, cost and lifetime of polymer solar cells. All these parameters can be used as an input for further development of roll-to-roll manufacturing.

Fig. 15. ITO-free organic photovoltaic devices produced in Holst Centre (15x15 cm^2, active area is 12x12 cm^2).

9. References

Aernouts, T., Aleksandrov, T., Girotto, C., Genoe, J., & Poortmans, J. 2008. Polymer based organic solar cells using ink-jet printed active layers. *Applied Physics Letters*, 92(3): 033306-3.

Aernouts, T., Vanlaeke, P., Geens, W., Poortmans, J., Heremans, P., Borghs, S., Mertens, R., Andriessen, R., & Leenders, L. 2004. Printable anodes for flexible organic solar cell modules. *Thin Solid Films*, 451-452: 22-25.

Ahlswede, E., Muhleisen, W., Wahi, M. W. b. M., Hanisch, J., & Powalla, M. 2008. Highly efficient organic solar cells with printable low-cost transparent contacts. *Applied Physics Letters*, 92(14): 143307.

Al-Ibrahim, M., Roth, H. K., & Sensfuss, S. 2004. Efficient large-area polymer solar cells on flexible substrates. *Applied Physics Letters*, 85(9): 1481-1483.

Baek, W. H., Yang, H., Yoon, T. S., Kang, C. J., Lee, H. H., & Kim, Y. S. 2009. Effect of P3HT:PCBM concentration in solvent on performances of organic solar cells. *Solar Energy Materials and Solar Cells*, 93(8): 1263-1267.

Berson, S., De Bettignies, R., Bailly, S., & Guillerez, S. 2007. Poly(3-hexylthiophene) Fibers for Photovoltaic Applications. *Advanced Functional Materials*, 17(8): 1377-1384.

Blankenburg, L., Schultheis, K., Schache, H., Sensfuss, S., & Schrödner, M. 2009. Reel-to-reel wet coating as an efficient up-scaling technique for the production of bulk-heterojunction polymer solar cells. *Solar Energy Materials and Solar Cells*, 93(4): 476-483.

Brabec, C., Scherf, U., Dyakonov, V., & (Ed(s).) 2008. *Organic Photovoltaics: Materials, Device Physics, and Manufacturing Technologies*, Wiley-VCH, ISBN 978-3527316755.

Chang, Y. M., Wang, L., & Su, W. F. 2008. Polymer solar cells with poly(3,4-ethylenedioxythiophene) as transparent anode. *Organic Electronics*, 9(6): 968-973.

Chang, Y. H., Tseng, S. R., Chen, C. Y., Meng, H. F., Chen, E. C., Horng, S. F., & Hsu, C. S. 2009. Polymer solar cell by blade coating. *Organic Electronics*, 10(5): 741-746.

Colsmann, A., Stenzel, F., Balthasar, G., Do, H., & Lemmer, U. 2009. Plasma patterning of Poly(3,4-ethylenedioxythiophene):Poly(styrenesulfonate) anodes for efficient polymer solar cells. *Thin Solid Films*, 517(5): 1750-1752.

Ding, J. M., de la Fuente Vornbrock, A., Ting, C., & Subramanian, V. 2009. Patternable polymer bulk heterojunction photovoltaic cells on plastic by rotogravure printing. *Solar Energy Materials and Solar Cells*, 93(4): 459-464.

Eom, S. H., Senthilarasu, S., Uthirakumar, P., Hong, C. H., Lee, Y. S., Lim, J., Yoon, S. C., Lee, C., & Lee, S. H. 2008. Preparation and characterization of nano-scale ZnO as a buffer layer for inkjet printing of silver cathode in polymer solar cells. *Solar Energy Materials and Solar Cells*, 92(5): 564-570.

Eom, S. H., Senthilarasu, S., Uthirakumar, P., Yoon, S. C., Lim, J., Lee, C., Lim, H. S., Lee, J., & Lee, S. H. 2009. Polymer solar cells based on inkjet-printed PEDOT:PSS layer. *Organic Electronics*, 10(3): 536-542.

Espinosa, N., García -Valverde, R., Urbina, A., & Krebs, F. C. 2011. A life cycle analysis of polymer solar cell modules prepared using roll-to-roll methods under ambient conditions. *Solar Energy Materials and Solar Cells*, 95(5): 1293-1302.

Galagan, Y., Rubingh, J. E., Andriessen, R., Fan, C. C., Blom, P., Veenstra, S. C., & Kroon, J. M. 2011. ITO-free flexible organic solar cells with printed current collecting grids. *Solar Energy Materials and Solar Cells*, 95(5): 1339-1343.

Galagan, Y., Andriessen, R., Rubingh, E., Grossiord, N., Blom, P., Veenstra, S., Verhees, W., & Kroon, J. 2010. Toward fully printed Organic Photovoltaics: Processing and Stability. *Proceedings of Lope-C*, 88-91, ISBN 978-3-00-029955-1.

Galagan, Y., de Vries, I. G., Langen, A. P., Andriessen, R., Verhees, W. J. H., Veenstra, S. C., & Kroon, J. M. 2011b. Technology development for roll-to-roll production of organic photovoltaics. *Chemical Engineering and Processing: Process Intensification*, 50(5-6): 454-461.

Gamota, D. R., Brazis, P., Kalyanasundaram, K., & Zhang, J. (Es(s.)) 2004. *Printed Organic and Molecular Electronics*, Springer, ISBN 978-1402077074.

Girotto, C., Moia, D., Rand, B. P., & Heremans, P. 2011. High-performance organic solar cells with spray-coated hole-transport and active layers. *Advanced Functional Materials*, 21(1): 64-72.

Glatthaar, M., Niggemann, M., Zimmermann, B., Lewer, P., Riede, M., Hinsch, A., & Luther, J. 2005. Organic solar cells using inverted layer sequence. *Thin Solid Films*, 491(1-2): 298-300.

Green, R., Morfa, A., Ferguson, A. J., Kopidakis, N., Rumbles, G., & Shaheen, S. E. 2008. Performance of bulk heterojunction photovoltaic devices prepared by airbrush spray deposition. *Applied Physics Letters*, 92(3): 033301-033303.

Gupta, D., Bag, M., & Narayan, K. S. 2008. Area dependent efficiency of organic solar cells. *Applied Physics Letters*, 93(16): 163301-163303.

Hagemann, O., Bjerring, M., Nielsen, N. C., & Krebs, F. C. 2008. All solution processed tandem polymer solar cells based on thermocleavable materials. *Solar Energy Materials and Solar Cells*, 92(11): 1327-1335.

Hau, S. K., Yip, H. L., Zou, J., & Jen, A. K. Y. 2009. Indium tin oxide-free semi-transparent inverted polymer solar cells using conducting polymer as both bottom and top electrodes. *Organic Electronics*, 10(7): 1401-1407.

He, C., Germack, D. S., Joseph Kline, R., Delongchamp, D. M., Fischer, D. A., Snyder, C. R., Toney, M. F., Kushmerick, J. G., & Richter, L. J. 2011. Influence of substrate on crystallization in polythiophene/fullerene blends. *Solar Energy Materials and Solar Cells*, 95(5): 1375-1381.

Hoppe, H. & Sariciftci, N. S. 2006. Morphology of polymer/fullerene bulk heterojunction solar cells. *Journal of Materials Chemistry*, 16(1): 45-61.

Hoth, C. N., Choulis, S. A., Schilinsky, P., & Brabec, C. J. 2007. High Photovoltaic Performance of Inkjet Printed Polymer:Fullerene Blends. *Advanced Materials*, 19(22): 3973-3978.

Hoth, C. N., Schilinsky, P., Choulis, S. A., & Brabec, C. J. 2008. Printing Highly Efficient Organic Solar Cells. *Nano Letters*, 8(9): 2806-2813.

Hoth, C. N., Steim, R., Schilinsky, P., Choulis, S. A., Tedde, S. F., Hayden, O., & Brabec, C. J. 2009. Topographical and morphological aspects of spray coated organic photovoltaics. *Organic Electronics*, 10(4): 587-593.

Hotta, S., Rughooputh, S. D. D. V., & Heeger, A. J. 1987. Conducting polymer composites of soluble polythiophenes in polystyrene. *Synthetic Metals*, 22(1): 79-87.

Hsiao, Y. S., Chen, C. P., Chao, C. H., & Whang, W. T. 2009. All-solution-processed inverted polymer solar cells on granular surface-nickelized polyimide. *Organic Electronics*, 10(4): 551-561.

Inganäs, O., Salaneck, W. R., Österholm, J. E., & Laakso, J. 1988. Thermochromic and solvatochromic effects in poly(3-hexylthiophene). *Synthetic Metals*, 22(4): 395-406.

Ishikawa, T., Nakamura, M., Fujita, K., & Tsutsui, T. 2004. Preparation of organic bulk heterojunction photovoltaic cells by evaporative spray deposition from ultradilute solution. *Applied Physics Letters*, 84(13): 2424-2426.

Jang, J., Seok-Soon, K., Seok-In, N., Byung-Kwan, Y., & Dong-Yu, K. 2009. Time-Dependent Morphology Evolution by Annealing Processes on Polymer:Fullerene Blend Solar Cells. *Advanced Functional Materials*, 19(6): 866-874.

Janssen, G., Aguirre, A., Goovaerts, E., Vanlaeke, P., Poortmans, J., & Manca, J. 2007. Optimization of morphology of P3HT/PCBM films for organic solar cells: effects of thermal treatments and spin coating solvents. *Eur.Phys.J.Appl.Phys.*, 37(3): 287-290.

Jørgensen, M., Norrman, K., & Krebs, F. C. 2008. Stability/degradation of polymer solar cells. *Solar Energy Materials and Solar Cells*, 92(7): 686-714.

Jun-Seok, Y. 2011. Variations of cell performance in ITO-free organic solar cells with increasing cell areas. *Semiconductor Science and Technology*, 26(3): 034010.

Kaihovirta, N. J., Tobjörk, D., Mäkelä, T., & Österbacka, R. 2008. Low-Voltage Organic Transistors Fabricated Using Reverse Gravure Coating on Prepatterned Substrates. *Advanced Engineering Materials*, 10(7): 640-643.

Kawano, K., Sakai, J., Yahiro, M., & Adachi, C. 2009. Effect of solvent on fabrication of active layers in organic solar cells based on poly(3-hexylthiophene) and fullerene derivatives. *Solar Energy Materials and Solar Cells*, 93(4): 514-518.

Kim, S. O., Chung, D. S., Cha, H., Hwang, M. C., Park, J. W., Kim, Y. H., Park, C. E., & Kwon, S. K. 2011. Efficient polymer solar cells based on dialkoxynaphthalene and benzo[c][1,2,5]thiadiazole: A new approach for simple donor-acceptor pair. *Solar Energy Materials and Solar Cells*, 95(7): 1678-1685.

Kopola, P., Aernouts, T., Guillerez, S., Jin, H., Tuomikoski, M., Maaninen, A., & Hast, J. 2010. High efficient plastic solar cells fabricated with a high-throughput gravure printing method. *Solar Energy Materials and Solar Cells*, 94(10): 1673-1680.

Kopola, P., Aernouts, T., Sliz, R., Guillerez, S., Ylikunnari, M., Cheyns, D., VSlimSki, M., Tuomikoski, M., Hast, J., Jabbour, G., MyllylS, R., & Maaninen, A. 2011. Gravure printed flexible organic photovoltaic modules. *Solar Energy Materials and Solar Cells*, 95(5): 1344-1347.

Koppe, M., Brabec, C. J., Heiml, S., Schausberger, A., Duffy, W., Heeney, M., & McCulloch, I. 2009. Influence of Molecular Weight Distribution on the Gelation of P3HT and Its Impact on the Photovoltaic Performance. *Macromolecules*, 42(13): 4661-4666.

Krebs, F. C., Tromholt, T., & Jørgensen, M. 2010. Upscaling of polymer solar cell fabrication using full roll-to-roll processing. *Nanoscale*, 2(6): 873-886.

Krebs, F. C. 2009b. Fabrication and processing of polymer solar cells: A review of printing and coating techniques. *Solar Energy Materials and Solar Cells*, 93(4): 394-412.

Krebs, F. C. 2009c. Pad printing as a film forming technique for polymer solar cells. *Solar Energy Materials and Solar Cells*, 93(4): 484-490.

Krebs, F. C. 2009a. All solution roll-to-roll processed polymer solar cells free from indium-tin-oxide and vacuum coating steps. *Organic Electronics*, 10(5): 761-768.

Krebs, F. C. 2009d. Polymer solar cell modules prepared using roll-to-roll methods: Knife-over-edge coating, slot-die coating and screen printing. *Solar Energy Materials and Solar Cells*, 93(4): 465-475.

Krebs, F. C. 2009e. Roll-to-roll fabrication of monolithic large-area polymer solar cells free from indium-tin-oxide. *Solar Energy Materials and Solar Cells*, 93(9): 1636-1641.

Krebs, F. C., Alstrup, J., Spanggaard , H., Larsen, K., & Kold, E. 2004. Production of large-area polymer solar cells by industrial silk screen printing, lifetime considerations and lamination with polyethyleneterephthalate. *Solar Energy Materials and Solar Cells*, 83(2-3): 293-300.

Krebs, F. C., Gevorgyan, S. A., & Alstrup, J. 2009. A roll-to-roll process to flexible polymer solar cells: model studies, manufacture and operational stability studies. *Journal of Materials Chemistry*, 19(30): 5442-5451.

Krebs, F. C., Jørgensen, M., Norrman, K., Hagemann, O., Alstrup, J., Nielsen, T. D., Fyenbo, J., Larsen, K., & Kristensen, J. 2009. A complete process for production of flexible large area polymer solar cells entirely using screen printing--First public demonstration. *Solar Energy Materials and Solar Cells*, 93(4): 422-441.

Krebs, F. C. & Norrman, K. 2010. Using Light-Induced Thermocleavage in a Roll-to-Roll Process for Polymer Solar Cells. *ACS Applied Materials & Interfaces*, 2(3): 877-887.

Krebs, F. C. & Norrman, K. 2007. Analysis of the failure mechanism for a stable organic photovoltaic during 10 000 h of testing. *Progress in Photovoltaics: Research and Applications*, 15(8): 697-712.

Krebs, F. C., Spanggard, H., Kjær, T., Biancardo, M., & Alstrup, J. 2007. Large area plastic solar cell modules. *Materials Science and Engineering*: B, 138(2): 106-111.

Krebs, F. C., Søndergaard, R., & Jørgensen, M. 2011. Printed metal back electrodes for R2R fabricated polymer solar cells studied using the LBIC technique. *Solar Energy Materials and Solar Cells*, 95(5): 1348-1353.

Lawrence, D., Kohler, J., Brollier, B., Claypole, T., & Burgin, T. 2004. Manufacturing platforms for Printing Organic Circuit. *In Printed organic and molecular electronics*, Gamota D. R., Brazis P., Kalyanasundaram K. & Zhang J. (Eds.), 161-346. Kluwer Academic Publishers, ISBN 978-1402077074.

Lim, Y. F., Lee, J. K., Zakhidov, A. A., DeFranco, J. A., Fong, H. H., Taylor, P. G., Ober, C. K., & Malliaras, G. G. 2009. High voltage polymer solar cell patterned with photolithography. *Journal of Materials Chemistry*, 19(30): 5394-5397.

Lungenschmied, C., Dennler, G., Neugebauer, H., Sariciftci, S. N., Glatthaar, M., Meyer, T., & Meyer, A. 2007. Flexible, long-lived, large-area, organic solar cells. *Solar Energy Materials and Solar Cells*, 91(5): 379-384.

Malik, S., Jana, T., & Nandi, A. K. 2001. Thermoreversible Gelation of Regioregular Poly(3-hexylthiophene) in Xylene. *Macromolecules*, 34(2): 275-282.

Manceau, M., Angmo, D., Jørgensen, M., & Krebs, F. C. 2011. ITO-free flexible polymer solar cells: From small model devices to roll-to-roll processed large modules. *Organic Electronics*, 12(4): 566-574.

Manor, A., Katz, E. A., Tromholt, T., Hirsch, B., & Krebs, F. C. 2011. Origin of size effect on efficiency of organic photovoltaics. *Journal of Applied Physics*, 109(7).

Medford, A. J., Lilliedal, M. R., Jørgensen, M., Aarø, D., Pakalski, H., Fyenbo, J., & Krebs, F. C. 2010. Grid-connected polymer solar panels: initial considerations of cost, lifetime, and practicality. *Opt.Express*, 18(S3): A272-A285.

Ming-Yi, Lin, Chun-Yu, Lee, Shu-Chia, Shiu, Jen-Yu, Sun, Yu-Hong, Lin, Wen-Hau, Wu, and Ching-Fuh, Lin. (2010). All-solution-processed-inverted polymer solar cells on PET substrates with CuOx thin film as an anode interlayer. *Proceedings of Photovoltaic Specialists Conference (PVSC), 35th IEEE.*, 001648-001649. ISBN/ISSN 0160-8371. June 2010.

Moulé, A. J. & Meerholz, K. 2007. Minimizing optical losses in bulk heterojunction polymer solar cells. *Applied Physics B: Lasers and Optics*, 86(4): 721-727.

Moule, A. J., Bonekamp, J. B., & Meerholz, K. 2006. The effect of active layer thickness and composition on the performance of bulk-heterojunction solar cells. *Journal of Applied Physics*, 100(9): 094503-094507.

Paetzold, R., Heuser, K., Henseler, D., Roeger, S., Wittmann, G., & Winnacker, A. 2003. Performance of flexible polymeric light-emitting diodes under bending conditions. *Applied Physics Letters*, 82(19): 3342-3344.

Park, S. Y., Kang, Y. J., Lee, S., Kim, D. G., Kim, J. K., Kim, J. H., & Kang, J. W. 2011. Spray-coated organic solar cells with large-area of 12.25 cm^2. *Solar Energy Materials and Solar Cells*, 95(3): 852-855.

Schilinsky, P., Waldauf, C., & Brabec, C. J. 2006. Performance Analysis of Printed Bulk Heterojunction Solar Cells. *Advanced Functional Materials*, 16(13): 1669-1672.

Servaites, J. D., Yeganeh, S., Marks, T. J., & Ratner, M. A. 2010. Efficiency Enhancement in Organic Photovoltaic Cells: Consequences of Optimizing Series Resistance. *Advanced Functional Materials*, 20(1): 97-104.

Shaheen, S. E., Radspinner, R., Peyghambarian, N., & Jabbour, G. E. 2001. Fabrication of bulk heterojunction plastic solar cells by screen printing. *Applied Physics Letters*, 79(18): 2996-2998.

Steirer, K. X., Berry, J. J., Reese, M. O., van Hest, M. F. A. M., Miedaner, A., Liberatore, M. W., Collins, R. T., & Ginley, D. S. 2009. Ultrasonically sprayed and inkjet printed thin film electrodes for organic solar cells. *Thin Solid Films*, 517(8): 2781-2786.

Susanna, G., Salamandra, L., Brown, T. M., Di Carlo, A., Brunetti, F., & Reale, A. 2011. Airbrush spray-coating of polymer bulk-heterojunction solar cells. *Solar Energy Materials and Solar Cells*, 95(7): 1775-1778.

Tvingstedt, K. & Inganäs, O. 2007. Electrode Grids for ITO Free Organic Photovoltaic Devices. *Advanced Materials*, 19(19): 2893-2897.

Vak, D., Kim, S. S., Jo, J., Oh, S. H., Na, S. I., Kim, J., & Kim, D. Y. 2007. Fabrication of organic bulk heterojunction solar cells by a spray deposition method for low-cost power generation. *Applied Physics Letters*, 91(8): 081102-081103.

Voigt, M. M., Guite, A., Chung, D. Y., Khan, R. U. A., Campbell, A. J., Bradley, D. D. C., Meng, F., Steinke, J. H. G., Tierney, S., McCulloch, I., Penxten, H., Lutsen, L., Douheret, O., Manca, J., Brokmann, U., Sönnichsen, K., Hülsenberg, D., Bock, W., Barron, C., Blanckaert, N., Springer, S., Grupp, J., & Mosley, A. 2010. Polymer Field-Effect Transistors Fabricated by the Sequential Gravure Printing of Polythiophene, Two Insulator Layers, and a Metal Ink Gate. *Advanced Functional Materials*, 20(2): 239-246.

Voigt, M. M., Mackenzie, R. C. I., Yau, C. P., Atienzar, P., Dane, J., Keivanidis, P. E., Bradley, D. D. C., & Nelson, J. 2011. Gravure printing for three subsequent solar cell layers of inverted structures on flexible substrates. *Solar Energy Materials and Solar Cells*, 95(2): 731-734.

Winther-Jensen, B. & Krebs, F. C. 2006. High-conductivity large-area semi-transparent electrodes for polymer photovoltaics by silk screen printing and vapour-phase deposition. *Solar Energy Materials and Solar Cells*, 90(2): 123-132.

Zhao, J., Swinnen, A., Van Assche, G., Manca, J., Vanderzande, D., & Mele, B. V. 2009. Phase Diagram of P3HT/PCBM Blends and Its Implication for the Stability of Morphology. *The Journal of Physical Chemistry B*, 113(6): 1587 1591.

Zhou, Y., Li, F., Barrau, S., Tian, W., Inganäs, O., & Zhang, F. 2009. Inverted and transparent polymer solar cells prepared with vacuum-free processing. *Solar Energy Materials and Solar Cells*, 93(4): 497-500.

Zimmermann, B., Glatthaar, M., Niggemann, M., Riede, M. K., Hinsch, A., & Gombert, A. 2007. ITO-free wrap through organic solar cells--A module concept for cost-efficient reel-to-reel production. *Solar Energy Materials and Solar Cells*, 91(5): 374-378.

Zimmermann, B., Schleiermacher, H. F., Niggemann, M., & Würfel, U. 2011. ITO-free flexible inverted organic solar cell modules with high fill factor prepared by slot die coating. *Solar Energy Materials and Solar Cells*, 95(7): 1587-1589.

Zou, J., Yip, H. L., Hau, S. K., & Jen, A. K. Y. 2010. Metal grid/conducting polymer hybrid transparent electrode for inverted polymer solar cells. *Applied Physics Letters*, 96(20): 203301-203303.

4

Life Cycle Assessment of Organic Photovoltaics

Annick Anctil[1] and Vasilis Fthenakis[1,2]
[1]*Photovoltaic Environmental Research Center, Brookhaven National Laboratory*
[2]*Center for Life Cycle Analysis, Columbia University*
USA

1. Introduction

The unlimited abundance of solar resources ensures that photovoltaic technologies have the potential to supply a significant amount of the energy required to fulfill current- and future-energy demands while reducing greenhouse- gases emissions. So far, the high cost of photovoltaics compared to other energy sources has limited their use. However, emerging technologies, such as organic photovoltaics (OPV), which take advantage of man-made materials and solution processing, hold the promise for inexpensive devices. While solar cells could be an alternative to energy produced from fossil fuels, it is necessary to ensure that, in doing so, new environmental issues are not created. For this reason, life cycle assessments (LCAs) can be undertaken on emerging organic technologies to evaluate a priori the environmental impact of large-scale production and identify pathways toward sustainable energy production. In this chapter, the methodology of life-cycle assessment is presented, emphasizing photovoltaics usages, and is applied to emerging organic photovoltaics.

2. Life cycle assessment methodology

Life-cycle assessment (LCA) is a "cradle-to-grave" approach that begins with the extraction of the raw material (cradle), and ends when the material returns to the earth (grave). The cumulative environmental impacts from all stages in a product's life cycle are included, so affording a comprehensive view of the environmental impacts, and allowing evaluations of trade-offs in product and process selection. By undertaking this comprehensive analysis, encompassing all product life cycle stages and multiple metrics, the LCA helps to avoid environmental problems.

LCA is a ISO standardized method that requires specific methodology. It consists of four stages, as illustrated in Fig. 1(b) (International Organization for Standardization 2006; EPA 2006), which are summarized as follows:

1. **Goal definition and scoping:** The product, process, or activity of interest is defined and described by establishing clear system boundaries and metrics. During this stage, the type of information required for the analysis, and how the results of the LCA should be interpreted and used must be included. Furthermore, the distinction between foreground and background data must be established. The foreground system is the system of primary concern, while the background system generally uses aggregated datasets that are similar for all the various scenarios being considered.

2. **Inventory Analysis:** All relevant data are collected and organized. The level of accuracy and detail will influence the quality of the analysis. The assumptions and limits of data collection must be clearly defined, such as cut-off rules. The goal and scope stage has defined general system boundaries that need to be further detailed and analyzed using the following four stages:
   a. Develop a flow diagram of the processes being evaluated
   b. Formulate a data-collection plan
   c. Collect data
   d. Evaluate and report results
3. **Impact Assessment:** The inventory information is used in combination with appropriate metrics to predict the potential human- and ecological-impacts.
4. **Interpretation:** Evaluate the results of the inventory analysis and impact assessment to select the preferred product, process or service.

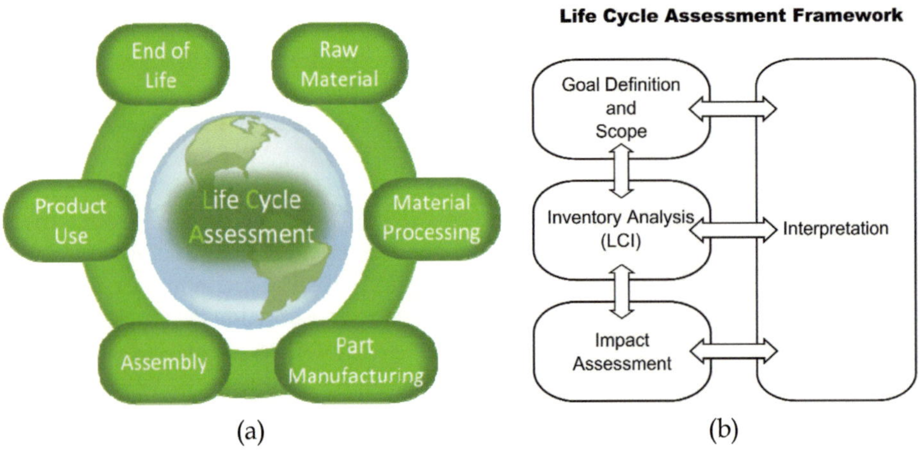

Fig. 1. (a) Cradle-to grave Life Cycle Assessment and (b) LCA Framework according to ISO standards 14040 and 14044 (International Organization for Standardization 2006).

As illustrated in Fig. 1(b), the LCA process is iterative. For example, interpretation is undertaken throughout the process and each stage is reevaluated, based on the results obtained. There is a variety of commercial software that can be used for life-cycle assessment to simplify the compilation and analysis of data. The software also contains life-cycle-inventory (LCI) databases for various chemicals and processes that generally are used to model the background system.

The ISO- and EPA-standards afford a framework for an LCA. However, this framework offers the individual practitioner a range of choices that can affect the validity and reliability of the results. To ensure consistency between photovoltaics (PV) LCA studies, the IEA recently published guidelines (Fthenakis et al., 2011). They represent a consensus among PV-LCA experts in North America, Europe, and Asia, for assumptions about PV performance, process input, and emissions allocation, methods of analysis, and reporting of the findings. The document offers guidance on photovoltaic-specific parameters (e.g., life expectancy, irradiation, performance ratio, degradation) that are the inputs in LCA, on choices and assumptions in analyzing data on life-cycle inventory (LCI) data, and on implementing modeling approaches.

The IEA Guidelines specially emphasize reporting transparency. This is of the utmost importance as parameters vary with geographical zones, and a system's boundary conditions and modeling approach can affect the findings significantly. At a minimum, the following parameters should be reported (Fthenakis et al., 2011): 1) On-plane irradiation-level and location; 2) module-rated efficiency; 3) system's performance ratio; 4) time-frame of data; 5)type of system (e.g., roof-top, ground mount fixed tilt or tracker); 6) expected lifetime for PV and BOS; 7) system's boundaries (whether capital goods, installation, maintenance, disposal, the transportation- and recycling-stages are included for both PV modules and balance-of-system (frame, mounting, cabling, inverter; for utility applications the transformer, site preparation, and maintenance)); 8) the place/country/region of production modeled, and, 9) the study's explicit goal.

These parameters should be listed in the captions of figures showing the results of the LCA. In addition, the report should identify the following: The LCA method used, especially if is not process-based; the LCA tool; databases used; the method of calculating the energy-payback time; the commercial representativeness of the study; and, assumptions on the production of major input materials.

2.1 Metrics

The following metrics are suggested for undertaking an impact assessment of PV LCA.

- **Greenhouse gas emissions:** The greenhouse gas (GHG) emissions during the life cycle stages of a PV system are estimated as an equivalent of CO_2, using an integrated 100-year time-horizon that incorporates the ICPP's most recently published global-warming potential factors.
- **Cumulative Energy Demand (CED):** The CED describes the consumption of fossil-, nuclear-, and renewable-energy sources throughout the life cycle of goods or services. This includes the direct uses and as the indirect (grey) consumption of energy due to employing materials (e.g. plastic or wood in construction), the consumables necessary in manufacturing (e.g., solvents, gloves, packaging), and the raw materials. The simplicity of concept of a CED assessment assures that it is a good starting point for an assessment, along with and its comparability with CED results in other studies. The following two CED indicators are well known: CED, non-renewable [MJ-eq.] – fossil and nuclear and CED; and, renewable [MJ-eq.] – hydro, solar, wind, geothermal, biomass.
- **Acidification potential (AP):** Acidification describes a change in acidity in the soil due to atmospheric deposition of sulphates, nitrates, and phosphates. Major acidifying substances are NO_X, NH_3, and SO_2.
- **Ozone-depletion potential (ODP):** The thinning of the stratospheric ozone layer as a result of anthropogenic emissions is described by the stratospheric ozone-depletion indicator. The impacts potentially are harmful to human health, animal health, terrestrial- and aquatic-ecosystems, biochemical cycles, and materials.
- **Human toxicity:** This indicator overs the impacts of toxic substances on human health. The health risks of exposure in the workplace also can be included.
- **Ecotoxicity:** This indicator includes the impacts of toxic substances on aquatic-, terrestrial-, and sedimentary-ecosystems.
- **Land use and water use:** These are environmental impacts are of growing importance. Recommendations are to list separately the withdrawal and consumption of water, and the occupation and transformation of land.

2.2 LCA interpretation

Considering that the main objective of using solar PV is to reduce the environmental impact associated with the employing fossil-fuel energy resources, the recommendations (Fthenakis et al., 2011) are to include specific indicators, such as the energy payback time (EPBT), the nonrenewable energy payback time (NREPBT), the energy return on investment (EROI), and the impact mitigation potentials (IMP).

2.2.1 Energy Payback Time (EPBT)

The EPBT denotes the time needed to compensate for the total renewable- and non-renewable- primary energy required during the life cycle of a PV system. The annual electricity generation (E_{agen}) is converted into its equivalent primary-energy, based on the efficiency of electricity conversion at the demand side, using the current average (in attributional LCAs), or the long-term marginal grid mix (in decisional/consequential LCAs) where the PV plant is being installed.

$$EPBT = (E_{mat}+E_{manuf}+E_{trans}+E_{inst}+E_{EOL}) \;/\; ((E_{agen} \;/\; n_G)- E_{O\&M}) \qquad (1)$$

where,

E_{mat} : Primary energy demand to produce materials comprising PV system
E_{manuf}: Primary energy demand to manufacture PV system
E_{trans}: Primary energy demand to transport materials used during the life cycle
E_{inst} : Primary energy demand to install the system
E_{EOL} : Primary energy demand for end-of-life management
E_{agen} : Annual electricity generation
$E_{O\&M}$: Annual primary energy demand for operation and maintenance
n_G : Grid efficiency, the average primary energy to electricity conversion efficiency at the demand side

2.2.2 Non-renewable Energy Payback Time (NREPBT)

The NREPBT denotes the time needed to compensate for the non-renewable energy required during the life cycle of a PV system. It accounts for only the non-renewable primary energy; renewable primary energy is *not* accounted for, neither on the demand side, nor during the operation phase. The annual electricity generation (E_{agen}) likewise is converted to the primary energy equivalent, considering the efficiency of the non-renewable primary energy to electricity conversion of the grid mix where the PV plant is being installed. The formula of NREPBT is identical to that of the EPBT given above except replacing "primary energy" with "non-renewable primary energy". Accordingly, grid efficiency, η_G, accounts for only non-renewable primary energy.

Both EPBT and NREPBT depend on the grid mix; however, excluding the renewable primary energy makes NREPBT more sensitive to local- or regional- conditions (e.g., product-specific use of hydro-power) that which may not be extrapolated to large global scales. On the other hand, the EPBT metric with an average large-scale grid-conversion efficiency (e.g., EU, or US, or World) might not capture the conditions of local- or regional- conditions. The calculated EPBT and NREPBT do not differ significantly in cases wherein the power plant mix of a country or region is dominated by non-renewable power generation. However, as an increasing share of renewable energies is expected in future power-grid mixes as well as within the PV supply- chain, the two opposing effects of a reduction in the CED of PV, and an increase in grid efficiency will require careful consideration, hence, the numerical values of EPBT or NREPBT may vary considerably according to the chosen approach.

2.2.3 Energy Return on Investment (EROI)

The traditional way of calculating the EROI of PV is as follows (Reich-weiser, Dornfeld, and Horne 2008):

$$EROI = lifetime / EPBT = T \cdot ((E_{agen}/n_G) - E_{O\&Mr}) / (E_{mat}+E_{manuf}+E_{trans}+E_{inst}+E_{EOL}) \quad (2)$$

2.2.4 Impact Mitigation Potentials (IMP)

This may comprise the mitigation potentials for climate change and high-level nuclear waste (Jungbluth et al., 2008). Clearly reference the impact assessment method applied, and specify the reference system, e.g., today's European electricity mix, or the national electricity supply mix.

3. Organic photovoltaics

While 20 years of research were necessary to increase device power efficiencies from 0.1 to 3.5% in 2005 (Janssen, Hummelen, and Sariciftci 2005), the last five year have seen unprecedented interest in the technology, so resulting in a rapid increase in efficiency up to 8.3% for both small molecule- and polymer-photovoltaics (Green, Emery, Hishikawa, Warta, et al. 2011). Compared to other photovoltaics technologies, commercialization occurred earlier than expected by Konarka (polymers) and Heliatek (small molecules) since the potential for inexpensive new applications counterbalance their low efficiency. For example, incorporating low-efficiency organic solar cells in electronic products has the advantage of increasing the lifetime of the battery since it is being charged by ambient lighting while not in use. Smart fabrics (also referred to as wearable technology) add new functionality to traditional applications, for example, in clothing, tents, and military uniforms. Applications are numerous for off-grid remote locations, in particular to provide lighting at night, or play a double role in shading an area while supplying electricity for another usage, for example, shade structures. The area where organic photovoltaic is expected to have the largest impact is in building-integrated photovoltaics (BIPV). The low absorption of organic films supports light reduction while providing power for the building during the day when demand is at its peak. Compared to traditional semiconductors, organic molecules offer a considerable improvement in design since they s can be tuned to the desirable colors by slightly changing their chemical properties, thereby allowing solar- cells to be an integral part of the design.

The most common device structure for OPV uses a mixture of donor- and acceptor-materials referred to as a bulk heterojunction (BHJ) that resides between two electrodes. Fig. 2 (a) illustrates a typical BHJ made of Poly (3-hexylthiophene) (P3HT) and Phenyl-C_{61}-butyric- acid methyl ester (C_{60}PCBM). As depicted, the photovoltaic effect follows the following steps: the illumination of an organic semiconductor donor (1) generates excitons (2) with a binding energy of about 0.4 eV instead of free charges (Hadipour, de Boer, and Blom 2008). To separate into free charges, the exciton must diffuse until it reaches a donor/acceptor interface (3) with a difference in electron affinities and an ionization potential large enough to overcome the binding energy. The energy cascade required for charge extraction is illustrated in Fig. 2(b). The free charges then can travel (4) through either the donor- or acceptor- material (5), and then are collected at the electrodes (6). The overall efficiency of the device therefore is determined by the optical absorption and the efficiency of each of those steps.

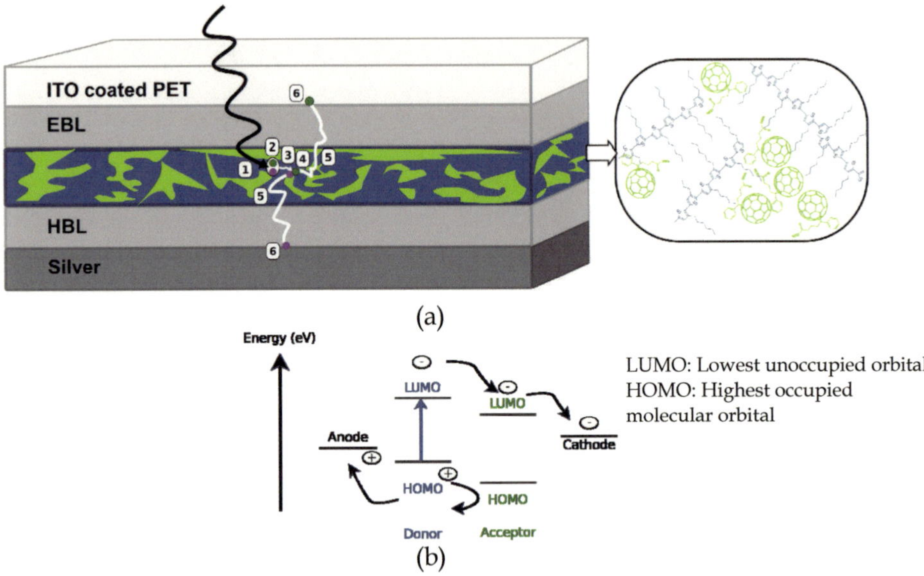

Fig. 2. (a) Photovoltaic effect in a bulk heterojunction organic solar cell with details about the chemical structure of P3HT:C_{60}PCBM material used in the active layer; (b) conditions for charge transfer in a donor/acceptor photovoltaic device where the colors correspond to the donor and acceptor molecules illustrated in (a).

Various challenges have limited the use of OPV; in particular, the large bandgap of most organic polymers is responsible for low power-conversion efficiency because a large portion of the solar spectrum is unabsorbed. In theory, in an optimal solar cell, the acceptor bandgap should be around 1.4 eV, wherein the maximum efficiency would be 31% under 1 sun AM1.5 (Gregg 2011). Most early- generation semiconducting polymers, such as polythiophenes, have bandgaps higher than 2- eV (620 nm), so limiting their maximum efficiency. There are two main alternatives to increase the devices' efficiency: Lowering the bandgap to absorb a maximum of photons in one layer, or using a multi-junction approach where two different materials absorb in a different region of the solar spectrum.

The low bandgap approach has received considerable interest in the last few years and allowed the current record efficiency for polymer devices. By lowering the bandgap from 2eV to 1.5 eV, the maximum theoretical efficiency increases from 8% to 13% (Siddiki et al. 2010). To increase efficiencies further, the multi-junction approach, illustrated in Fig. 3 is necessary. For OPV this approach is not only advantageous to capture a broader range of the solar spectrum, but it also helps overcome the poor charge carrier mobility and lifetime of carriers which prevents the fabrication of a thick absorbing layer (Ameri and Al. 2009). In comparison with inorganic material, organic semi-conductors absorb only a narrow portion of the spectrum as illustrated in Fig. 3 and therefore a combination of multiple materials is necessary to absorb a larger portion of the spectrum.

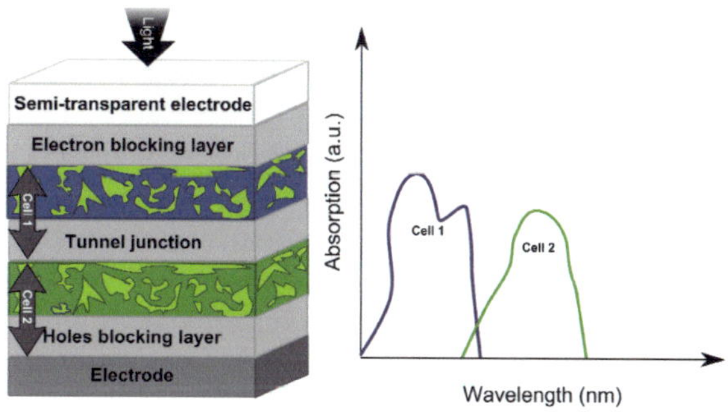

Fig. 3. Organic photovoltaic with two sub-cells having different complementary absorption-spectra.

Multi-junction devices either can be connected in series or in parallel depending on the nature of the intermediate layer. The most common type of connection is in series wherein the voltage across the whole device is equal to the sum of the voltage of each sub-cell, and the current is limited by the lowest sub-cell current. While multi-junction devices increase efficiency, they also increase the device's complexity and require additional processing steps and materials.

In addition to the problem of photon absorption, another issue for organic photovoltaics is charge recombination where free charges recombine before reaching the electrodes (step 4 in Fig. 2). One option that was shown to be effective is to use a hole-blocking layer, such as titanium dioxide (TiOx) before depositing the metallic electrode (Hayakawa et al. 2007).

4. Life-cycle assessment of organic solar cells

There are some early evaluations of the life-cycle impact of producing organic photovoltaics, but all recent studies were limited to single-junction polymer devices (Roes et al. 2009; García-Valverde, Cherni, and Urbina 2010; Espinosa et al. 2010), i.e., the best types of devices five years ago, but since, have been surpassed. As described in the previous section, there are various alternatives types of photovoltaics devices, including small molecule- and multi-junction- ones that have not been rigorously studied using LCA.

Furthermore, the few existing LCAs of organic PV were constrained by the unavailability of life-cycle inventory data, (i.e., materials and energy inputs and emissions and waste outputs over the life cycle of the cells and processes) because these 3rd-generation PV technologies are in R&D or in pilot-scale status. When direct, detailed inventory data y from industrial production are unavailable, such studies often adopt data from laboratory-scale or prototype-production, and from literature on similar process products, or their own modeling; sometimes this information is extrapolated or adjusted to describe the nanotechnologies studied. Accordingly, the LCAs of organic technologies inherently carry large uncertainties. For example, some of then include in their system boundary a large amount of solvents used in synthesizing and purifying nanomaterials, which is common under R&D conditions. However, fully scaled-up nanotechnologies would utilize them much more efficiently through recycling, energy recovery, or improved processes. We can expect that process yields in the former conditions will be much poorer than in scaled-up production lines.

Therefore, there is a need to develop realistic LCI data to study the life-cycle impacts of OPV technologies The first step is to compile an LCI data-base for characterizing the impact of

individual new materials, including various donor materials (polymers and small molecules) and acceptors (fullerenes and derivatives) as well as materials used as electron- and hole-blocking layers (PEDOT:PSS and TiOx, respectively). Using the results from these material inventories, the impact of various configurations of devices can be calculated.
The LCA results are obtained from analyses using SimaPro® based upon existing inventory data obtained from available databases and previously published primary literature. Inventory data for chemicals not available in databases are estimated using default values and stoichiometric reactions according to previously published guidelines (Frischknecht et al. 2007; Geisler, Hofstetter, and Hungerbiihler 2004). The life-cycle impact of OPV was assessed using specific data from the life cycle of fullerene production, semiconductor polymer-, small molecule-, and interfacial material-processing. Different donor/acceptor combinations are examined in conjunction with the reported efficiencies and specific processing conditions. The life-cycle assessment is characterized via the cumulative energy demand (CED) method, using the Ecoinvent electricity profile for the United States. The CED was shown to be correlated with most environmental indicators (Capello et al. 2009; Huijbregts et al. 2006), and, since comprehensive data on environmental impact are scarce in current inventories , in particular on emissions in terms of their toxicology and releases, this approach often yields a better estimate of the environmental impact.
The Life Cycle of PV typically consists of materials production, solar cell- and PV module-manufacturing, installation, operation and maintenance, recycling and end-of-life (Fig. 4). The LCA of PV systems will be based on the life-cycle inventories of materials and energy for the solar cells, PV modules, and the balance of system (e.g., wiring, frames, inverters, support structures).

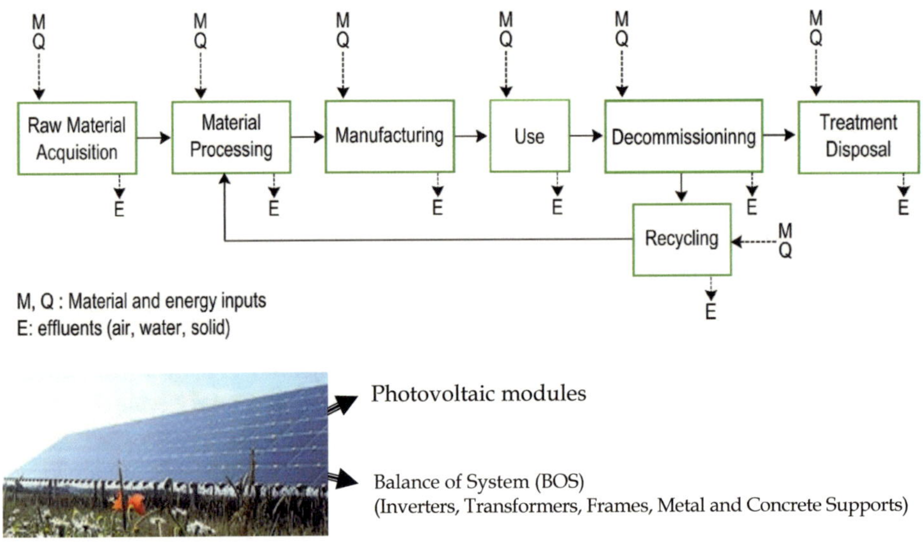

Fig. 4. The Life Cycle of Photovoltaic Systems.

LCI data on the cell/module manufacturing stages when possible, are obtained directly from the PV manufacturers and are documented with actual records of material inventories and energy use. The energy/emissions background data for the U.S. electricity mixture are obtained from established databases.

4.1 Life cycle inventory of materials for organic photovoltaics
4.1.1 Fullerenes and fullerene-derivatives

For OPV, the most common acceptor molecules are fullerenes for which there exists no detailed inventory information in particular with regard to the purity and chemical modifications required for application in organic photovoltaics. The methano-fullerene derivatives such as those illustrated in Fig. 2(a) are commonly used as an acceptor molecule when mixed with a conducting polymer to create a bulk heterojunction of donor/acceptor material in organic solar cells. The contribution from nanomaterials in developing products has generally not been thoroughly studied since there is a belief that a small amount of nanomaterials in a product will result in minimal environmental impact (Meyer, Curran, and Gonzalez 2009; Som et al. 2010). Whereas previous studies have limited the LCA scope to direct energy input (e.g. electricity usage) during fullerene synthesis (Roes et al. 2009; Kushnir and Sanden 2008) this study quantifies total material intensity and embodied energy with a scope that includes all direct and upstream feedstock and fuel energy inputs for modern production methods as illustrated in Fig. 5.

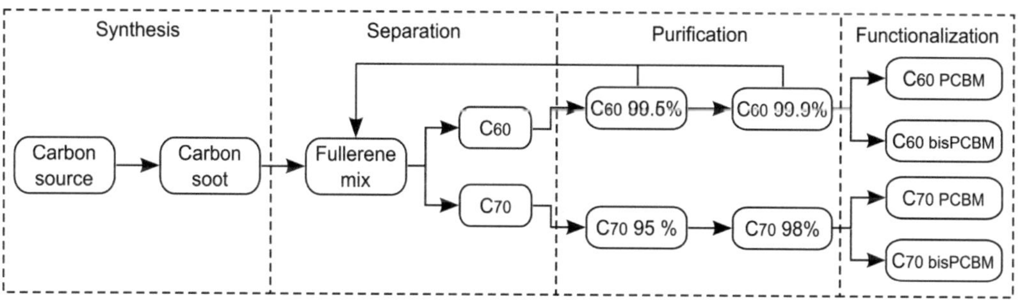

Fig. 5. Overview of the process flow for producing modified fullerene compounds for use as functional materials in organic solar cells (Anctil et al. 2011).

We specify the variation in synthesis methods and purification processes to more accurately quantify the impact from factors as the synthesis reaction, the purity of reactants and solvents, low reaction yields, repeated purification steps, and toxic chemicals or solvents. Four different methods are compared, using the conditions for synthesis summarized in Table 1. In general, pyrolysis techniques produce large amounts of various sized fullerenes, while the plasma techniques generate smaller quantities of mainly C_{60}. Plasma methods result an incomplete combustion process, and therefore, the main product is gaseous emissions while the soot containing the fullerenes is a secondary product per mass.

From this analysis, pyrolysis methods seemingly have an order-of-magnitude lower impact than does plasma, as illustrated for C_{60} in Fig. 6. Detailed contributions for each method are given in Fig. 6(b). Electricity is the dominant contributor for plasma methods, while the carbon feedstock (either toluene or tetralin) are most important ones for pyrolysis. RF plasma has was developed to increase the production rate of fullerenes compared to arc plasma: accordingly, a larger amount of gas is required to provide sufficient energy to convert the carbon precursor into fullerenes.

Process	Carbon source	Fullerene yield in soot (%)	Production rate (g fullerene/hr)	Ratio $C_{60}/C_{70}/$ higher fullerenes	Ref.
Arc Plasma	High purity graphite electrodes	13.1	1.2	69 / 24 / 7	(Marković et al. 2007)
Pyrolysis	Toluene	17.5	44	43 / 28.5 / 28.5	(Alford et al. 2008)
	Pyrolysis 1,4-tetrahydro-naphtalene (tetralin)	30	70	39 / 30.5 / 30.5	(Alford et al. 2008)
RF Plasma	Graphite powder	5.9	2.7	70 / 23 / 7	(Szépvölgyi et al. 2006)

Table 1. Reported values for fullerene production under different conditions of synthesis used to perform the LCA of fullerenes.

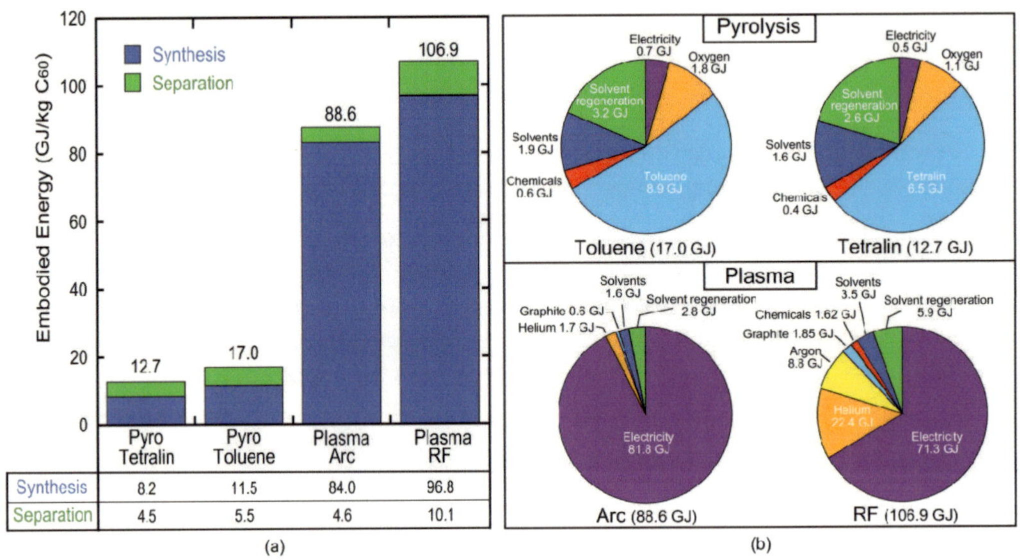

Fig. 6. (a) Embodied energy of 1 kg of C_{60} after synthesis and separation, as produced by pyrolysis (tetralin and toluene) and plasma methods and (b) contribution of various components for the total embodied energy for each type of synthesis methods after fullerenes separation (Anctil et al. 2011).

The production of fullerenes is energy intensive because they account for less than 30% of the material generated during the synthesis stage, the rest being soot. Since fullerenes have low solubility, a large amount of solvent is required during separation and purification stages as illustrated in Fig. 7 for both C_{60} and C_{70} , considering the best-case scenario (pyro-tetralin). The electronic-grade material (grey region) in Fig. 7 corresponds to the material purity required for electronic applications.

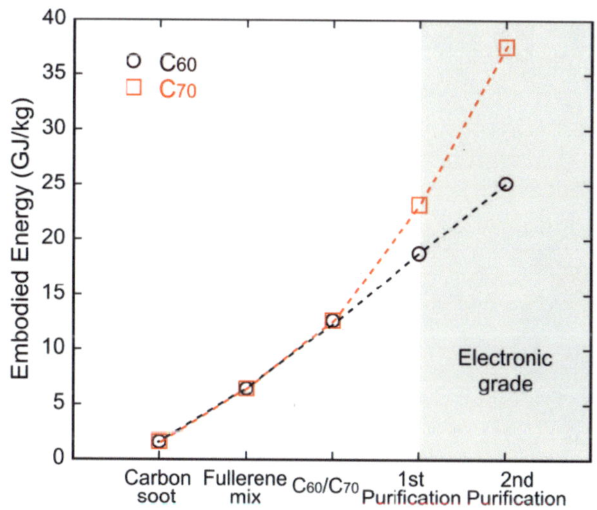

Fig. 7. Embodied energy for C_{60} and C_{70} products as a function of product stage (Anctil et al. 2011).

C_{70} is produced along with C_{60} in the carbon soot for each of the synthesis methods, albeit in ratios varying relative to the synthesis technique. Since C_{70} is more difficult to separate from the other higher order fullerenes than C_{60}, about 50% more C_{70} is required after separation to produce the same amount of high purity material.

While a high degree of purity is required for most electronic applications, in certain cases there is an additional need to modify the fullerene structure. This is the case for fullerenes used in organic solar cells which need to have high purity to avoid impurity trap states, but also need to be modified to PCBM to increase solubility for solution processing and device viability. The additional embodied energy associated with each of the various chemicals, solvents, and solvent regeneration steps further increases the embodied energy up to 64.7 GJ/kg for C_{60}-PCBM, more than five-fold greater than embodied energy calculated for C_{60} produced by pyro-tetralin. Although considerable research has utilized C_{60}-PCBM for organic solar cells, recent results show that C_{70}-PCBM produces higher efficiency devices (Liang et al. 2009) thereby increasing interest in the larger fullerenes. Fig. 8 illustrates the contribution from each component along the complete process required to produce either 1 kg of C_{60}-PCBM or C_{70}-PCBM. Due to the increased energy intensity associated with purification and the functionalization reactions, the embodied energy of C_{70}-PCBM is 90.2 GJ/kg as compared to 64.7 GJ/kg for C_{60}-PCBM. Therefore, each modified fullerene has a significantly different embodied energy, which in turn influences the embodied energy of the organic solar cell.

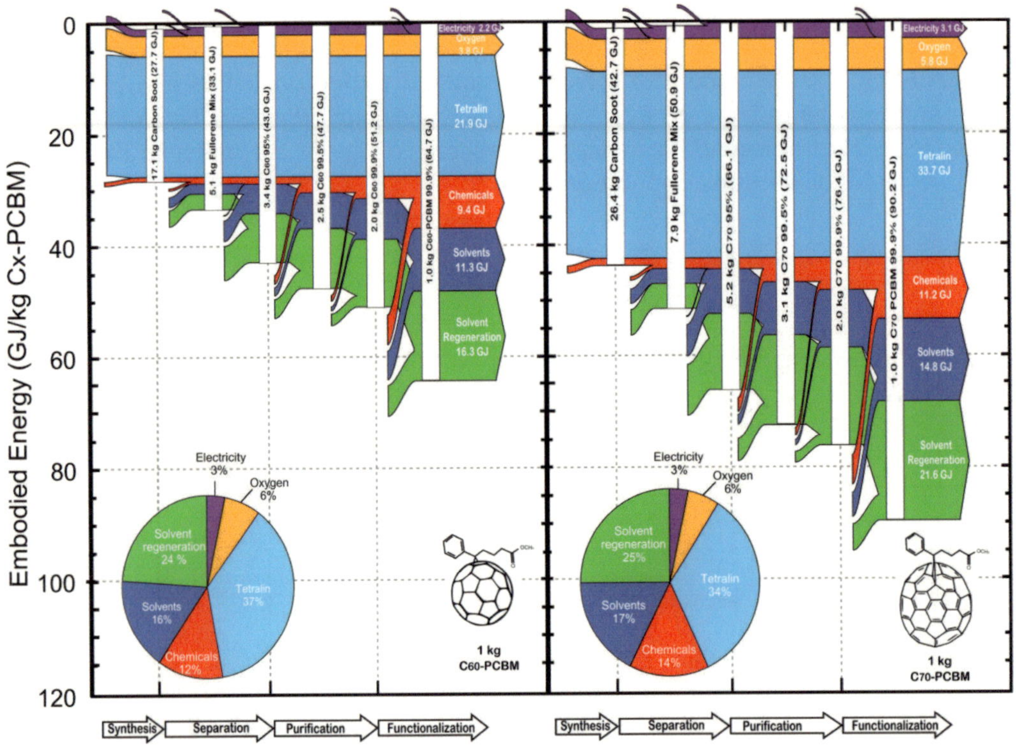

Fig. 8. Energy flow diagram for production using the pyro-tetralin synthesis method of (a) 1 kg of C_{60}PCBM, compared to (b) 1kg of C_{70}PCBM (Anctil et al. 2011).

4.1.2 Polymers and small molecules

Organic semiconductors require alternating single- and double-carbon bonds where each carbon binds only to three adjacent atoms, leaving one electron in the P_z orbital. The overlap of these P_z orbitals causes the formation of π bonds along the backbone, and the delocalization of the π electrons. Molecules with a delocalized π electron system can absorb sunlight, photo-generate charge carriers, and transport these charge carriers. The organization of the semi-conducting material is critical as the π-π bonds must be close enough to allow charge transport. Organic semiconductors fall into two categories: Polymers that can be solution-processed; and, small molecules that are generally evaporated under high vacuum.

Current interest in OPV can be attributed to regioregular poly(3-hexylthiophene) (P3HT) that allowed the efficiencies of devices to reach 5% for the first time. Morphology was critical, requiring precise control over the the donor :acceptor ratio, the choice of solvent , and post-treatment annealing to induce the reorganization and crystallization of the polymer. In addition to P3HT, there is a significant interest in small molecules, in particular phthalocyanine molecules that already produced inexpensively in large amounts, mainly for pigments in paints and other products (Lambourne and Strivens 1999). In Fig. 9 depicts the energy required to produce 1 mol of various polymers and small molecules. . For the polymers, increasing the number of steps from 4 for P3HT, to 9 for the low bandgap PCDTBT significantly impact the final embodied energy. Even if only 3 steps required for small molecules, there is a rapid increase in embodied energy as 4 moles of the starting material is required to produce 1 mole of phthalocyanine.

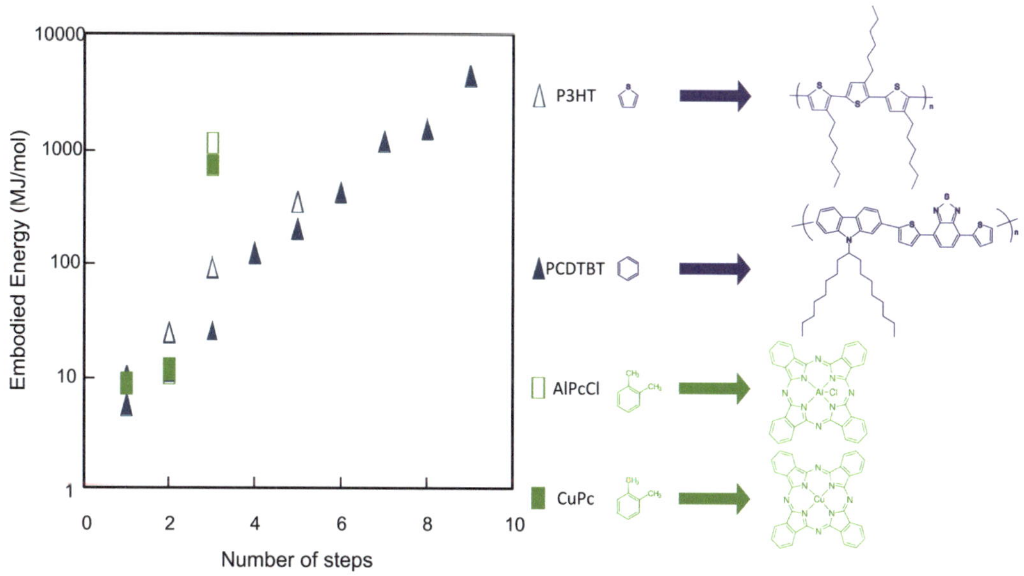

Fig. 9. Comparison of embodied energy in synthesizing of two types of phthalocyanines small molecules (green) and polymers (blue).

In addition to the active layer material (fullerenes, polymer, and small molecules) the impact of blocking layers used to improve charge collection is calculated. The results from all LCIs of materials used for organic photovoltaics considered in this work are illustrated in Fig. 10.

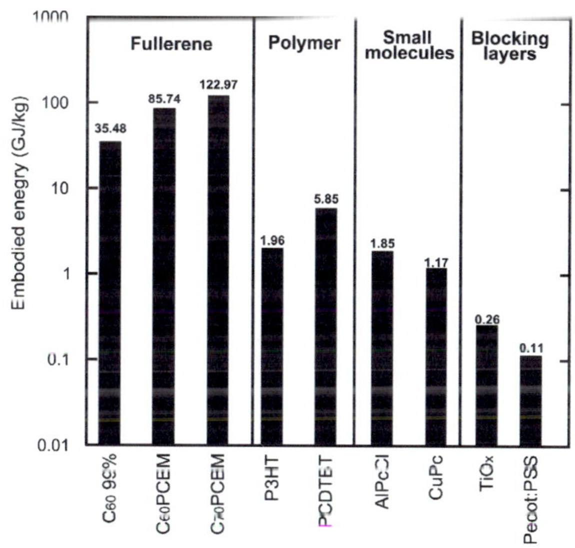

Fig. 10. Cumulative energy demand (CED) for various materials commonly used in organic photovoltaics.

As illustrated, C_{70} PCBM has the highest embodied energy, followed by both functionalized- and unmodified-C_{60}. The embodied energy of all fullerenes are an order-of-magnitude higher than most common chemicals, and therefore, are likely to influence the embodied energy of the product they will be used in, even though they might only represent a small fraction of its total mass. While low bandgap polymer PCDTBT requires larger amount of energy compared to AlPCCl or CuPc, this is significantly lower than other materials. Finally, the blocking layers are the least energy-intensive materials to produce.

4.2 Life-cycle impact of polymer, small molecules, and multi-junction organic photovoltaics

The primary goal of this LCA is to compute the CED of various organic photovoltaic technologies, including single-junction small molecule- and polymer-photovoltaics as well as their multi-junction counterparts that are responsible for the rapid increase in the devices' efficiency. A secondary goal is to examine the effect of specific processing conditions, such as using thermal treatments, interface layers, low bandgap polymers, and the type of heterojunction. Fig. 11 illustrated the systems considered. The functional unit of this study is the CED to produce a power of 1 watt-peak (CED/Wp).

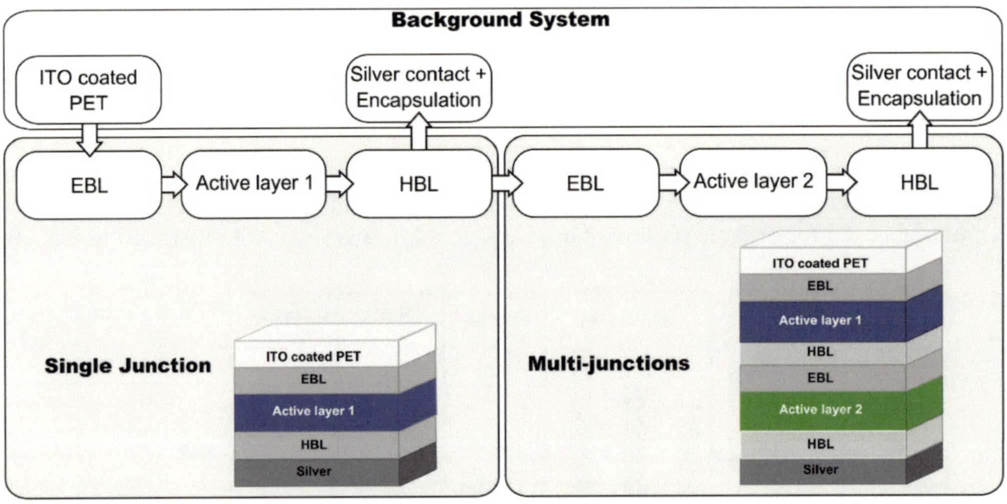

Fig. 11. System boundaries for both types of organic photovoltaic devices considered in this work.

Since the focus of this work is the active and interfacial layer material as well as the process energy associated with the deposition and annealing of those layers, the substrate (ITO coated PET), silver contact, and encapsulation constitute the background system (Fig. 11). The properties of those different components are described in Table 1.

Using the results from the previous section, the life cycle assessment (LCA) of various polymers, small molecules, single- and multi-junction-photovoltaics, representing various types of organic photovoltaics devices developed recently is calculated according to the fabrication conditions described in Table 2, including material ratio, layer thicknesses, and annealing conditions.

Stage	Details	Embodied Energy (MJ/m^2)	Reference
ITO-coated PET	Polyethylene terephthalate film (130 um thick)	16.9	(PlasticsEurope 2010)
	Sputtering 180 nm ITO (including production ITO)	68.3	(Hischier et al. 2007)
Contacts Printing	Silver	4.9	(Espinosa et al. 2010)
	Screen-printing	9.9	
Encapsulation	PET Covers + epoxy	11.7	
	Lamination energy	0.1	

Table 2. Values used for materials and energy commodities for the background system used in the inventory analysis.

Figure 12 summarizes the results of the LCA for both polymer- and small molecule-photovoltaics as a function of materials, processing, and other components. In general, the CED spans the same magnitude of 3-6 MJ/Wp for both cases. Also, the CED shows a trend of decreasing value with the device's increasing efficiency. The CED/Wp of small molecule devices is on average slightly higher than that of polymer devices, but this is largely due to the lower efficiency of the latter for reported measurements.

Device name	Efficiency (%)	Type	Reference
P3HT:C_{60} PCBM	5.0	Single junction polymer	(Ma et al. 2005)
PCDTBT: C_{70}PCBM	5.5	Single junction low-bandgap polymer	(Park et al. 2009)
PCDTBT:C_{70}PCBM - TiOx	6.1	Single junction low-bandgap polymer with electron blocking layer	(Park et al. 2009)
P3HT:C_{70}PCBM+ PCDTBT: C_{60}PCBM	6.5	Multi-junction polymer	(Jin Young Kim et al. 2007)
AlPcCl:C_{60}	3.0	Small molecules	(Cheyns, Rand, and Heremans 2010)
CuPc:C_{60}	5.0	Small molecules	(Xue et al. 2005)

Table 3. Organic photovoltaics-devices considered herein, representing a variety of organic photovoltaic technologies, including single-junction small molecule- and polymer-photovoltaics, as well as their multi-junction counterparts.

Fig. 12 also illustrates a difference between polymer- and small molecule-photovoltaics in terms of the relative contribution from materials, processing, and other components. For the polymer devices, the material contributions dominate the processing conditions, whereas the opposite trend applies to small-molecule devices where they are more balanced. For example, the total thickness of a small-molecule photovoltaic is lower (<60 nm) than that of the polymer photovoltaic, and these structures utilize pure C_{60}, which has a dramatically lower embodied energy than the modified fullerenes in the polymer devices. Polymer photovoltaics need a thicker layer (e.g., 80-250 nm), requiring a larger amount of material of higher energy.

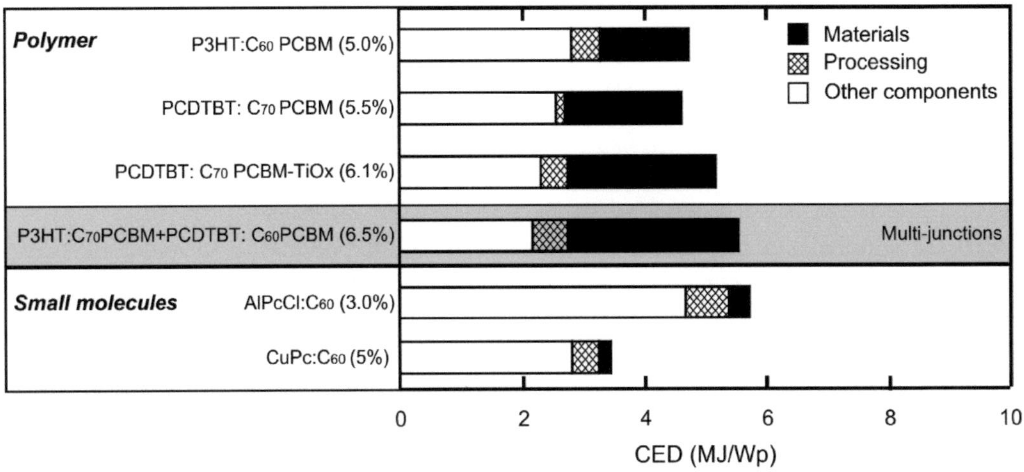

Fig. 12. Cumulative energy demand (CED) for various types of organic photovoltaics, including low bandgap polymer-, small molecule- and multi-junction-polymer devices.

While multi-junction devices increase efficiency, they also increase the devices' complexity and require additional processing steps and materials. In polymer photovoltaics, the major material contributor is the electron acceptor, the importance of which rises with the increasing efficiency of the device due to a higher concentration of larger fullerenes needed to attain the gain in efficiency. There is a small increase in the direct contribution from a block-copolymer compared to the P3HT, but it is actually an improvement considering all other factors. The block copolymer has a lower bandgap with higher absorption coefficient that allows the use of a thinner layer (80-100 nm, compared to 150-200 nm for P3HT) and higher efficiency, so lowering the amount of material required. In addition, the polymer chain can be optimally organized by optimization of the solvent with the copolymer, so eliminating the need for post-processing annealing compared to P3HT, thereby further reducing the contribution from annealing.

In addition, it is apparent that the impact of processing of small-molecule photovoltaics is greater than that of the materials. The relative contribution of the electron acceptor in the small-molecule devices, i.e., unmodified C_{60}, is nearby 30% lower than in polymer photovoltaics, which require the high-energy functionalized fullerenes the fabrication of multi-junction photovoltaics requires additional processing and material to obtain higher efficiency, but presently, they are the predominant option to increase organic photovoltaic efficiencies above 13%. There are few publications on multi-junction photovoltaics, although efficiencies of 8.3% were reported for double-junction small-molecule photovoltaics (Green, Emery, Hishikawa, and Warta 2011). As illustrated in Fig. 12, the efficiency of multi-junction polymer photovoltaics increases from 6.1% to 6.5%, but the overall impact is an increase in CED, mostly due to the total amount of active material required. The tunnel junction uses materials such as TiO_x and Pedot:PSS that negligibly impact the CED based upon the comparison in Figure 10 with the other materials. The contribution from processing, while markedly similar, actually has a lower impact from annealing and greater impact from solution processing because six layers have to be deposited rather than three. In previous studies, it was assumed that, for future organic photovoltaics, the embodied energy of a photovoltaic with twice the device's efficiency would be half of that of the initial device (García-Valverde, Cherni, and Urbina 2010). This assumption is questionable, as we have shown from this work that increases in the devices' efficiency results from additional processing and new materials that are more energy intensive to produce. Therefore the total embodied energy likely will be higher for the multi-junction- photovoltaics compared to single-junction ones, unless the efficiency is significantly higher to compensate for the extra processing and material requirement. For small molecules, the largest energy contribution is associated with the evaporation process, which requires creating a vacuum. Adding additional layers for a small molecule multi-junction does not significantly raise the CED, so the multi-junction approach might be attractive if the device's efficiency surpasses that of polymer-based devices.

5. Conclusions

This chapter illustrates the use of life cycle assessment for emerging photovoltaics technologies, such as organic photovoltaics. We directly compared small molecule- and polymer-photovoltaics, with new calculations for the emerging specialty materials being used as the donor, acceptor, and interface layers in both single- and multi-junction-designs. An indirect outcome from this work is the creation of life cycle inventory data for new semiconductor materials that feature prominently in organic photovoltaics, but also have tremendous potential in other organic electronic devices (e.g., transistors, light emitting diode). Fullerenes used as acceptors are the most energy-intensive component of organic photovoltaics. While the polymer energy requirement is much lower than that of fullerenes, the embodied energy in low bandgap polymer materials rises with increasing number of steps in their fabrication. Small molecule- and polymer-photovoltaics were found to have a similar CED. For polymer photovoltaics, there is a trend towards using a higher quantity of larger fullerenes, which increases the CED/Wp. Although the differing interface layers are essential for enhanced efficiency, especially for the multi-junction approach, these materials have negligible direct impact , but the deposition and annealing processes are significant.

6. Acknowledgments

The work has been financially supported by the Department of Energy under project DE-FG36-08GO88110 and the Golisano Institute for Sustainability at RIT.

7. References

Alford, J.M. et al., 2008. Fullerene production in sooting flames from 1,2,3,4-tetrahydronaphthalene. Carbon, 46, pp.1623-1625.

Ameri, T. et al., 2009. Organic tandem solar cells: A review. Energy & Environmental Science, 2, pp.347-363.

Anctil, A. et al., 2011. Material and Energy Intensity of Fullerene Production. Environmental Science and Technology, 45(6), pp.2353-2359.

Capello, C. et al., 2009. A comprehensive environmental assessment of petrochemical solvent production. International Journal of Life Cycle Assessment, 14, pp.467-479.

Cheyns, D., Rand, B.P. & Heremans, P., 2010. Organic tandem solar cells with complementary absorbing layers and a high open-circuit voltage. Applied Physics Letters, 97, p.033301.

EPA, 2006. Life Cycle Assessment: Principles and Practice, Available at: http://www.epa.gov/nrmrl/lcaccess/pdfs/600r06060.pdf.

Espinosa, N. et al., 2011. A life cycle analysis of polymer solar cell modules prepared using roll-to-roll methods under ambient conditions. Solar Energy Materials & Solar Cells, 95, pp.1293-1302.

Frischknecht, R. et al., 2007. Overview and Methodology Final report ecoinvent Data v2.0, Duebendorf, CH.

Fthenakis, V. et al., 2011. Life Cycle Inventories and Life Cycle Assessments of Photovoltaic Systems, International Energy Agency (IEA) PVPS Task 12, Report T12-02:2011,

García-Valverde, R., Cherni, J.A. & Urbina, A., 2010. Life cycle analysis of organic photovoltaic technologies. Progress in Photovoltaics: Research and Applications, 18(7), pp.535-558.

Geisler, G., Hofstetter, T.B. & Hungerbiihler, K., 2004. Production of Fine and Speciality Chemicals: Procedure for the Estimation of LCIs. International Journal of Life Cycle Analysis, 9(2), pp.101-113.

Green, M.A. et al., 2011. Solar cell efficiency tables (Version 38). Progress in Photovoltaics: Research and Applications, 19, pp.565-572.

Gregg, B.A., 2011. The Photoconversion Mechanism of Excitonic Solar Cells. MRS Bulletin, 30(1), pp.20-22.

Hadipour, A., de Boer, B. & Blom, P.W.M., 2008. Organic tandem and multi-junction solar cells. Advanced Functional Materials, 18(2), pp.169-181.

Hayakawa, A. et al., 2007. High performance polythiophene/fullerene bulk-heterojunction solar cell with a TiOx hole blocking layer. Applied Physics Letters, 90(16), pp.163517/1-163517/3.

Hischier, R. et al., 2007. Life Cycle Inventories of Electric and Electronic Equipment - Production, Use & Disposal F. report ecoinvent D. v2.0, ed., Duebendorf and St. Gallen.

Huijbregts, M.A.J. et al., 2006. Is cumulative fossil energy demand a useful indicator for the environmental performance of products? Environmental Science & Technology, 40, pp.641-648.

International Organization for Standardization, 2006. Environmental management-Life cycle assessment-Requirements and guidelines, Switzerland.

Janssen, R.A.J., Hummelen, J.C. & Sariciftci, N.S., 2005. Polymer-fullerene bulk heterojunction solar cells. MRS Bulletin, 30(1), pp.33-36.

Kim, J.Y. et al., 2007. Efficient tandem polymer solar cells fabricated by all-solution processing. Science (New York, N.Y.), 317(5835), pp.222-5. Available at: http://www.ncbi.nlm.nih.gov/pubmed/17626879 [Accessed June 10, 2011].

Kushnir, D. & Sanden, B.A., 2008. Energy Requirements of Carbon Nanoparticle Production. Journal of Industrial Ecology, 12(3), pp.360-375.

Lambourne, R. & Strivens, T.A., 1999. Paint and Surface Coatings. Theory and Practice,

Liang, Y. et al., 2009. Development of New Semiconducting Polymers for High Performance Solar Cells. Journal of American Chemical Society, 131, pp.56-57.

Ma, W. et al., 2005. Thermally stable efficient polymer solar cells with nanoscale control of the interpenetrating network morphology. Advanced Functional Materials, 15(10), pp.1617-1622.

Marković, Z et al., 2007. Comparative Process Analysis of Fullerene Production by the Arc and the Radio-Frequency Discharge Methods. Journal of Nanoscience and Nanotechnology, 7, pp.1357-1369.

Meyer, D.E., Curran, M.A. & Gonzalez, M.A., 2009. An examination of existing data for the industrial manufacture and use of nanocomponents and their role in the life cycle impact of nanoproducts. Environmental Science & Technology, 43(5), pp.1256-1263.

Park, S.H. et al., 2009. Bulk heterojunction solar cells with internal quantum efficiency approaching 100%. Nature Photonics, 3(May), pp.297-303.

PlasticsEurope, 2010. PlasticsEurope's Eco-profiles. Available at: http://www.plasticseurope.org/.

Reich-weiser, C., Dornfeld, D. & Horne, S., 2008. Greenhouse Gas Return on Investment: A New Metric for Energy Technology. In 15th International CIRP Conference on Life Cycle Engineering.

Roes, A.L. et al., 2009. Ex-ante Environmental and Economic Evaluation of Polymer Photovoltaics. Progress in Photovoltaics: Research and Applications, 17, pp.372-393.

Siddiki, M.K. et al., 2010. A review of polymer multijunction solar cells. Energy & Environmental Science, 3, pp.867-883.

Som, C. et al., 2010. The importance of life cycle concepts for the development of safe nanoproducts. Toxicology, 269, pp.160-169.

Szépvölgyi, J. et al., 2006. Effects of Precursors and Plasma Parameters on Fullerene Synthesis in RF Thermal Plasma Reactor. Plasma Chemistry and Plasma Processing, 26(6), pp.597-608.

Takehara, H. et al., 2005. Experimental study of industrial scale fullerene production by combustion synthesis. Carbon, 43, pp.311-319.

Xue, J. et al., 2005. A Hybrid Planar-Mixed Molecular Heterojunction Photovoltaic Cell. Advanced Materials, 17(1), pp.66-70.

5

Investigating New Materials and Architectures for Grätzel Cells

Alex Polizzotti, Jacob Schual-Berke, Erika Falsgraf and Malkiat Johal*
Pomona College
USA

1. Introduction

A third-generation photovoltaic cell known as a Gratzel or Dye-Sensitized Solar Cell (DSSC) has been gaining attention due to its high lab efficiencies, ease of manufacture, and earth-abundant, nontoxic composition. These cells promise to be marketable in the near future, but several hurdles must still be overcome before Gratzel Cells can be mass-produced. This chapter outlines the technological theory behind Gratzel cells, summarizes the state-of-the-art of DSSC research, and identifies the major challenges that still must be addressed to reduce recombination and increase voltage, current density, and cell lifetime.

2. Grätzel cell overview

Some of the most promising photovoltaic technologies are the emerging third-generation technologies. These range from polymer cells to cells made from proteins extracted from jellyfish (Chiragwandi *et al*). One variety in particular, known as the Grätzel cell, has gained prominence due to its very low cost and ease of manufacture. Developed in 1992 by Michael Grätzel at the Ecole Polytechnique in Switzerland (O'Regan & Grätzel, 1991), these biomimetic devices represent a significant departure from typical silicon solar cells. Unlike p-n junction cells, Grätzel cells, also known as dye-sensitized solar cells (DSSCs), work much like a photosynthetic plant cell (O'Regan & Grätzel, 1991).

The heart of a DSSC is the dye, which operates much like chlorophyll in a photosynthetic plant cell. This dye, usually a ruthenium-based organometallic complex, is responsible for harvesting photons. When a photon hits the dye, an electron is excited from the ground-state (HOMO) into the first excited state (LUMO), leaving behind a hole. In other words, the dye is responsible for generating an exciton (O'Regan & Grätzel, 1991).

Just as with a silicon cell, the hole and excited electron will recombine if they are not separated. Thus, in order to ensure an efficient design the electron and hole must be quickly separated from each other to prevent recombination. This separation has two parts: first, the excited electron must be brought to one side of the cell – the anode. Second, the hole must be transported to the opposite side of the cell – the cathode.

The first part – transporting the electron – is typically performed by a titania (TiO_2) nanostructure and is known as the electron transport layer (ETL). This titania nanostructure is mesoporous, meaning that it has a pore structure on the order of 2-50 nanometers, and is annealed directly onto the anode. The dye molecules are introduced into this titania matrix via immersion in a dye solution, where they chelate to the TiO_2 directly. When electrons are excited in the dye molecule, they are able to flow very quickly from the dye molecule LUMO into the lower-energy TiO_2 conduction band, where they flow down to the anode's conduction band, which is still lower in energy (Duffy *et al.*, 2000). The importance of the nanoporous semiconductor layers, not only on electron transport, but also on increasing photon absorption are described in section 3.

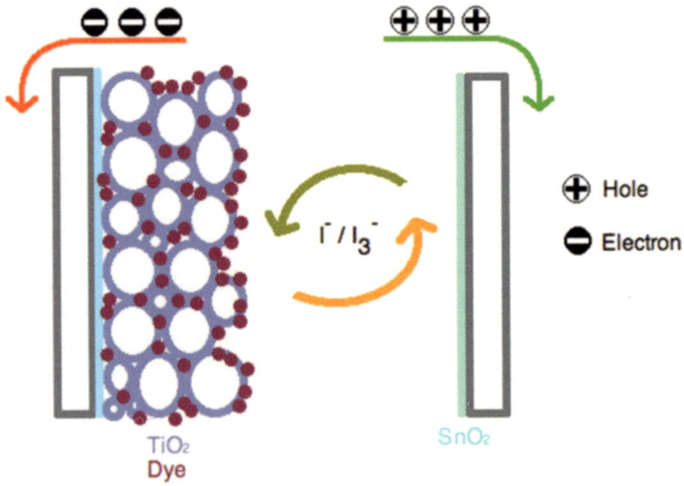

Fig. 1a. Structure of a Grätzel Cell. A mesoporous titania nanoscaffolding houses small dye molecules, which harvest light and generate excitons. Electrons flow through the titania to an anode mounted on glass substrate, and the dye's electrons are regenerated by the cathode via a redox couple such as I^-/I_3^-.

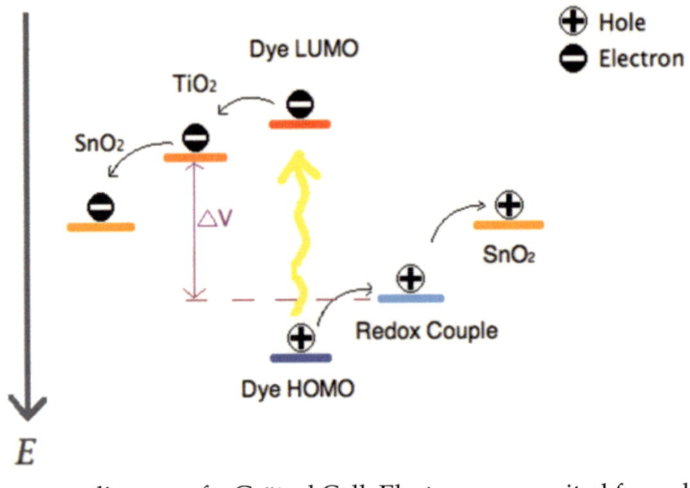

Fig. 1b. Typical energy diagram of a Grätzel Cell. Electrons are excited from dye HOMO to LUMO, where they transfer to the conduction band of the TiO_2 nanostructure, and on to the anode. Meanwhile, holes left in dye LUMO travel up in energy through the redox couple, and then up to the cathode. The maximum attainable voltage is believed to be approximately determined by the energy difference between the Fermi energy of illuminated electrons in the semiconductor and the redox energy of the HTL.

The second task – transporting the hole – is performed by the hole transport layer (HTL) and can be accomplished in a variety of ways. The first dye-sensitized cells created by Grätzel *et al.* utilized an iodide / triiodide redox couple, where iodide regenerates the oxidized dye molecules. The resulting oxidized species, mostly triiodide, are transported to the cathode via diffusion, where they are reduced back to iodide. In this way, positive charge carriers (holes) are transported from dye molecules to the cathode (O'Regan & Grätzel, 1991). A variety of new HTL materials have emerged since the first Grätzel cell, ranging from new liquid electrolytes through solid-state conjugated polymers to quasi-solid polymer-electrolyte mixes. The function of the HTL and ways to improve it are described in section 4. Regardless of materials, the electrical characteristics of these cells are still not completely understood. The voltage, specifically, is not well explained in the literature (Li *et al.*, 2011; Johansson *et al.*, 2005). Voltage is, ultimately, the difference in work function under illumination of the two electrodes from which voltage measurements are taken. In other words, this voltage is a result of difference in electrochemical potential of mobile electrons under illumination at the electrodes (Cahen *et al.*, 2000). In a DSSC, the electrochemical potential at the electrodes is primarily determined by differences in energy between the HTL and ETL - the maximum possible photovoltage is widely agreed to be the difference in energy between the Fermi energy of electrons in illuminated TiO_2 (roughly approximated as the conduction band edge) (Duffy *et al.*, 2000), and the redox potential of the liquid electrolyte (O'Regan & Grätzel, 1991). In the case of a solid-state device, the difference between TiO_2 conduction band and the HOMO level of the HTL material determines the maximum voltage instead (Johansson *et al.*, 2005).
However, the observed photovoltage is rarely this maximum value – for example, similar cells using Poly(3,4-ethylenedioxythiophene) poly(styrenesulfonate) (PEDOT:PSS) and poly(4-undecyl-2,20-bithiophene) (P3PUT) differ in photovoltage by 0.7 V, even though HOMO levels of these polymers lie 1 eV apart (Johansson *et al* 2005; Smestad *et al.*, 2003). One of the major detractors of voltage is recombination, the process by which excited electrons and holes recombine, giving off energy as photons. This not only detracts from current, since it means fewer electrons doing useful work, but it also lowers the observed voltage (Ghadiri *et al.*, 2010; Ito *et al.*, 2008; Saito *et al.*, 2004). Because recombination causes fewer electrons to reach the anode, the Fermi energy of electrons at the anode will be decreased. Similarly, fewer holes reaching the cathode will increase the energy of electrons at the cathode. Photovoltage, which is determined by the difference between the two electrodes, is therefore decreased. This and other factors can seriously affect the photovoltage, thus complicating the mechanism for establishing a photovoltage.
Though these cells are nascent and their properties still being explored and optimized, DSSCs have achieved moderate efficiencies of over 11%, which is encouraging given the extremely low cost of these cells. Although these cells are not yet marketable, improvements in each of the three main components (Dye, ETL and HTL), as well as improved reliability and lifetime, will help Grätzel cells to achieve their full industrial potential.

3. The sensitizer

The dye in a Dye Sensitized Solar Cell has one of the most fundamental tasks. This molecule is responsible for photon harvesting and generation of an exciton. The dye, in other words, uses light to create the charge separation that is the goal of any electrochemical cell. This dye, though a complex molecule, is very simple functionally: it must only satisfy a few key requirements. The first of these is that it must absorb radiation across a broad visible spectrum. The second is that it must be small enough to enter the pores in the nanoporous semiconductor lattice. The third is that once inside the lattice, it must be able to chelate to the TiO_2 or other semiconductor. The last, and potentially the most challenging requirement, is that the dye must be able to accomplish the other three requirements while being cheap and environmentally benign.

The majority of incoming solar power is due to visible wavelengths. Thus, a dye must be able to absorb across these wavelengths - it must appear black. Porphyrin-based sensitizers have been shown to exhibit such effects. Figure 2 demonstrates the broad-spectrum absorption properties of N-719 and Z-907. Note that the wavelengths at which these dyes absorb photons are determined by the band gap properties of the dye, while the magnitude of the absorption is due to the molar absorption coefficient, ε, of the dye. In order for a dye to be effective, it must have a high value for ε in addition to exciting over a wide spectrum.

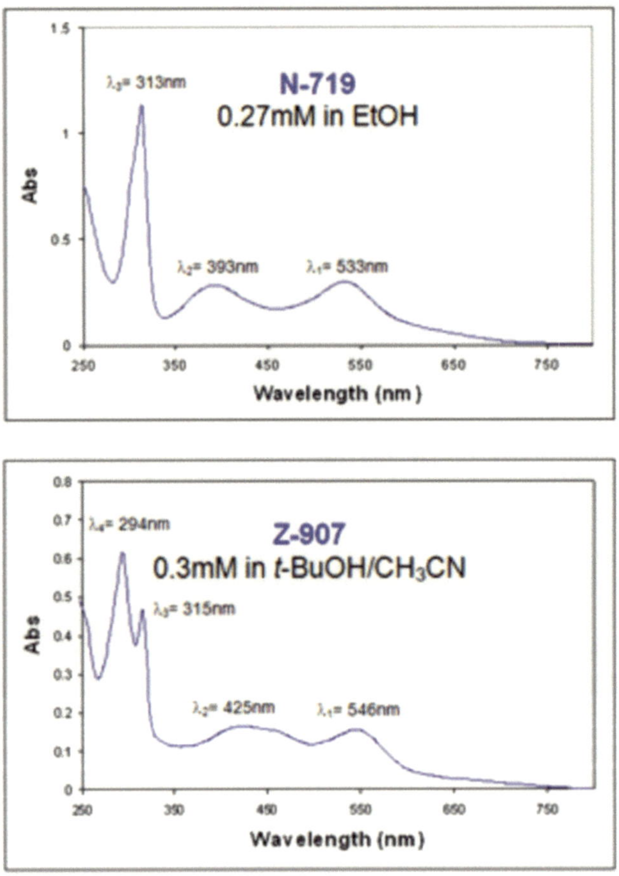

Fig. 2. UV/Vis absorption spectra of Ru-based dyes N719 and Z907. Reproduced from (Desilvestro & Hebting, 2011), with permission.

The second requirement of the dyes, that they must be able to enter the anatase structure, is relatively easy to satisfy. Though in the original Grätzel article the large size of the dye molecule prevented complete saturation of chelation sites (O'Regan & Grätzel, 1991), dye molecules are typically on the order of 1 nm in diameter (Katoh *et al.*, 2010), and thus are easily able to penetrate into the 2-50 nm pores characteristic of mesoporous materials. Moreover, further work is being done into alternative semiconductor morphologies such as anatase nanotubes that will increase dye access to adsorption sites. This research is further discussed in section 3.

Third, the dye must be able to chelate to the semiconductor structure. This is usually accomplished via carboxylate or carboxylic acid groups, with the deprotonated species being far more soluble (N3 and N719 dyes, two of the most common, are identical except for this difference). Chelation can also be accomplished via phosphate and sulfonate groups (Johannsson *et al.*, 2005).

3.1 Ruthenium dyes

Ruthenium-based dyes have met the described requirements (i.e., photon absorption, spreading in porous media and chelation to the semiconductor) superbly. Dyes made of ruthenium-based organo-metallic complexes have achieved overall conversion efficiencies of greater than 11% (O'Regan & Grätzel, 2004). The carboxyl moieties on the molecules allow them to chelate to the TiO_2 surface, they have broad-spectrum absorption, and exhibit high molar extinction coefficients. N3 (fig. 3), for instance, has absorption maxima at 518 and 380 nm with a high value of $\varepsilon = 1.35*10^4$ M^{-1} cm^{-1} in alkaline solution at 500nm (Katoh *et al.*, 2010). N719, the partially deprotonated salt of N3, has near-identical properties (ε=$1.42*10^{-4}$ M^{-1} cm^{-1}) (G.W. Lee *et al.*, 2010) plus enhanced solubility, while Z907 is a hydrophobic variant with two long photoactive carbon chains that shows broad absorption across the visible range. The structures of these dyes, along with that of N749, are shown in figure 3.

Further advancements in dye engineering have produced even more functional compounds. Many dyes are unable to absorb in the red wavelengths, but black dye (N749), with its three thiocyanate functional groups, absorbs wavelengths in the red and near IR range, giving it a dark brown-black color. Figure 4 shows that the photocurrent response of black dye extends to about 900 nm while that of N3 is attenuated in the range of 750-900 nm. Despite a lower value for ε of $7.7*10^3$ M^{-1} cm^{-1} (G.W. Lee *et al.*, 2010), cells using black dye have achieved a conversion efficiency of 10.4% in full sunlight, approaching the maximum efficiencies of N3 (Grätzel, 2004), because of this spectral advantage. As well, Chiba et al. have built cells with black dye that yielded efficiencies of 11.1% when scattering particles were added to TiO_2 to increase haze (ratio of diffused transmittance to total optical transmittance) (Chiba *et al*, 2006). Other emerging ruthenium sensitizers include HY2, which has absorption peaks at 437 and 550nm with a molar absorption coefficient of 1.98 x 10^4 $M^{-1}cm^{-1}$ at the latter wavelength, significantly higher than that of N3 at its peak wavelength of 530nm. DSSCs with HY2 have attained efficiencies of 8.07% (Chen *et al*, 2009). A ruthenium sensitizer called K20 has recently been submitted for a patent with the remarkably high molar absorption coefficient of 2.2 x 10^4 $M^{-1}cm^{-1}$ at 520nm, a significant improvement over N3, N719, and black dye. Characterization of DSSCs sensitized with K20 has not yet been pursued (Ryan, 2009).

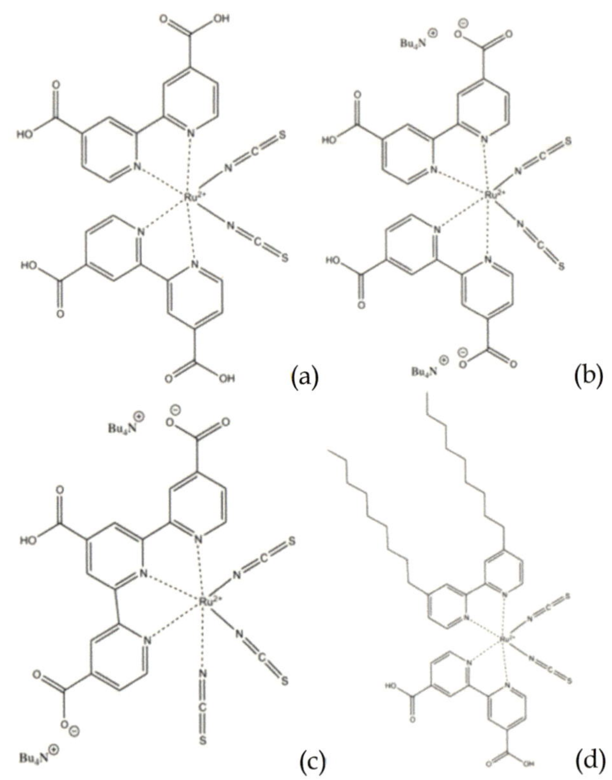

Fig. 3. Molecular structures of ruthenium complex dyes: (a) N3, IUPAC name cis-bis(isothiocyanato)bis(2,2'-bipyridyl-4,4'-dicarboxylato)-ruthenium(II); (b) N719, IUPAC name cis-diisothiocyanato-bis(2,2'-bipyridyl-4,4'-dicarboxylato) ruthenium(II) bis(tetrabutylammonium); (c) Black dye (N749), IUPAC name triisothiocyanato-(2,2':6',6"-terpyridyl-4,4',4"-tricarboxylato) ruthenium(II) tris(tetra-butylammonium); and (d) Z907, IUPAC name cis-disothiocyanato-(2,2'-bipyridyl-4,4'-dicarboxylic acid)-(2,2'-bipyridyl-4,4'-dinonyl) ruthenium(II).

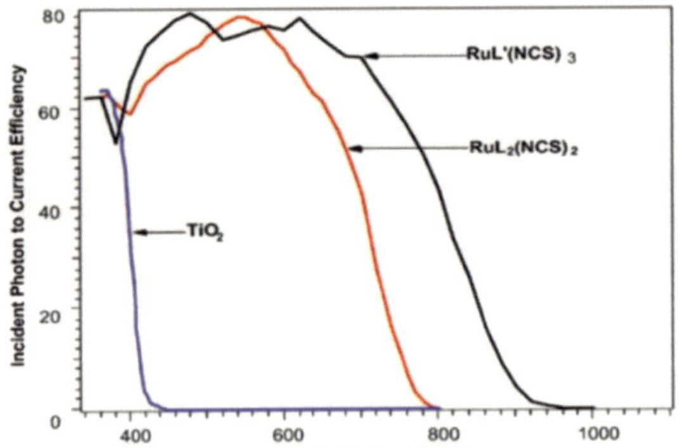

Fig. 4. Photocurrent action spectra of N3 (ligand L), black dye (ligand L'), and bare TiO_2. Reproduced from (Grätzel, 2004), with permission.

All of these dyes have several beneficial shared properties. Ruthenium-based complexes exhibit metal-to-ligand charge transfer, a process in which the excited electron in a d-orbital of Ru is transferred to the π^* orbital of the carboxyl ligand, from which it is injected into the conduction band of TiO_2. This transfer is fast and irreversible, and will be discussed further in section 3. Moreover, they are remarkably stable over long periods of time. In the 1991 article by Grätzel and O'Regan, dyes sustained $5*10^6$ turnovers without serious decomposition (O'Regan & Grätzel, 1991). More modern dyes are even more stable - in its solid state, N3 can withstand temperatures up to 280°C, and maintains high performance over 10^8 redox cycles, the equivalent of 20 years of use in sunlight (Grätzel, 2004).

If pure functionality were the only requirement for these dyes, then perhaps alternatives to these existing dyes would not be needed. However, the last and most difficult requirement is that these dyes must be cheap, nontoxic, and environmentally benign. The greatest barrier to such a dye is the use of ruthenium as a metal center in these metal complex dye molecules. World production of ruthenium is approximately 12 metric tons (Lenntech), and though it is found in the rare minerals laurite, ruarsite, and ruthenarsenite, its commercial recovery is limited almost entirely to trace elemental amounts in nickel deposits in Africa and the Americas. Due to its rarity in difficulty to obtain, Ruthenium is expensive even in small quantities. For instance, a 10 mg sample of N-3 from Solaronix currently costs approximately $244 USD (Solaronix.com as Ruthenizer 535). Moreover, ruthenium complexes are potentially hazardous to health and the environment. Though not yet fully studied for toxicity in their own right, when heated in air ruthenium complexes form ruthenium tetroxide, a highly volatile and toxic compound that damages the eyes and the upper respiratory system (Dierks, 2006). For all these reasons, a variety of entirely organic dyes as well as dyes using different metal centers are being developed for use in DSSCs.

3.2 Ru-free porphyrin dyes

Several main approaches have emerged to bypass the need for ruthenium dyes. One of the most common is to modify the dye structure to allow for a different metal center such as zinc or magnesium. Porphyrins, a class of naturally occurring biological compounds that include chlorophyll and hemoglobin, are known to absorb radiation in the visible range due to a conjugated system of pyrolles, frequently complexed around a metal ion center (iron in the case of hemoglobin, and magnesium in the case of chlorophyll). As chlorophyll is the biological inspiration for many of the dyes in DSSCs today, much work is being done currently to produce porphyrin-based, ruthenium-free organometallic dyes. The Grätzel lab recently developed a porphyrin dye called YD-2 that achieved an 11% power conversion efficiency in a 16μm TiO_2 layer, the highest efficiency demonstrated by a ruthenium-free sensitizer. The structure of YD-2 is shown in figure 5.

Moreover, chlorophyll itself is relatively simple to modify, which allows researchers to create highly biomimetic devices while manipulating properties such as excited-state lifetime and LUMO energy level as well as enhance adsorption to TiO_2 (Wang *et al.*, 2010b). Chlorophyll (Chl) comes in several distinct forms, most commonly Chl *a*, Chl *b*, and Chl *c*; the latter two are typically found as accessory pigments in light-harvesting complexes, which serve to absorb additional solar radiation and transport excited electrons to the reaction center, which contains Chl *a*. (Figure 6a). In an effort to mimic the multimolecular approach used by plants to harvest light, researchers constructed DSSCs with co-sensitized Chl derivatives, which were labeled *a*-type, *b*-type, or *c*-type depending on which Chl they structurally resembled. Combinations of *a*-type Chl sensitizers with *c*-type Chl sensitizers resulted in augmented power conversion efficiencies of 5.4% (Wang *et al.*, 2010a). This group subsequently developed a Chl-*a* derivative, chlorin-3, with a dodecyl ester group at C17, where a carboxyl group would normally be. DSSCs with chlorin-3 obtained a conversion efficiency of 8% with an AcCN redox electrolyte HTL, the highest efficiency for any chlorophyllous sensitizer to date. (Wang *et al.*, 2010b).

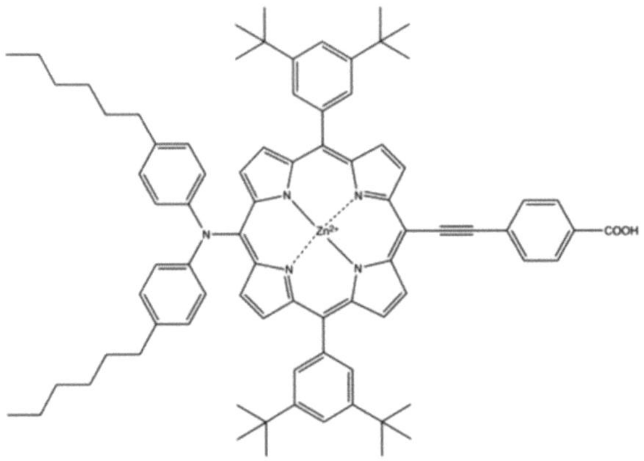

Fig. 5. Molecular structure of YD-2.

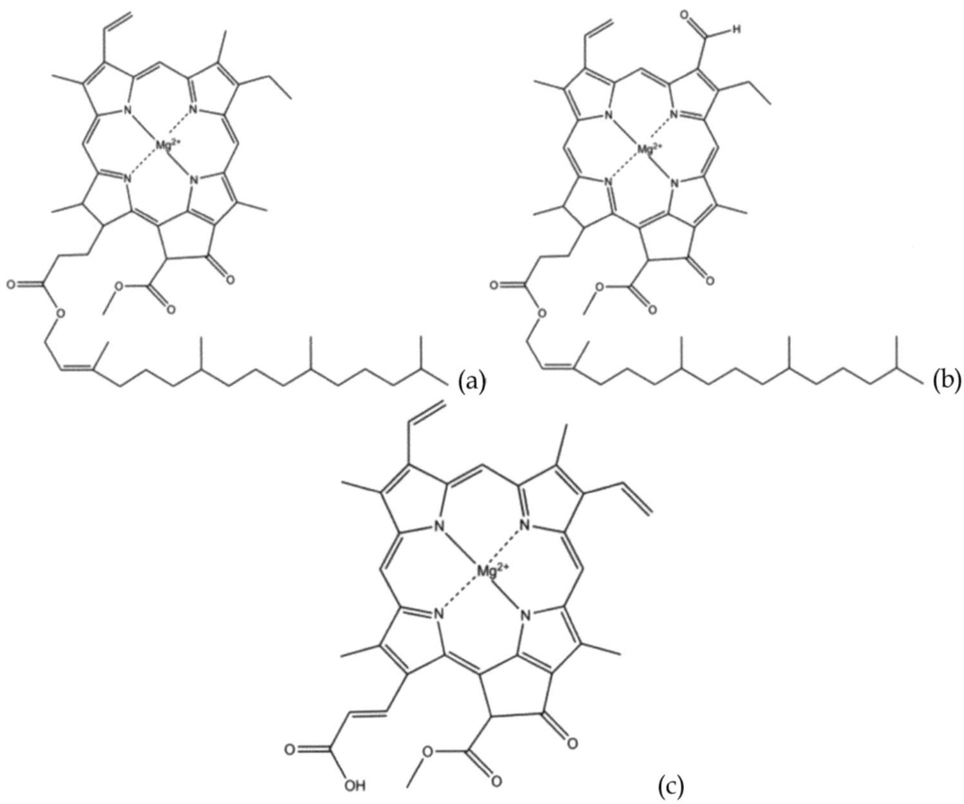

Fig. 6. Molecular structures of some chlorophyll: (a) Chlorophyll *a*; (b) Chlorophyll *b*; and Chlorophyll *c*2.

3.3 Fully organic dyes

Another fast-growing method for doing away with ruthenium in sensitizers is to create completely organic sensitizers without the need for a metal center at all. Indoline sensitizers such as D149 show promise for their extremely high molar absorption coefficient of 6.87 x 10^4 $M^{-1}cm^{-1}$ at 526 nm. DSSCs with D149 have achieved efficiencies of 9.0% and 6.7% in cells with acetonitrile-based electrolytes and cells with ionic liquid (IL) electrolytes, respectively (Ito *et al.*, 2006). Subsequent work using the indoline sensitizer D205 with IL electrolytes achieved an efficiency of 7.2%, the highest efficiency obtained in DSSCs with organic dyes and an IL HTL (Kuang *et al.*, 2008).

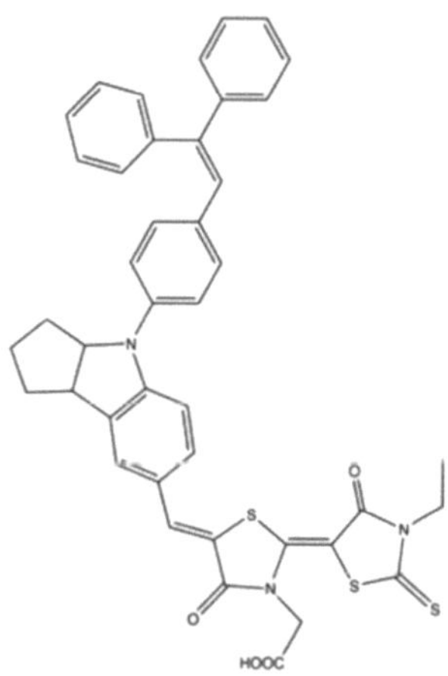

(a)

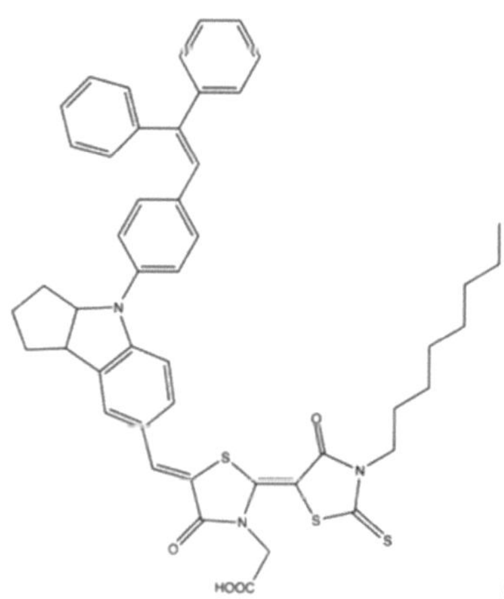

(b)

Fig. 7. Molecular structures of indoline sensitizers: (a) D149; and (b) D205.

Yum *et al.* have demonstrated efficiencies of 7.43% and respectable absorbances in the red/near IR region using cells cosensitized with JK2 and SQ1, both organic, non-porphyrin molecules (Figure 8). JK2 and SQ1 have impressive molar absorption coefficients of 4.2 x 10^4 $M^{-1}cm^{-1}$ at 452 nm and 15.9 x 10^4 $M^{-1}cm^{-1}$ at 636 nm, respectively (Yum et al., 2007).

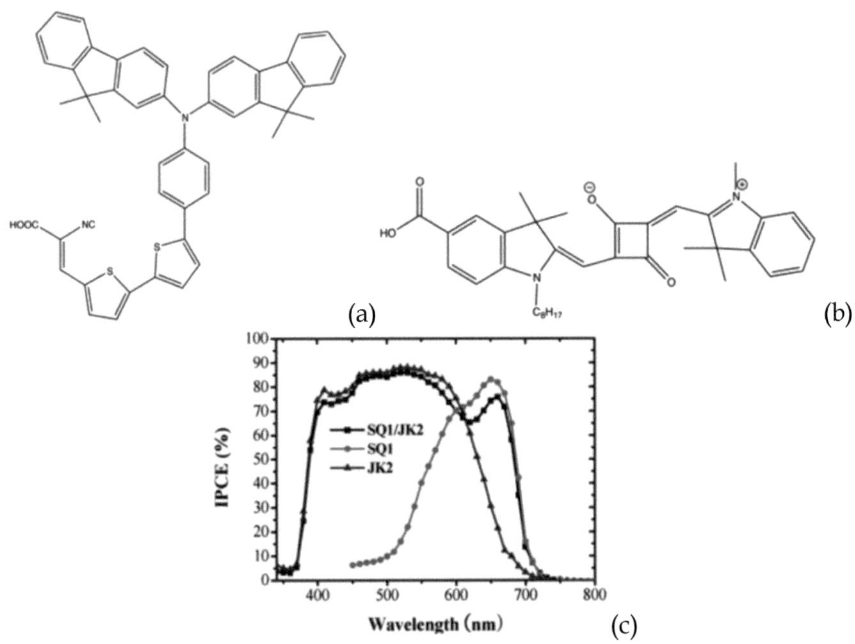

Fig. 8. (a) Structure of JK2; (b) structure of SQ1; and (c) IPCE profile of cells sensitized with SQ1 (light gray), JK2 (dark gray), and SQ1/JK2 (black). Graph reproduced from (Yum *et al.*, 2007), with permission.

Advanced dye engineering has produced superior organic dyes. However, the simplest of these molecules can be found as the colorful dyes in berries and flowers. Anthocyanins and betalains are two classes of biological molecules being investigated for their photoactive properties. Anthocyanins are colorful flavonoids that are found in fruits such as blackberries and raspberries. Though they can be modified - their energy absorption is pH sensitive with acidic solutions turning many anthocyanins into a red dye (Calogero *et al.*, 2009) – their overall conversion efficiencies are quite low. DSSCs sensitized with juice from red Sicilian oranges (*Citrus Sinensis*) achieved an efficiency of 0.66%(Calogero & Di Marco, 2008). Nevertheless, anthocyanins are easily extracted from plants and are both abundant and diverse. Betalains are found in Caryophyllales plants, which include red turnip, wild purple sicilian prickly pear, and bougainvillea flowers. Betalain pigments include betanine, which has a red-purple coloration, and betaxanthins, which are yellow-orange; both compounds have carboxyl functional groups, allowing them to readily chelate to TiO_2. DSSCs sensitized with betalains from red turnip yielded an overall conversion efficiency of 1.75%, a remarkable value for nonsynthetic, biological dyes (Calogero *et al.*, 2009). These biological sensitizers are in the initial stages of development and may be a nontoxic and inexpensive option for DSSCs of the future.

Long-term stability is a factor to consider for all dyes, and the stability of ruthenium complexes is the main reason that they are still the standard sensitizers in high-performance DSSCs. However organic dyes are also gaining the potential to exhibit long-term stability; a sensitizer called D21L6 sustained 1000 hours of light soaking at 60°C and maintained an efficiency of 90% of the initial value, an unprecedented degree of stability for organic sensitizers. Furthermore, D21L6 has a molar absorption coefficient of $3.7 \times 10^4\ M^{-1}cm^{-1}$ at 458 nm, superior to ruthenium-based dyes, and have achieved overall conversion efficiencies of 7.25% and 4.44% in cells with AcCN electrolytes and solid-state HTL material, respectively (Yum et al., 2009). This combination of stability and efficiency indicates that D21L6 and other organic dyes may be able to replace ruthenium complexes as the preferred dyes in DSSCs.

The high price and limited availability of ruthenium dyes is one of the main drawbacks to DSSCs. However, new advancements in dye fabrication promise to take the ruthenium out of Grätzel cells and significantly improve their watt-to-cost ratio. Though ruthenium dyes are still the most commonly used dyes, both organic and porphyrin-based organometallic dyes are poised to become the cheap and nontoxic alternatives for future cells.

4. The Electron Transport Layer (ETL)

Sensitizers generate excitons, thus creating the initial charge separation that is vital to the function of any solar cell. However, these excitons are very short lived, and excited electrons tend to recombine with the holes to which they are still coulombically (i.e. electrostatically) bound. Thus, an effective semiconductor electron-transport matrix is essential in a DSSC to efficiently separate charge.

Optically transparent microfilms of anatase TiO_2 (and some other metal oxide semiconductors such as ZnO) have been found to effectively transport charge from an excited dye to an electrode. However, charge transport alone is insufficient. A monolayer film of dye adsorbed onto a one-dimensional TiO_2 layer yields very low-efficiency solar cells, well below 1% (O'Regan & Grätzel, 1991). This low efficiency is due to the poor absorption properties of dyes in a monolayer - less than 1% of incident light is absorbed by a monolayer of dye on a flat TiO_2 surface (O'Regan & Grätzel, 1991).

Though Grätzel cells differed from previous dye-sensitized photoelectrical cells by introducing a more efficient dye, the true innovation that made the first Grätzel cell so significant was the use of a highly mesoporous (pores in the range of 2-50 nm) semiconductor layer. This layer, fabricated by the deposition of TiO_2 or another metal oxide semiconductor in a colloidal solution followed by drying and sintering, allows dye to be adsorbed in the pores of the semiconductor, thus vastly increasing the amount of dye adsorbed onto the material. Figure 9 shows scans of a typical anatase mesoporous film.

To give an idea of the impact of such a mesoporous layer, the 10 micrometer thick layer of 15 nm average diameter TiO_2 particles used in the original Grätzel cells, packed cubically, is expected to have a roughness factor of 2000. In other words, these films are expected to produce a 2000-fold increase in surface area as opposed to a perfectly flat film. In reality, Grätzel and O'Regan found that this first attempt at a mesoporous TiO_2 anode only produced a roughness factor of 780 (O'Regan & Grätzel, 1991). This discrepancy was attributed to "necking" - inter-particle fusions that decrease surface area - as well as the fact that some parts of the porous structure were too small for the relatively large dye molecules to penetrate. Regardless of these imperfections, a new method had been demonstrated for extremely high adsorption of dye molecules into a small area while still allowing contact between dye and liquid electrolyte. With this remarkably simple mechanism, which allowed for 7% efficiencies in ambient sunlight, and 46% harvesting of incident photons (O'Regan & Grätzel, 1991), dye-sensitized photoelectrical cells began to show real promise of future commercial viability.

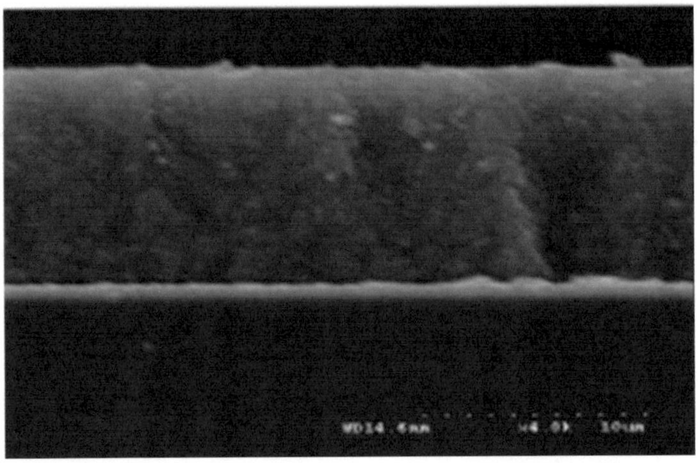

a)

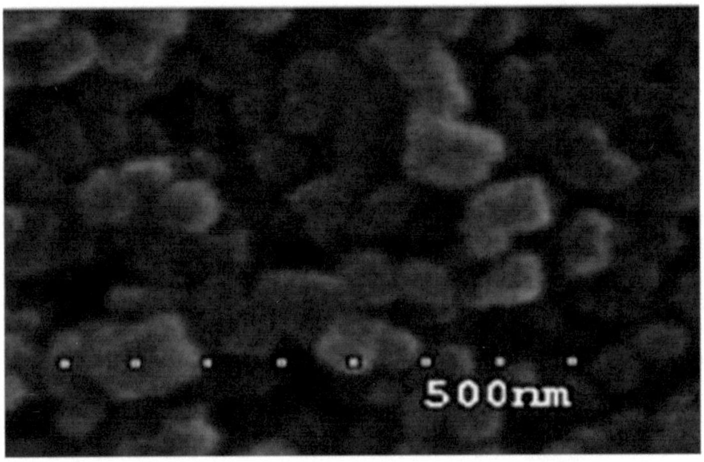

b)

Fig. 9. a) cross section view and b) top view of SEM scans of anatase TiO_2 films prepared via doctor blade method. Reproduced from (Singh *et al.*, 2008), with permission.

One great advantage to a mesoporous TiO_2 structure aside from higher dye packing is its ability to trap light within its pore structure. Instead of photons hitting a flat surface (and either being absorbed or reflecting away), unused photons are scattered by TiO_2 nanoparticles until they either escape or are absorbed by a dye molecule. This process repeats, immensely increasing a photon's chance of interacting with dye. As opposed to the 1% light harvesting efficiency seen on a flat ETL, recent papers have achieved over 95% harvesting of incoming photons (Ghadiri *et al.*, 2010).

Mesoporous TiO_2 layers are effective because they allow for very dense dye adsorption, and increase the path length of light inside their porous structure. These factors contribute to a sufficiently high rate of exciton generation upon illumination – in other words, photons have a high probability of reflecting around inside the crystal TiO_2 structure until they hit one of the densely adsorbed dye molecules. However once this occurs, the TiO_2 must then accept this excited electron from the dye and also transport the excited electron to the anode before it has a chance to recombine with holes in the oxidized dye or in the HTL.

The driving force behind electron injection into the TiO_2 from the dye is twofold. Firstly, there is an enthalpic impetus to electron injection from dye LUMO to the TiO_2 conduction band. Because the conduction band lies below the dye LUMO in energy, it is energetically favorable for an electron to transport from dye to TiO_2 immediately upon excitation (Cahen *et al.*, 2000; Liu *et al.*, 2002). The energy gap $E_{LUMO} - E_{CB}$ between the two materials determines the strength of the enthalpic driving force. For a typical dye such as N3, which has a LUMO energy of -3.74 eV vs. vacuum (Nazeeruddin *et al.*, 1993; Zafer *et al.*, 2005), injecting electrons into mesoporous anatase TiO_2 with conduction band edge at -4.1 eV vs vacuum (Cahen *et al.*, 2000; Liu *et al.*, 2002; Tang *et al.*, 1994), this energy difference is ~0.34 eV.

The second impetus to electron injection from dye to TiO_2 is entropic. Per unit area, anatase TiO_2 has a much higher density of possible electronic states than the dye molecule. Thus, an electron diffusing from dye to TiO_2 causes a gain in entropy, much like a gas molecule flowing from a small container to a large container. This entropy gain can result in an additional, substantial driving force of up to ~0.1 eV (Cahen *et al.*, 2000).

The advantage of these two driving forces - entropic and enthalpic favorability - is that they are essentially irreversible. Reverse transport of electrons from the TiO_2 conduction band edge to dye LUMO is unlikely (O'Regan & Grätzel, 1991; Cahen *et al.*, 2000). Moreover, since there are essentially no minority charge carriers in TiO_2 (i.e. there are no free holes in the TiO_2), recombination within the semiconductor is unlikely (O'Regan & Grätzel, 1991). The implication is that rough or broken TiO_2 films can still produce perfectly functional Grätzel cells. Limiting recombination is key for a solar cell - as discussed in the introduction, though open-circuit voltage (V_{oc}) is primarily determined by the energy difference between ETL Fermi energy upon illumination and HTL oxidation potential (Nernst potential in the case of a liquid electrolyte redox couple), recombination detracts from this value significantly (Ghadiri *et al.*, 2010; Li *et al.*, 2011; Peng *et al.*, 2004;).

Thus, the majority of recombination in a Grätzel cell is due to interactions between electrons in the TiO_2 semiconductor and anode material, and holes in the HTL. For a liquid iodine redox couple, this means that oxidized species of iodine, such as I_2^- and I_3^-, can absorb electrons in the TiO_2 matrix. However, this recombination requires electrons in the TiO_2 nanostructure to be within tunneling distance of the HTL (Cahen *et al.*, 2000), or within approximately 3 nm. In a system where particles have an average diameter of 20 nm, this means that only electrons on the surface of TiO_2 particles have the possibility of recombining at all. Moreover, this implies that the recombination rate is inversely proportional to particle size, and directly proportional to the specific surface area.

There are other mechanisms that prevent recombination in Grätzel cells. Among the most important of these is ionic screening in a liquid electrolyte cell. Iodine and counterions shield holes in the HTL and electrons in the ETL from their coulombic attraction, thus decreasing the likelihood that these particles will recombine (Cahen *et al.*, 2000).

These and other lesser mechanisms make for a very low rate of recombination in Grätzel cells, especially those employing a liquid electrolyte (O'Regan & Grätzel, 1991; Cahen *et al.*, 2000; Peng 2004). This low recombination rate results in a relatively high fill factor (i.e. the ratio of observed efficiency to theoretical max efficiency). The fill factor for the first DSSC was 0.684 (O'Regan & Grätzel, 1991), and recent cells attain values of up to 0.75 (Ghadiri *et al.*, 2010). To give a comparison, the fill factor for an average commercial silicon cell is ~0.83, while some advanced inorganic thin-film cells have achieved almost 0.90 (Green, 1981; PVEducation.org). Grätzel cells therefore compare very favorably with current commercial cells in this respect.

Thus, four important characteristics have emerged that work to make a TiO_2 mesoporous semiconductor layer more effective than a flat electrode. First, the TiO_2 nanostructure provides a very high effective surface area for dye molecule adsorption due to its very high roughness factor. Second, the TiO_2 provides very favorable and effectively irreversible entropic and enthalpic driving forces for charge injection. Third, low recombination within the semiconductor matrix as well as between the semiconductor and the HTL allows for very efficient electron transport to the anode. Fourth, light scattering within the porous crystalline structure increases the radiation path-length and thus improves photon harvesting. And, because of the porous nature of the ETL, this all happens while simultaneously allowing the HTL to come into direct contact with every single dye molecule just like for a flat electrode (O'Regan & Grätzel, 1991).

4.1 Improving the ETL through new morphologies

These four main requirements have guided efforts to improve ETL function further. If dense adsorption of dye onto TiO_2 is responsible for the efficiency of the original Grätzel cells, then increased-surface-area materials and more favorable dye-semiconductor adsorption interactions can make these cells even more effective. Likewise, finding ways to prevent recombination, increase light scattering, and increase the driving force for charge injection will further increase cell performance.

The use of light-scattering semiconductor particles is one of the simplest and easiest ways to improve device efficiency. Ito *et al.* used larger TiO_2 particles on top of a mesoporous TiO_2 layer to scatter photons into the semiconductor pores, increasing path-length of the photons and increasing cell efficiency substantially (Ito *et al.*, 2007).

Other methods have emerged to prevent recombination rates in Grätzel cells. One major problem is contact between the hole-rich HTL and the electron-rich photoanode. In solid-state devices, which use a solid HTL, this is an even larger problem. In the original Grätzel cells, liquid electrolyte provided ionic screening to prevent charge recombination. In a solid device, this phenomenon does not occur. Moreover, some solid HTLs cause a short circuit by providing direct electrical contact between anode and cathode, thus rendering the device non-functional (Peng *et al.*, 2004). A simple solution to this problem is the use of a nonporous semiconductor layer in between the anode and the mesoporous semiconductor layer, shown in figure 10. While providing electron transport in the same fashion as the mesoporous layer, this nonporous "blocking" layer prevents HTL material, whether liquid electrolyte or solid material, from directly contacting the anode (Peng *et al.*, 2004; Saito *et al*, 2004). Work by Peng *et al.* has shown that a blocking layer of TiO_2 can significantly prevent charge recombination at the anode, and that the ideal thickness for a blocking TiO_2 layer is between 120 and 200 nm (Peng *et al.*, 2004). Since the porous layer of titania is on the microscale - typically 10 or more micrometers, this study shows that a significant increase of efficiency is possible using only minimal extra amounts of material.

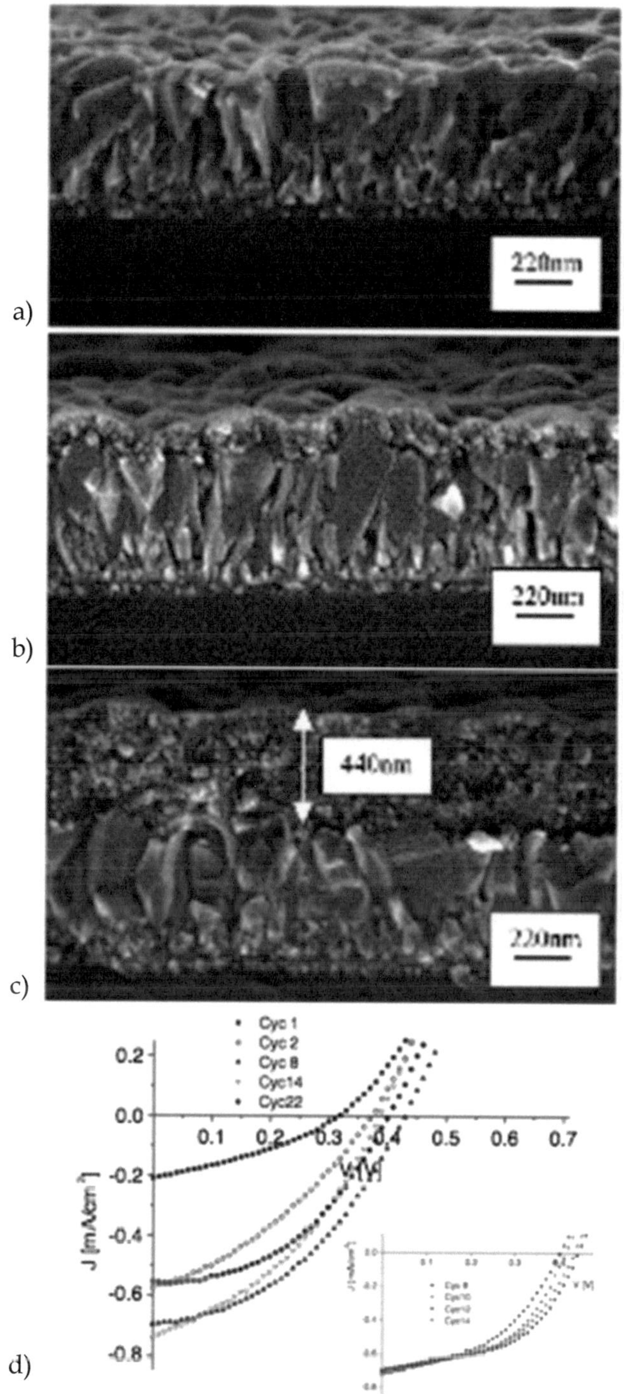

Fig. 10. a) Bare FTO electrode compared to b) 8 and c) 30 spraying cycles of blocking TiO_2. d) I-V curves of cells made with blocking TiO_2 films. Performance is optimized between 8 and 14 spraying cycles. Reproduced from (Peng *et al.*, 2004), with permission.

4.2 Semiconductor nanotubes as ETL

One ETL technology that has been gathering interest recently is the use of semiconductor nanotubes. One of the problems with mesoporous films is that cracks and gaps between TiO_2 particles in the anatase layer create electron traps (Ghadiri *et al.*, 2010; Li *et al.*, 2011). In other words, gaps between nanoparticles in the mesoporous structure make it difficult for electrons to tunnel from particle to particle, thus retarding electron movement through the semiconductor (Mor *et al.*, 2006). This slowing of electrons gives rise to a higher probability of recombination with holes in the HTL. One-dimensional electron transport via semiconductor nanotubes seeks to remedy this problem. Adachi *et al.* reported efficiencies of 4.88% using disordered nanotube arrays in 2003 (Adachi *et al.*, 2003). Mor *et al.* reported the use of highly-ordered TiO_2 nanotubes for improved electron transport in 2006. Though their devices exhibited only 2.9% efficiencies, this was for a nanotube layer only 360 nm thick, or under 1/30 the thickness of the original 10 micrometer films used by Grätzel and O'Regan (Mor *et al.*, 2006; O'Regan & Grätzel, 1991). They noted that efficiency scaled linearly with nanotube length, meaning that dye adsorption appears to be the limiting factor. They hypothesized that with several-micrometer-long nanotubes this system has an ideal limit of ~31% efficiency (Mor *et al.*, 2006). SEM scans of nanotube systems are shown in figure 11.

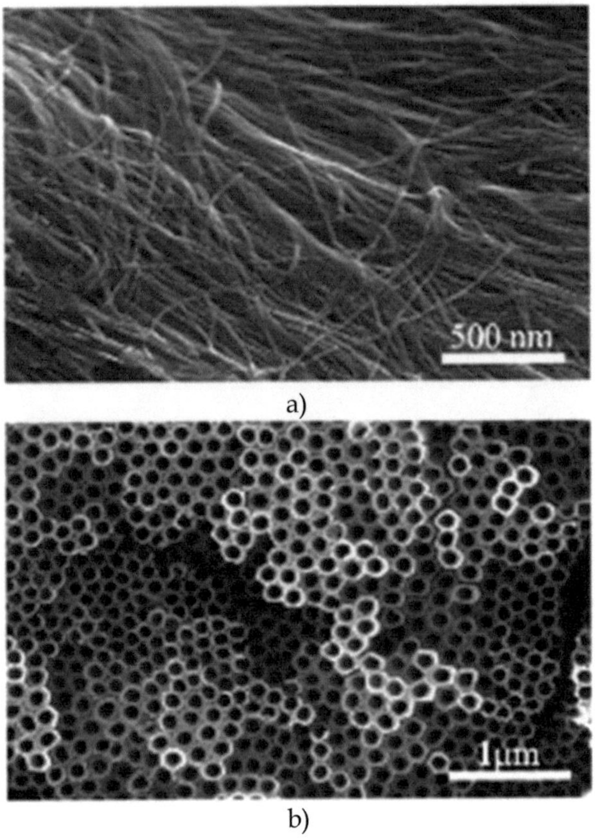

b)

Fig. 11. SEM scans of TiO_2 nanotubes a) without treatment and covered with a layer of nanowires and b) after ultrasonic treatment in ethanol to remove nanowires. Reproduced from (Li *et al.*, 2011), with permission.

The reason for this high ideal efficiency with nanotubes is due to their simple linear architecture. A nanotube, unlike a mesoporous TiO_2 layer, provides very direct and one-dimensional electron transport (Bendall *et al.*, 2011; Ghadiri *et al.*, 2010; Li *et al.*, 2011). The small boundary between TiO_2 particles **in** highly-ordered nanotubes largely bypasses the problem of electron traps caused by gaps in mesoporous structures. Because of this, the average length an electron can travel in these tubes before recombining, known as the diffusion length, is much longer (Ghadiri *et al.*, 2010; Li *et al.*, 2011). It has been estimated by Jennings *et al.* that the diffusion length in a nanotube cell is approximately 100 micrometers (Jennings *et al.*, 2008). As many nanotube layers in this kind of study have only reached 20 micrometers or less, there is great potential for scaling up tube size, and thus effective surface area, without significantly increasing the rate of recombination.

Additionally, nanotubes are more "open" than mesoporous layers. This means that oxidized species in the liquid electrolyte can quickly escape from the tube structure as opposed to being trapped in the pore structure. Once the electrolyte accepts a hole from the dye, it can quickly move to the cathode and be reduced, and thus has less chance of recombining with electrons in the ETL (Ghadiri *et al.*, 2010, Li *et al.*, 2011).

One of the downsides to carbon nanotubes is their high manufacturing cost. They usually require much more energy input than mesoporous layers, and are more time-consuming to fabricate (Bendall *et al.*, 2011; Li *et al.*, 2011). Thus, efforts are being made to decrease cost of these cells while increasing efficiency.

Hollow TiO_2 fibers were recently fabricated by Ghadiri *et al.* through a nanotemplating method, shown in figure 12. By creating cellulose nanofibers, coating them in TiO_2, and then removing the fibers via heat treatment, this group produced hollow TiO_2 fibers similar to nanotubes via a very cheap and effective method (Ghadiri *et al.*, 2010), outlined in figure 13. Moreover, these structures were shown to have 2-3 times faster electron transport than mesoporous films, higher voltage due to lower recombination (despite nanofiber energy actually being *closer* to electrolyte Nernst potential than mesoporous films), and high efficiencies. In fact, cells made with fibers only 20 micrometers long exhibited 7.2% efficiencies, nearly identical to similar devices made with TiO_2 nanoparticles (Ghadiri *et al.*, 2010). There are downsides to this technology, including lower roughness factors and less light trapping than mesoporous TiO_2. These problems can be mitigated by more efficient dyes (thus reducing the need for high-density adsorption), and decreasing fiber diameter to increase surface area. Nevertheless, despite these drawbacks nanotube cells performed similarly to mesoporous ETL cells. Thus, this technology has great potential for improvement and has promise to play a role in future generations of DSSCs.

Nanowires are a variation on the theme of highly-ordered nanostructures for the ETL. ZnO nanowire and nanotube fabrication require much milder conditions than for TiO_2 nanowires and nanotubes, and thus are an attractive option for their low cost of manufacture (Bendall *et al.*, 2011). ZnO also has improved electron mobility as compared to TiO_2, which would lead to potentially longer diffusion length. Generally, ZnO exhibits lower affinity for dye adsorption, and tubes made from ZnO are less stable. However, Bendall *et al.* reported 2.53% efficiency using a mix of ZnO and TiO_2 nanowires, and reported effective dye adsorption with use of a novel organic dye (Bendall *et al.*, 2011). Thus, ZnO remains a potentially attractive alternative to TiO_2 as the ETL material.

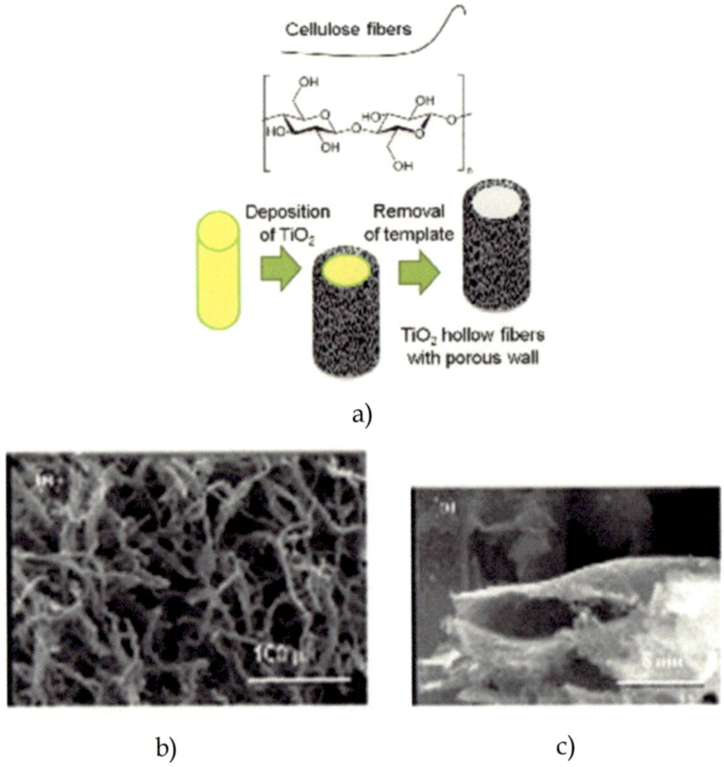

Fig. 12. a) Creation of porous hollow anatase fibers using a template method. b) SEM scan of hollow fibers immediately after template removal. c) cross section SEM scan of one hollow fiber. Reproduced from (Ghadiri *et al.*, 2010), with permission.

The ETL of a Grätzel cell, though usually made of just one material, is as vital as it is simple. Superior surface area allows for hundreds of times more dye to adsorb to surfaces than a flat ETL. Light trapping increases the likelihood that photons will interact with dyes. Low recombination rates due to unfavorable back-reactions combined with protective surface morphologies result in very high fill factors. And while performing all these tasks, anatase and ZnO semiconductor matrices allow light and HTL materials to contact the dye fairly unhindered. New developments in nanotubes and nanowires take these properties to the next level. They allow ultrafast, one-dimensional electron transport from dye to anode while still providing very efficient dye packing and easy access to HTL materials. New breakthroughs in low-cost nanotubes, higher-efficiency mesoporous layers, and nanowires, along with alternative ETL materials such as ZnO, continue to raise the efficiency of Grätzel cells, lower their cost, and further elucidate their inner workings.

5. The Hole Transport Layer (HTL)

The HTL in a DSSC plays a critical role in determining the device's efficiency, closed-circuit voltage, photocurrent and physical architecture. An effective HTL must be able to transfer charge rapidly so as to be able to regenerate the dye quickly and mitigate losses due to charge recombination between oxidized dyes and injected electrons (Hao *et al.*, 2011). In addition, the material must be stable over long periods of time and elevated temperatures in the cell's operating range, around 80-85 °C (Wang *et al.*, 2003). Though the most common and efficient HTL materials to this date are liquid electrolytes, a wealth of new and innovative research is being done on a wide variety of materials to make hole transport cheaper, more effective, and more robust (Helgesen *et al.*, 2009).

5.1 Electrolyte redox couples as HTL

In the original DSSC design described in Grätzel's seminal 1991 publication, the cell utilized an I^-/I_3^- redox couple composed of tetrapropylammonium iodide, iodine, ethylene carbonate and acetonitrile. While the iodine is the active component of the redox shuttle, the other components are necessary to create a liquid electrolyte. Using this formulation, even without any of the significant improvements that have been developed in recent years, the cell exhibited an impressive efficiency in the range of 7-8% (O'Regan & Grätzel, 1991). This relatively high efficiency is due to the excellent mobility of the liquid electrolyte and its ability to make contact with, and rapidly regenerate, the dye inside the ETL. In a follow-up paper, Grätzel also demonstrated the significance of the electrolyte composition by showing that the I^-/I_3^- counterion (e.g. Li^+, tetrapropylammonium) had a significant effect on the photocurrent of the system (Kay & Grätzel, 1996) due to the improved ionic screening provided by these counterions. Among the iodide salts that form room-temperature ionic liquids, 1-propyl-3-methylimidazolium iodide (PMII) has the lowest viscosity, making it the most effective and widely used liquid electrolyte HTL (*vide infra*) (Bai *et al.*, 2008).
The role of iodide in the electrolyte is to regenerate oxidized dye according to the following reduction reactions:

$$Dye^+ + 2I^- \rightarrow I^-_2 + Dye \quad (1)$$

$$2I^-_2 \rightarrow I^-_3 + I^- \quad (2)$$

The resulting triiodide must diffuse away to the cathode where it is reduced back to iodide according to the equation

$$I_3^- + 2e^- \rightarrow 3I^- \quad (3)$$

Thus, the diffusion coefficients of iodine and triiodide are very important in determining the rate at which the electrolyte can regenerate the dye and prevent recombination (Hao *et al.*, 2011). The diffusion coefficients are a function of physical diffusion and exchange reaction constants. Physical diffusion is dictated by viscosity, which increases with electrolyte concentration due to ionic and van Der Waals interactions. A more viscous solution inhibits I^-/I_3^- diffusion, which reduces its ability to regenerate the dye. However, a high electrolyte concentration is necessary to yield sufficient ionic conductivity and enhances the rate of exchange reactions, so there is some optimal electrolyte concentration that balances these factors. Using a binary electrolyte system consisting of 1-ethyl-3-methylimidazolium dicyanamide (EMIDCA) / PMII with a fixed iodine concentration and a varying EMIDCA volume fraction, Lin *et. al.* concluded that a volume fraction of 10 vol% EMIDCA produced the maximal balance between physical diffusion and exchange reactions, leading to the optimal electrolyte composition. Indeed, when fabricated into a working cell, this electrolyte led to the highest efficiency of 5.20% (Figure 13) (Hao *et al.*, 2011).

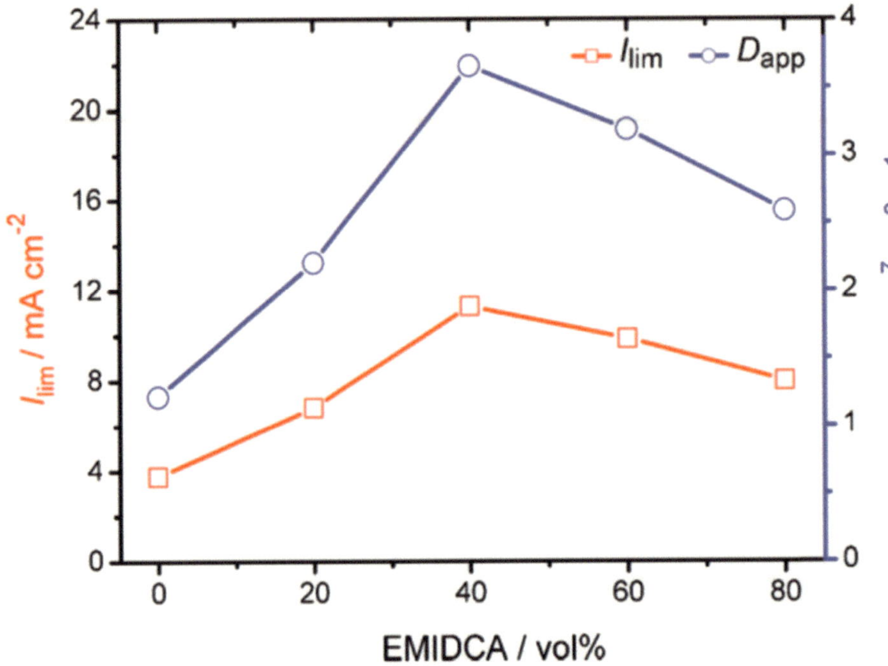

Fig. 13. Dependence of the limiting current density and the apparent diffusion coefficient on the EMIDCA volume fraction in the binary IL electrolyte. Reproduced from (Hao *et al.*, 2011), with permission.

Since 1991, numerous liquid electrolytes have been tested in DSSCs (often by varying the counterion), with the most efficient yielding cells with about 11% efficiency (W. Peng *et al.*, 2003; Hao *et al.*, 2011). However, Grätzel noted that the presence of a liquid solvent for the electrolyte requires "hermetic" sealing of the module in order to prevent evaporation of the volatile solvent (Kay *et al.*, 1996). Such processes are difficult and costly to perform, and mitigate the ease of fabrication that makes DSSCs an attractive alternative to silicon solar cells. The necessity to create a vacuum within the cell and backfill with liquid electrolyte requires slow heating and cooling in order to avoid cracking the glass, and large modules would require significant mechanical strength in order to hold up against bending under wind load. It has also been shown that the I^-/I_3^- redox couple is highly reactive and corrosive with the other components of the DSSCs, especially the sealing material (I. Lee *et al.*, 2010). Finally, the disproportionation reaction of I_2^- has a redox energy that is much higher than the reaction between dye and iodide - this reaction lies closer in energy to the TiO_2 conduction band edge. Thus, the energy difference between HTL and ETL is made smaller by this reaction, and it in fact accounts for an approximately 0.75V loss in maximum photovoltage (Boschloo *et al.*, 2009).

5.2 Ionic liquids as HTL

To address these problems, non-volatile, room-temperature ionic liquid electrolytes have been investigated as hole transporters in DSSCs. These electrolytes have virtually zero vapor pressure in the temperature range experienced inside the cell, although sealing is still required to contain the liquid (Welton *et al.*, 1999). Although these non-volatile liquid electrolytes obviate the need for hermetic sealing, they are by nature very viscous and incapable of achieving the rapid physical diffusion rates of the solvent-based liquid electrolytes. Using 1-alkyl-3-methylimidazolium iodide as a room-temperature ionic liquid electrolyte (ILE), Yanagida et. al. were able to achieve a device efficiency of 5.0%, as compared to 7.9% when using the redox couple and organic liquids (OGEs). In order to achieve a quasi-solid state device, which is more robust and resistant to leakage over time, they added a gelator to form an ionic gel electrolyte (IGE) (Figure 14), and they observed the same device efficiency of 5.0%. Furthermore, they found that at high temperatures, the IGE cells outperformed the OGEs, due to the thermal stability of the ionic gel electrolyte (Kubo *et al.*, 2003). Finally, it was determined that the viscosity of the liquid could be manipulated by varying the alkyl chain length. Longer chains lowered the viscosity of the solution (by interrupting ionic interactions) and led to the most efficient device. Along similar lines, Zakeeruddin et. al. used a mixture of carbon black and PMII-based ionic liquid and obtained a device efficiency of 6.37% (Lei *et al.*, 2010). Lianos et. al. cleverly designed a sol-gel process whereby 1-methylimidazole is reacted with 3-iodopropyltrimethoxysilane to form the alkoxysilane derivitized electrolyte and I^-, giving an almost solid state electrolyte (Figure 15). By virtue of its solidity, albeit desirable, this electrolyte resulted in a mediocre device efficiency of only 3.2% (Jovanovski *et al.* 2006).

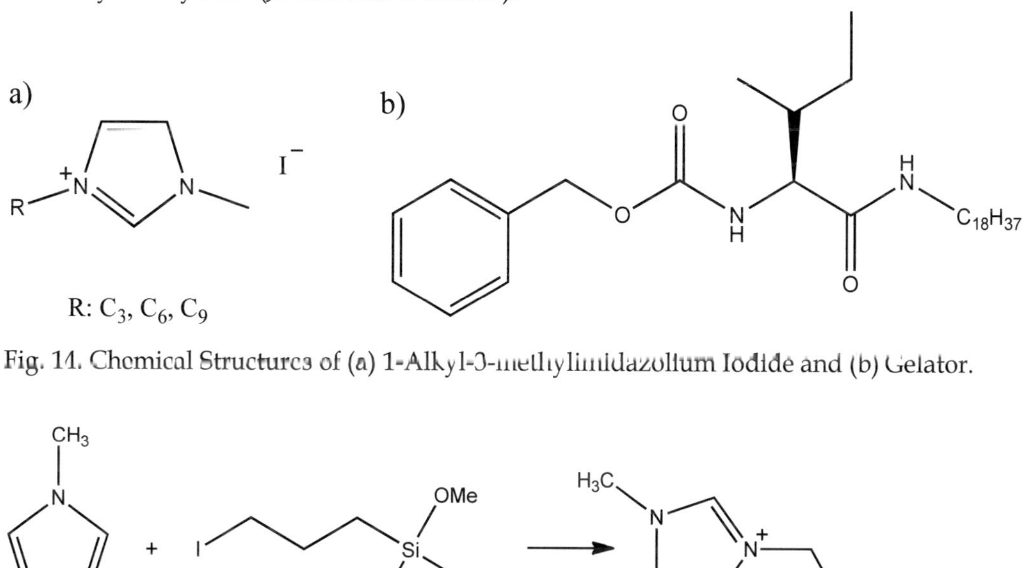

Fig. 14. Chemical Structures of (a) 1-Alkyl-3-methylimidazolium Iodide and (b) Gelator.

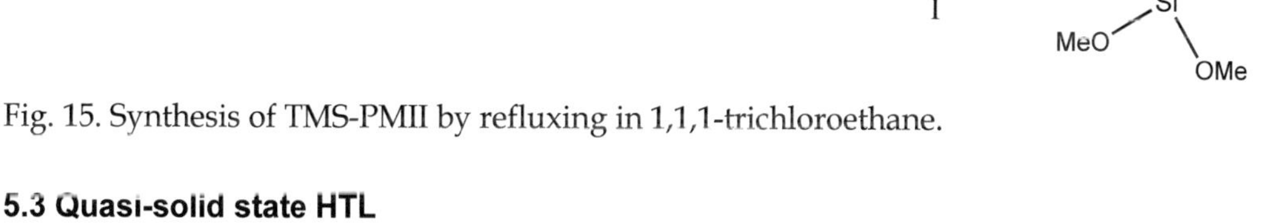

Fig. 15. Synthesis of TMS-PMII by refluxing in 1,1,1-trichloroethane.

5.3 Quasi-solid state HTL

Besides adding gelators to ionic liquids, it is possible to create quasi-solid state electrolytes by incorporating polymers into the electrolyte in a number of creative ways. One such method is to use a polymer to form a host matrix that traps the liquid electrolyte. These devices maintain the high ionic conductivities of liquid electrolytes, but protect against leakage and physical stress from thermal expansion. This concept was pioneered in 1995 by Searson et. al., who simply polymerized polyacrylonitrile in a mixture containing ethylene carbonate, propylene carbonate, acetonitrile and the desired concentration of NaI. The resulting gel was applied to the counter-electrode of a DSSC and efficiencies of 3-5% were achieved (Cao *et al.*, 1995). Since then and continuing today, a variety of polymers and electrolyte combinations have been explored, with reported efficiencies ranging from 2% to 8.2% (Bai *et al.*, 2008; Jin *et al.*, 2010). This remarkably high efficiency was achieved by Grätzel et. al., who reasoned that a eutectic melt of different polymers would have a lower viscosity than a binary melt or other simple formulations. Yet, despite this result, more recently published papers continue to report much lower efficiencies for polymer-based quasi-solid state devices, suggesting that there is a relatively poor understanding of the factors that lead to a highly effective electrolyte. Continued research in this field looks to be promising, and may eventually lead to devices that surpass the efficiency of liquid electrolyte systems at operating temperatures.

5.4 Solid state HTL

Even in quasi-solid-state devices, sealing is necessary to contain the liquid electrolyte. In order to completely preclude this necessity, it is possible to create solid-state DSSCs (ss-DSSCs) with completely solid-state materials, namely small molecules and conjugated polymers. Small molecules can be inorganic, such as the copper salts CuI and CuSCN, or entirely organic, such as SpiroMeOTAD (2,2′,7,7′-tetrakis(N,N-di-p-methoxyphenyl-amine)9,9′-spirobifluorene) (Bach *et al.*, 1998). Solutions of these molecules are given time to permeate the TiO_2, and then the solvent is evaporated, leaving only the solute molecules. However, device efficiencies with these materials are generally poor - devices using SpiroMeOTAD hover around 5% efficiency (Ding *et al.*, 2009). Although hole transport in ss-DSSCs is not fully understood, it is postulated that these materials function poorly because they transport charge electronically, as opposed to ionically. Studies suggest that the ionic nature of liquid electrolytes is critical to their high efficiency (Nogueira *et al.*, 2001).

Polyelectrolytes are a relatively recent innovation, in principle able to exploit both the efficiency advantages of ionic conduction and the stability of solid-state materials. The polyelectrolytes used are typically non-conjugated polymers containing covalently bound charged groups. Like other solid-state devices, they still suffer from diminished efficiencies because charge transport, while ionic, relies entirely on exchange reactions and not diffusion. Unfortunately, polymers are large and often unable to penetrate the TiO_2 matrix and contact the dye, so they must often be synthesized *in situ* for contact with the dye, and afterwards there is little or no further diffusion into the pore spaces. In 2001, De Paoli *et. al.* introduced the first solid state device using polyelectrolytes, achieving a maximum efficiency of only 2.6% under low-light conditions (Nogueira *et al.*, 2001). Since then, despite numerous attempts, the efficiency of such devices has generally remained low. In 2010, Jin et. al. achieved a device efficiency of only 2.42% (Jin *et al.*, 2010). These materials, however, still provide much promise for future improvement. Their performance is limited primarily by the decreased diffusion rates of the electrolytes they contain. With the discovery of new, more intensely absorbing dyes, the required thickness of DSSCs will decrease, mitigating the significance of electrolyte viscosity (Kubo *et al.*,

2003). As this happens, quasi-solid state electrolytes may become increasingly competitive with liquid electrolytes.

Conjugated polymers are a widely-used material in solid-state hole transport devices, though like all solid-state devices they remain fairly inefficient compared to liquid electrolyte cells. However, they hold promise for being stable, efficient solid-state hole transport-devices. These materials, which are characterized by a long, delocalized pi-bond system along their carbon backbone, are able to transport holes through their valence band. Many polymers have emerged, most notably polythiophenes such as Poly(3,4-ethylenedioxythiophene) poly(styrenesulfonate) (PEDOT:PSS). Recent work with graphene / PEDOT:PSS - coated cathodes has yielded cells with 4.5% efficiencies (Hong *et al.*, 2008). As well, these materials are often able to catalyze hole transport in liquid devices by providing an extra electronic step in between liquid electrolyte redox potential and cathode. In 2009, Cheng et. al. fabricated a counter-electrode from PEDOT:PSS, which catalyzed the regeneration of iodide at the counter-electrode and achieved an efficiency of 5.25% (Sirimanne *et al.*, 2010).

Though still a relatively low-efficiency component in DSSCs compared to liquid electrolytes, conjugated polymers provide a wealth of opportunities for improvement and experimentation. Work has been done in the Johal lab at Pomona College on the tuning of photoactive conjugated polyelectrolytes. These materials are adept at transferring holes through their HOMO, but are also excited in the visible range, and have charged side-groups that allow them to be tuned via surfactants. Work by Treger *et al* has shown that adding surfactant lowers the optical band gap of these polymers (presumably by increasing the effective conjugation length), improves internal quantum efficiency, and decreases photooxidation and degradation (Treger *et al*, 2008a, 2008b).

Recent work by Polizzotti, DeJong, Schual-Berke, and Johal has attempted to examine the effects on efficiency of adding the conjugated polyelectrolyte Poly[5-methoxy-2-(3-sulfopropoxy)-1,4-phenylenevinylene] (MPS-PPV) to a liquid electrolyte based cell. Though a marked increase in current and cell performance is observed (especially upon the addition of surfactant), voltammetric and spectroscopic analyses of these materials show hole transport from iodine redox couple to polymer to be forbidden (see figure 16a). To rectify this discrepancy, a mechanism in which the photoactive polyelectrolyte is generating a second exciton to facilitate hole transport is proposed and shown in figure 16b. As of yet, little to no work has been done on the subject of dual exciton generation in DSSCs. It is hoped, however, that this study will inspire further investigation into this matter, and prove to be yet another way of tuning and optimizing hole transport in DSSCs.

Liquid electrolytes remain the most efficient HTL to date. However, from a mechanical standpoint, solid-state electrolytes are the ideal option for DSSCs due to their improved stability. The main drawback of these materials is that they have yet to prove their viability as effective charge transporters, and a wealth of new research is based on creation of an optimal solid-state DSSC. Conjugated polymers, small-molecule solid HTLs, and others all show promise to provide the next breakthrough in ss-DSSCs, while quasi-solid-state gel electrolyte systems combine the benefits of polymeric materials and liquid electrolytes, also provide promise of high-efficiency, long-lived DSSCs. It is unclear which type of HTL will become the dominant technology, but it will surely have a drastic impact on the efficiency of DSSCs.

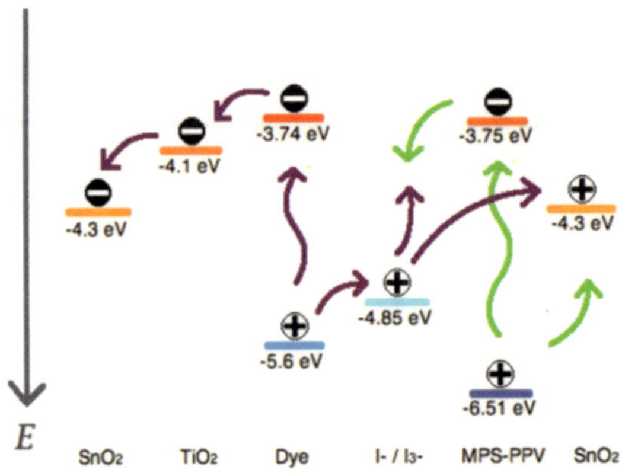

(a)

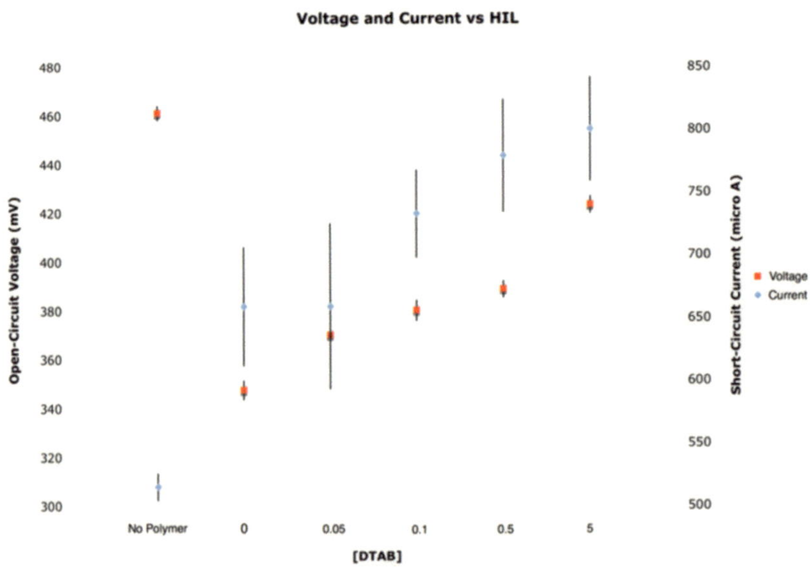

(b)

Fig. 16. Cells fabricated using both liquid iodine and a photoactive polymer HTL. (a) Current and voltage data for cells using MPS-PPV with varying concentrations of the surfactant DTAB. Current increases upon addition of polymer, and a 42+/- 12% increase in max power ($V_{oc}xJ_{sc}$) is observed when films of MPS-PPV with 5 mM DTAB are added to the cell. (b) Proposed electron flow mechanism involving secondary exciton generation by photoactive polymer HTL. Energy values are vs. vacuum and refer to MPS-PPV with [DTAB] = 0.

6. Conclusion

Grätzel cells are poised to become an important player in photovoltaic energy generation. They have already achieved over 11% efficiencies in prototype champion cells. Their watt-to-cost ratio promises to be very favorable, especially with the recent introduction of ruthenium-free dyes. As advances are made in more effective semiconductor morphologies, lower-cost, higher-efficiency dyes, and more effective and long-lasting hole conduction materials, these devices grow ever closer to commercial viability. However, there is still work to be done before these cells can start producing cheap, clean energy on a larger scale. In this chapter we described the functions and challenges for each of the components of a Grätzel cell, and outlined research needed for new and innovative ways of separating and transporting charges within these devices. By having a full understanding of the drawbacks facing DSSCs and the ways that current research is attempting to address them, can future experimenters be poised to discover the next groundbreaking solutions.

7. Acknowledgments

The authors would like to thank the Pomona College chemistry department for resources to conduct our research as well as their unending support. They would also like to thank the other members of the Johal lab for advice and assistance in and out of the lab.

8. References

Adachi, M., Murata, Y., Okada, I., & Yoshikawa, S.J. (2003). *Electrochem. Soc.* Vol. 150, No. 8, (June 2003), pp. G488-G493. ISSN 0013-4651

Bach, U., Lupo, D., Comte, P., Moser, J., Weissortel, F., Salbeck, J., Spreitzer, H., & Grätzel, M. (1998). Solid-state dye-sensitized mesoporous TiO_2 solar cells with high photon-to-electron conversion efficiencies. *Nature*, Vol. 395, No. 6702, (October 1998), pp. 583-585

Bai, Y., Cao, Y., Zhang, J., Wang, M., Li, R., Wang, P., Zakeeruddin, S., & Grätzel, M. (2008). High-performance dye-sensitized solar cells based on solvent-free electrolytes produced from eutectic melts. *Nat. Mater.* Vol. 7, No. 8, (June 2008), 626-630

Bendall, J.S., Etgar, L., Tan, S.C., Cai, N., Wang, P., Zakeeruddin, S.M., Grätzel, M., & Welland, M.E. (2011). An Efficient DSSC Based on ZnO Nanowire Photo-anodes and a New D-p-A Organic Dye. *Royal Soc. Chem.*, (May 2011)

Bessho, T., Zakeeruddin, S., Yeh, C., Diau, E., & Grätzel, M. (2010). Highly Efficient Mesoscopic Dye-Sensitized Solar Cells Based on Donor - Acceptor-Substituted Porphyrins. *Angewandte Chemie.* Vol.122, (2010) pp. 6796-6799

Cahen, D., Hodes, G., Grätzel, M., Guillemoles, J.F., & Riess, I. (2000). Nature of Photovoltaic Action in Dye-Sensitized Solar Cells. *J Phys Chem B*, Vol. 104, No. 9 (February 2000), pp. 2053-2059.

Calogero, G. & Di Marco, G. (2008). Red Sicilian orange and purple eggplant fruits as natural sensitizers for dye-sensitized solar cells. *Solar Energy Materials & Solar Cells*, Vol.92, No.11, (November 2008), pp. 1341-1346. ISSN 0927-0248

Calogero, G., Di Marco, G., Caramori, S., Cazzanti, S., Argazzi, R., & Bignozzi, C. (2009). Natural dye sensitizers for photoelectrochemical cells. *Energy & Environmental Science*, Vol.2, No.11, (August 2009), pp. 1162-1172

Cao, F., Oskam, G., & Searson, P. A solid state, dye sensitized photoelectrochemical cell. *J. Phys. Chem.* Vol. 99, No. 47, (October 1995), pp. 17071-17073. ISSN 0022-3654

Chen, K., Hong, Y., Chi, Y., Liu, W., Chen, B., & Chou, P. (2009). Strategic design and synthesis of novel tridentate bipyridine pyrazolate coupled Ru(II) complexes to achieve superior solar conversion efficiency. *Journals of Materials Chemistry*, Vol.19, No.30, (April 2009), pp. 5329-5335

Chiba, Y., Islam, A., Watanabe, Y., Komiya, R., Koide, N., & Han, L. (2006). Dye-sensitized solar cells with conversion efficiency of 11.1. *Japanese Journal of Applied Physics*, Vol.45, No.25, (June 2006), pp. 638-640, ISSN 0021-4922

Chiragwandi, Z.G., Gillespie, K., Zhao, Q.X., & Willander, M. (2006). Ultraviolet driven negative current and rectifier effect in self-assembled green fluorescent protein device. *Applied Physical Letters*, Vol. 89, (October 2006), pp. 162909, ISSN 0003-6951

Artificial Photosynthesis in Solar Cells, In: *Connect-Green*, 02.06.2011, available from: <http://www.connect-green.com/artificial-photosynthesis-in-solar-cells/>

Desilvestro, H. & Hebting, Y. Ruthenium-based dyes for Dye Solar Cells, In: *Sigma-Aldrich: Analytical, Biology, Chemistry & Materials Science Products and Services*, 02.06.2011, Available from: <http://www.sigmaaldrich.com/materials-science/organic-electronics/dye-solar-cells.html>

Dierks, S. (May 2006). Ruthenium Material Safety Data Sheet, In: E*SPI Metals*, 03.06.2011, Available from: http://www.espimetals.com/index.php/online-catalog/237-ruthenium

Ding, I.K., Tétreault, N., Brillet, J., Hardin, B., Smith, E., Rosenthal, S., Sauvage, F., Grätzel, M., & McGehee, M. (2009). Pore-filling of spiro-OMeTAD in solid-state dye sensitized solar cells: quantification, mechanism, and consequences for device performance. *Adv. Funct. Mater.* Vol. 19, No. 15, (2009), pp. 1-6

Duffy, N. W., Peter, L.M., Rajapakse, R.M.G., & Wijayantha, K.G.U. (2000). A Novel Charge Extraction Method for the Study of Electron Transport and Interfacial Transfer in Dye Sensitized Nanocrystalline Solar Cells. *Electrochemistry Communications* Vol. 2, (June 2000), pp. 658-662. ISSN 1388-2481

de Freitas, J.N., Nogueira, A., & De Paoli. M. (2009). New insights into dye-sensitized solar cells with polymer electrolytes. *J. Mater. Chem.*, Vol. 19, No. 30, (May 2009), pp. 5279-5294

Ghadiri, E., Taghavinia, N., Zakeeruddin, S.M., Grätzel, M., & Moser, J.E. (2010) Enhanced Electron Collection Efficiency in Dye-Sensitized Solar Cells Based on Nanostructured TiO2 Hollow Fibers. *Nano Letters*, Vol. 10, (April 2010), pp. 1632-638.

Grätzel, M. (2004). Conversion of sunlight to electric power by nanocrystalline dye-sensitized solar cells. *Journal of Photochemistry and Photobiology A*, Vol.164, No.1-3, (June 2004), pp. 3-14. ISSN 1010-6030

Grätzel, M. (2005). Solar Energy Conversion by Dye-Sensitized Photovoltaic Cells. *Inorganic Chemistry*, Vol.44, No.20, (October 2005), pp. 6841-6851

Green, M.A. (1981) Solar cell fill factors: General graph and empirical expressions. *Solid-State Electronics*, Vol. 24, No. 8, (August 1981), pp. 788 - 789. ISSN 0038-1101

Hao, F., Lin, H., Zhang, J., & Li, J. (2011). Balance between the physical diffusion and the exchange reaction on binary ionic liquid electrolyte for dye-sensitized solar cells. *J. Power Sources*. Vol. 196, No. 3, (February 2011), pp. 1645-1650. ISSN 0378 7753

Helgesen, M., Søndergaard, R., & Krebs, F. (2009). Advanced materials and processes for polymer solar cell devices. *J. Mater. Chem.* Vol. 20, No. 1, (October 2009), pp. 36-60

Hong, W., Xu, Y., Lu, G., Li, C., & Shi, G. (2008). Transparent graphene / PEDOT-PSS composite films as counter electrodes of dye-sensitized solar cells, *Electrochem. Communications*, Vol. 10, No. 10, (October 2008), pp. 1555-1558. ISSN 1388-2481

Ito, S., Zakeeruddin, S., Humphry-Baker, R., Liska, P., Charvet, R., Comte, P., Nazeeruddin, M., Péchy, P., Takata, M., Miura, H., Uchida, S., & Grätzel, M. (2006). High-efficiency organic dye-sensitized solar cells controlled by nanocrystalline-TiO_2 electrode thickness. *Advanced Materials*, Vol.18, No.9, (May 2006), pp. 1202-1205

Ito, S., Chen, P., Comte, P., Nazeeruddin, M.K., Liska, P., Péchy P., & Grätzel, M. (2007). Fabrication of screen-printing pastes from TiO_2 powders for dye-sensitized solar cells. *Prog. Photovoltaics* Vol. 15, No. 7, (March 2007), pp. 603–612.

Ito, S., Miura, H., Uchida, S., Takata, M., Sumioka, K., Liska, P., Comte, P., Pechy, P., & Grätzel, M. (2008). High-conversion-efficiency organic dye-sensitized solar cells with a novel inoline dye. *Chemistry Communications*, (September 2008), pp. 5194-5196

Jennings, J.R., Ghicov, A., Peter, L.M., Schmuki, P., & Walker, A.B. (2008). Dye-Sensitized Solar Cells based on Oriented TiO_2 Nanotube Arrays: Transport, Trapping, and Transfer of Electrons. *J. Am. Chem. Soc.*, Vol. 130, (2008), p. 13364.

Jin, X, Tao, J, & Yang, Y. (2010). Synthesis and characterization of poly(1-vinyl-3-propylimidazolium) iodide for quasi-solid polymer electrolyte in dye-sensitized solar cells. *J. Appl. Polym. Sci.* Vol. 118, No. 3, (June 2010), pp. 1455-1461

Johansson, E.M.J., Sandell, A.; Siegbahn, H., Rensmo, H., Mahrov, B., Boschloo, G., Figgemeier, E., Hagfeldt, A., Jonsson, S.K.M., & Fahlman, M. (2005). Interfacial Properties of Photovoltaic TiO2 / Dye / PEDOT-PSS Heterojunctions. *Synthetic Metals*, Vol. 149, (January 2005), pp. 157-167. ISSN 0379-6779

Jovanovski, V., Stathatos, E., Orel, B., & Lianos, P. (2006). Dye-sensitized solar cells with electrolyte based on a trimethoxysilane-derivatized ionic liquid. *Thin Solid Films*, Vol. 511-512, (January 2006), pp. 634-637. ISSN 0040-6090

Katoh, R., Yaguchi, K., Murai, M., Watanabe, S., & Furube, A. (2010). Differences in adsorption behavior of N3 dye on flat and nanoporous TiO_2 surfaces, *Chem. Phys. Letters*, Vol. 497, No. 1-3, (July 2010), pp. 48-51. ISSN 0009-2614

Kay, A., & Grätzel, M. (1996). Low cost photovoltaic modules based on dye sensitized nanocrystalline titanium dioxide and carbon powder. *Sol. Energ. Mat. Sol. C.* Vol. 44, No. 1, (October 1996), pp. 99-117. ISSN 0927-0248

Kuang, D., Uchida, S., Humphry-Baker, R., Zakeeruddin, S.M., & Grätzel, M. (2008). Organic Dye-Sensitized Ionic Liquid Based Solar Cells: Remarkable Enhancement in Performance through Molecular Design of Indoline Sensitizers, *Angewandte Chemie International Edition*, Vol. 47, No. 10, (February 2008), pp. 1923-1927

Kubo, W., Kambe, S., Nakade, S., Kitamura, T., Hanabusa, K., Wada, Y., & Yanagida, S. (2003). Photocurrent-determining processes in quasi-solid-state dye-sensitized solar cells using ionic gel electrolytes. *J. Phys. Chem. B.* Vol. 107, No. 18, (April 2003), pp. 4374-4381

Lee, G.W., Kim, D., Ko, M.J., Kim, K., & Park, N.G. (2010). Evaluation on over photocurrents measured from unmasked dye-sensitized solar cells, *Solar Energy*, Vol. 84, (January 2010), pp. 418-425. ISSN 0038-092X

Lee, I., Hwang, S., & Kim, H. (2010). Reaction between oxide sealant and liquid electrolyte in dye-sensitized solar cells. *Sol. Energ. Sol. Mat. C.* Vol. 95, No. 1, (2011), pp. 315-317. ISSN 0927-0248

Lei, B.X.., Fang, W., Hou, Y., Liao, J., Kuang, D., & Su, C. (2010). All-solid-state electrolytes consisting of ionic liquid and carbon black for efficient dye-sensitized solar cells. *J. Photoch. Photobio. A.* Vol. 216, No. 1, (September 2010), pp. 8-14. ISSN 1010-6030

Ruthenium (Ru) - Chemical properties, health and environmental effects, In: *Water Treatment Solutions - Lenntech*, 03.06.2011, Available from: http://www.lenntech.com/periodic/elements/ru.htm

Li, B., Wang, L., Kang, B., Wang, P., & Qiu, Y. (2006). Review of recent progress in solid-state dye-sensitized solar cells. *Sol. Energ. Mat. Sol. C.* Vol. 90, No. 5, (2006), pp. 549-573. ISSN 0927-0248

Li, S., Liu, Y., Zhang, G., Zhao, X., & Yin, J. (2011). The role of the TiO2 nanotube array morphologies in the dye-sensitized solar cells, *Thin Solid Films* (2011). ISSN 0040-6090

Liu, G., Jaegermann, W., He, J., Sundstrom, V., & Sun, L. (2002). XPS and UPS Characterization of the TiO2/ZnPcGly Heterointerface: Alignment of Energy Levels. *J. Phys. Chem. B* Vol. 106, (May 2002), pp. 5814-5819.

Mor, G.K., Shankar, K., Paulose, M., Varghese, O.K., & Grimes, C.A. (2006). Use of Highly-Ordered TiONanotube Arrays in Dye-Sensitized Solar Cells. *Nano Letters*, Vol. 6, No. 2, (2006), pp. 215-218.

Nazeeruddin, M.K., Kay, A., Rodicio, I., Humphry-Baker, R., Muller, E., Liska, P., Vlachopouslos, N., & Grätzel, M. (1993). Conversion of Light to Eelectricity by *cis*-X_2Bis(2,2′-bipyridyl-4,4′-dicarboxylate)ruthenium(II) Charge-Transfer Sensitizers (X = Cl^-, Br^-, I^-, CN^-, and SCN^-) on Nanocrystalline TiO_2 Electrodes. *J. Am. Chem. Soc.* Vol. 115, (July 1993), pp. 6382-6390. ISSN 0002-7863

Neild, B. (2010). Jellyfish' Smoothies Offer Solar Solutions. In: *CNN.com*, May 2011, Available from: <http://www.cnn.com/2010/TECH/innovation/09/27/jellyfish.solar.power/index.html>

Nogueira, A., Durrant, J., & De Paoli, M. (2001). Dye-sensitized nanocrystalline solar cells employing a polymer electrolyte. *Adv. Mater.* Vol. 13, No. 11, (June 2001), pp. 826 – 830. ISSN 0935-9648

O'Regan, B., & Grätzel, M. (1991). A Low-cost, High-efficiency Solar Cell Based on Dye-sensitized Colloidal TiO2 Films. *Nature* Vol. 353, No. 6346, (October 1991), pp. 737-740

Peng, B., Jungmann, G., Jager, C., Haarer, D., Schmidt, H.W., & Thelakkat, M. (2004). Systematic Investigation of the Role of Compact TiO2 Layer in Solid State Dye-sensitized TiO2 Solar Cells. *Coordination Chemistry Reviews*, Vol. 248, No. 13-14, (June 2004), pp. 479-489. ISSN 0010-8545

PVEducation. (2010). Fill Factor. In: *Pveducation.org*. Web. 07 June 2011. Available from: <http://www.pveducation.org/pvcdrom/solar-cell-operation/fill-factor>

Ryan, M. (2009). Progress in ruthenium complexes for dye sensitised solar cells. *Platinum Metals Review*, Vol.53, No.4, (October 2009), pp. 216-218

Saito, Y. (2004). Solid State Dye Sensitized Solar Cells Using in Situ Polymerized PEDOTs as Hole Conductor. *Electrochemistry Communications* Vol. 6, No. 1, (2004), pp. 71-74. ISSN 1388-2481

Schoonman, J. (2005). Nano-structured materials for the conversion of sustainable energy, In: *Nanostructured and advanced materials for applications in sensor, optoelectronic and photovoltaic technology,* Vaseashta, A., Dimova-Malinovska, D., & Marshall, J., pp. 270-280

Singh, P.K., Kim, K.W., Park, N.G., & Rhee, H.W. (2008). Mesoporous nanocrystalline TiO_2 electrode with ionic liquid-based solid polymer electrolyte for dye-sensitized solar cell application, *Synthetic Materials,* Vol. 158, No. 14, (August 2008), pp. 590-593. ISSN 0379-6779

Sirimanne, P.M., Winther-Jensen, B., Weerasinghe, H.C., & Cheng, Y.B. (2010). Towards an all-polymer cathode for dye sensitized photovoltaic cells, *Thin Solid Films,* Vol. 518, No. 10, (March 2010), pp. 2871-2875. ISSN 0040-6090

Smestad, G.P. (2003). A Technique to Compare Polythiophene Solid-State Dye Sensitized TiO_2 Solar Cells to Liquid Junction Devices. *Solar Energy Materials and Solar Cells,* Vol. 76, (April 2003), pp. 85-105. ISSN 0927-0248

Ruthenium Dyes, In: *Solaronix Products,* 03.06.2011, Available from: <http://www.solaronix.com/products/dyes/>

Tang, H., Berger, H., Schmid, P.E., & Levy, F. (1994). Optical Properties of Anatase (TiO2). *Solid State Communications,* Vol. 92, No. 3, (June 1994), pp. 267-271. ISSN 0038 1098

Treger, J.S., Ma, V.Y., Gao, Y., Wang, C.C., Wang, H.L., & Johal, M.S. (2008a). Tuning the optical properties of a water-soluble cationic poly(*p*-phenylenvinylene): surfactant complexation with a conjugated polyelectrolyte, *J.Phys. Chem. B.,* Vol. 112, (January 2008), pp. 760-763.

Treger, J.S., Ma, V.Y., Gao, Y., Wang, C.C., Jeon, S., Robinson, J.M., Wang, H.L., & Johal, M.S. (2008b). Controlling layer thickness and photostability of water-soluble cationic poly(*p*-phenylenevinylene) in miltilayer thin films by surfactant complexation, *Langmuir,* Vol. 24, (August 2008), pp. 13127-13131

Wang, P., Zakeeruddin, S., Moser, J., Nazeeruddin, M., Sekiguchi, T., & Grätzel, M. (2003). A stable quasi-solid-state dye-sensitized solar cell with an amphiphilic ruthenium sensitizer and polymer gel electrolyte. *Nat. Mater.* Vol. 2, No. 7, (June 2003), pp. 402-407

Wang, P., Zakeeruddin, S, Moser, J, & Grätzel, M. (2003). A new ionic liquid electrolyte enhances the conversion efficiency of dye-sensitized solar cells. *J. Phys. Chem. B.* Vol. 107, No. 48, (September 2003), pp. 13280-13285

Wang, X., Koyama, Y., Kitao, O., Wada, Y., Sasaki, S., Tamiaki, H., & Zhou, H. (2010a). Significant enhancement in the power-conversion efficiency of chlorophyll co-sensitized solar cells by mimicking the principles of natural photosynthetic light-harvesting complexes. *Biosensors and Bioelectrics,* Vol.25, No.8, (April 2010), pp. 1970-1976.

Wang, X., Tamiaki, H., Wang, L., Tamai, N., Kitao, O., Zhou, H., & Sasaki, S. (2010b) Chlorophyll-*a* derivatives with various hydrocarbon ester groups for efficient dye-sensitized solar cells: static and ultrafast evaluations on electron injection and charge collection processes. *Langmuir,* Vol.26, No.9, (May 2010), pp. 6320-6327

Welton, T. (1999). Room-temperature ionic liquids. Solvents for synthesis and catalysis. *Chem. Rev.* Vol. 99, No. 8, (April 1999), pp. 2071-2083

Yanagida, S., & Yu, Y. (2009). Manseki, K. Iodine/iodide-free dye-sensitized solar cells. *Accounts Chem. Res.* Vol. 42, No. 11, (November 2009), pp. 1827-1838

Yum, J., Jang, S., Walter, P., Geiger, T., Nüesch, F., Kim S., Ko, J., Grätzel, M., & Nazeeruddin, M. (2007). Efficient co-sensitization of nanocrystalline TiO_2 films by organic sensitizers, *Chemical Communications,* Vol.44, (November 2007), pp. 4680-4682

Yum, J., Hagberg, D., Moon, S., Karlsson, K., Marinado, T., Sun, L., Hagfeldt, A., Nazeeruddin, M., & Grätzel, M. (2009). A light-resistant organic sensitizer for solar-cell applications, *Angewandte Chemie International Edition*, Vol.48, No.9, (February 2009), pp. 1576-1580

Zafer, C., Karapire, C., Sariciftci, S.N., & Icli, S. (2005). Characterization of N, N0-bis-2-(1-hydoxy-4- Methylpentyl)-3, 4, 9, 10-perylene Bis (dicarboximide) Sensitized Nanocrystalline TiO2 Solar Cells with Polythiophene Hole Conductors. *Solar Energy Materials & Solar Cells* Vol. 88, (October 2005), pp. 11-21. ISSN 0927-0248

6

Nonisovalent Alloys for Photovoltaics Applications: Modelling IV-Doped III-V Alloys

Giacomo Giorgi, Hiroki Kawai and Koichi Yamashita
Department of Chemical System Engineering,
School of Engineering, The University of Tokyo
Japan

1. Introduction

In the next thirty years the annual global consumption of energy will rise by more than 50% (Hochbaum & Yang, 2010). Currently most of the energy production comes from the combustion of fossil fuels; nonetheless the deriving CO_2 emissions represent a real risk for the safety of the environment and for human health. Not secondarily, the prompt availability of fossil fuels is extremely influenced by geo-political factors. In the very last decades much attention has been devoted to the development of green renewable energetic sources as possible viable alternative. Owing to its almost ubiquitous availability, solar energy seems to be the most promising way to produce alternative energetic sources. The optimal choice of the materials for the device assembling, according to their prompt availability, in conjunction with recycling spent modules and thinning the semiconductor layers (these two latter are the key-points in reducing the material related sustainability deficits (Fthenakis, 2009)) will help passing from the gigawatt to the terawatt level, that required by the global consumption (Feltrin & Freundlich, 2008). The idea of getting energy from the sun is almost 150 years old. It derives from the discovery of the photovoltaics effect (1839) observed for the first time by the French physicist Edmond Becquerel in an experiment lead with an electrolytic cell made up of two metal electrodes: a weak electrical current was detected by exposing to sunlight a silver coated platinum electrode immersed in electrolyte. From the initial discovery many years have passed until the very first practical application, due to the celebrated Russian physicist Abram Ioffe, consisting in low efficient rectified thallium sulfide cells, has been realized.

The very basic components of a modern solar cell consists in a *p-n* junction, an N-semiconductor, and a P-one, in conjunction with the two electrical contact metal layers that provide the current flow internally and externally to the cell. The main physical quantity that describes a solar cell is the energy *conversion efficiency* (η), defined as the quantity of sunlight converted into electric power. Materials like crystalline Si and GaAs can reach efficiency of about 25%, although efficiencies for most commercially available multicrystalline Si solar cells are below 20%. Solar cells constituted by only one semiconductor material are characterized by a maximum energy efficiency for a bandgap (E_G) between 1.4 and 1.6 eV (Green, 1982).

The other key performance characteristics are represented by the *photocurrent density* (J_{sc}), the *fill factor* (*FF*), and the *open circuit voltage* (V_{oc}), related (also with the incident light power density, P_s) by the general expression:

$$\eta = J_{sc}\,FF\,V_{oc}/P_s \quad (1)$$

Solar cells can operate in a wide range of current (*I*) and voltage (*V*); the *maximum power point* of the cell is defined as the point that maximizes the product between *I* and *V*, in a process that involves the continuous increase of the resistive load on an irradiated cell from *short circuit* to *open circuit* regime. Modern cells can track the power by measuring continously *V* and *I*, equilibrating the load, in order to achieve always the maximum power, regardless of the lighting conditions.

The initial technology of the modern solar cells based on the well-assessed single and multi-crystalline Si cells (the latter cheaper but with reduced quality due to the presence of grain boundaries) has been progressively replaced by the *thin film* based one, the so-called "Second-generation solar cells", usually constituted by GaAs, CuInSe, CIGS (Copper Indium Gallium Selenide), or CdTe, which represent a cheap and simple alternative technology in photovoltaics (PV). A particular mention is deserved by the emerging technology based on the combination of dyes anchored on metal oxides (DSSC, dye sensitized solar cells, (O'Regan & Graetzel, 1991) consisting in porous layers of TiO_2 nanoparticles.

The usage of layered materials with different bandgaps is a desirable procedure for increasing the efficiency of the final PV device. Such assembling procedure results in the splitting of the solar spectrum in several parts. In this way, indeed, photons of different energies are absorbed depending on the used material: the stacking of higher bandgap material on the surface able to absorb high-energy photons with lower-energy photons absorbed by the lower bandgap material beneath is the key step for these *Multi-Junction* (*MJ*), "Third-generation" solar cells, characterized by the reduction of the transmission and that of the thermalization losses of hot carriers. According to the assembling procedure, MJs are *monolithic* or *stacked*. Owing to the requirement of electronic and lattice match of the constituents, the monolithic ones present additional intrinsic difficulties; they are epitaxially grown on Ge substrate and constituted by lattice matched (In,Ga)P and GaAs. The reader may understand the structure of similar devices observing Figures 1-2. In Figure 1, a sketch of (Ga,In)P/(Ga,In)As/Ge is reported, while Figure 2 shows its spectral irradiance of the solar AM1.5 spectrum in conjunction with the parts of the spectrum used. "AM"(Air-Mass) refers to the spectrum of the incident light and corresponds to the shortest, direct optical path length through the Earth atmosphere, that in conjunction with "1.5" represents the standard for the characterization of terrestrial power-generating panels. More precisely, AM is $1/\cos(z)$, where *z* represents the zenith angle. The choice of AM1.5 as standard stems from the fact that it corresponds to $z{\sim}43^{\circ}$, representing the yearly average at mid-latitudes. AM0 is the standard for space cells. In the mechanical stacked MJ the separate connections of top and bottom cells do not mandatorily require current matching, making the combination of bandgaps quite flexible. The only tight requirement is thus the transparent contacts. Anyway, the lattice mismatch may cause crystal dislocations introducing levels in the gap and thus mediating the Shockley-Read-Hall recombination.

Primary role in the rising success of monolithic tandem solar cells must be ascribed to the National Renewable Energy Laboratory (NREL) activity and in particular to the research conducted by J. M. Olson group (Olson et al., 1985). In details, the initial tandem solar cells based on GaAs and GaInP were made via a vapour phase epitaxy growth process and revealed problems related to the purity of source materials. The initial drawbacks were progressively solved during the nineties and subsequently due to their high efficiency and power-to-mass ratio, systems based on GaInP/GaInAs/Ge have represented and still represent the most reliable choice for communication satellites (Dimroth, 2006). In Figure 3, the calculated conversion efficiency of a triple-junction solar cell as function of the bandgap of the single junctions forming the stacking is reported.

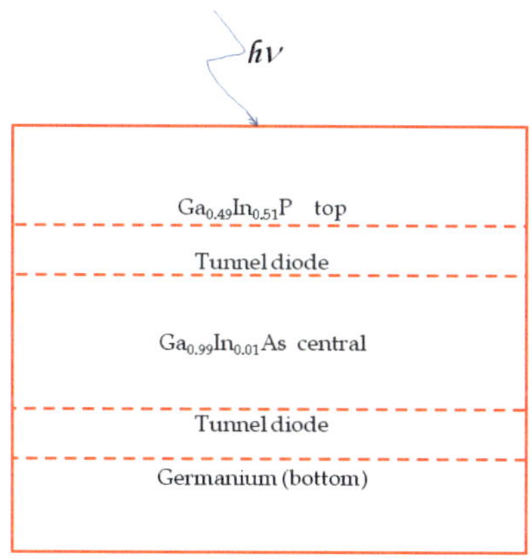

Fig. 1. General structure of a multi-junction solar cell.

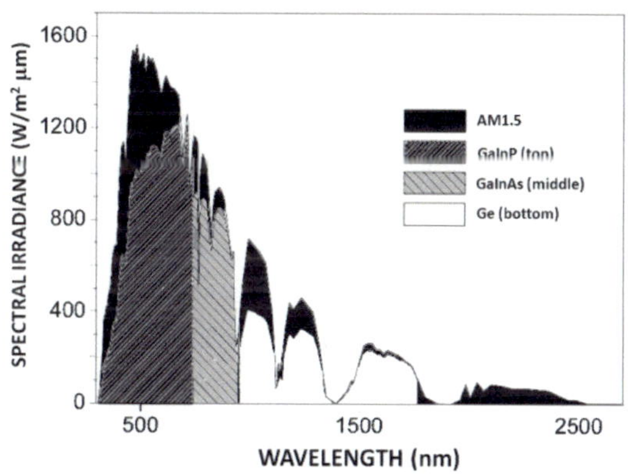

Fig. 2. Spectral irradiance of the solar AM1.5 spectrum with the regions of the spectrum used by a GaInP/GaInAs/Ge 3-junction solar cell.

MJ systems are usually constituted by three different possible substrate: Ge, GaAs, and InP. The introduction of chemical components in the different layers is beneficial (Yamaguchi et al., 2005): Al in the top cell has the property of increasing E_G to values that cover a larger part of the solar spectrum, while few amounts of In reduce the mismatch between layers.
In the context of the Japanese "New Sunshine Project", InGaP/InGaAs/Ge monolithic integrated cells have reached an efficiency of 31.7%, while for the stacked InGaP/GaAs//InGaAs ones an efficiency of 33.3% has been reported (Yamaguchi, 2003), representing at the time of their production the efficiency World Guinness for such cells.

Metamorphic triple-junction $Ga_{0.44}In_{0.56}P/Ga_{0.92}In_{0.08}As/Ge$ terrestrial concentrator solar cells (owing to the macroscopic differences between terrestrial spectrum and the conditions in space, the structure of the solar cell must be adjusted. Therefore, the optimum bandgap combination of materials is not the same) have been recently grown and with the usage of a buffer structure, able to minimize the dislocation formation (King et al., 2007), a record of 40.7% in efficiency has been established in 2007 at 240 suns and at AM1.5D. Even more recently (Guter et al., 2009) a new World Guinness has been established for the metamorphic triple-junction solar cell: an efficiency of 41.1% has been achieved under 454 suns and at same standard conditions: this latter cell combines $Ga_{0.35}In_{0.65}P$ (top cell), $Ga_{0.83}In_{0.17}As$ (middle cell) with a Ge bottom cell. An electrically inactive buffer is used in order to make inactive the formation of threading dislocations that in any case have densities below 10^6 cm^{-2}. Such highly efficient cells are integrated by the usage of a Fresnel lens placed ~10 cm over the cells, ensuring a concentration of the sunlight increased by a factor ranging between 400 and 500.

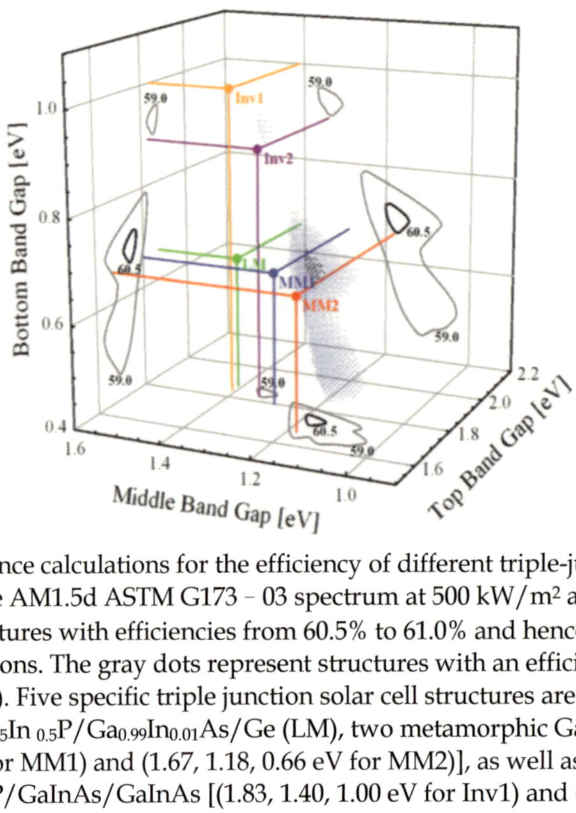

Fig. 3. Detailed balance calculations for the efficiency of different triple-junction solar cell structures under the AM1.5d ASTM G173 - 03 spectrum at 500 kW/m² and 298 K. The black dots represent structures with efficiencies from 60.5% to 61.0% and hence mark the optimal band gap combinations. The gray dots represent structures with an efficiency of 59.0%-60.5% (see contours). Five specific triple junction solar cell structures are also shown. The lattice-matched $Ga_{0.5}In_{0.5}P/Ga_{0.99}In_{0.01}As/Ge$ (LM), two metamorphic GaInP/GaInAs/Ge [(1.8, 1.29, 0.66 eV for MM1) and (1.67, 1.18, 0.66 eV for MM2)], as well as two inverted metamorphic GaInP/GaInAs/GaInAs [(1.83, 1.40, 1.00 eV for Inv1) and (1.83, 1.34, 0.89 eV for Inv2)] devices. [Reprinted figure 3 with permission from: Guter, W.; Schöne, J.; Philipps, S. P.; Steiner, M.; Siefer, G.; Wekkeli, A.; Welser, E.; Oliva, E.; Bett, A. W. & Dimroth, F., *Applied Physics Letters*, Vol. 94, No. 22, pp. 223504-3 ISSN 1077-3118. (2009). Copyright (2011) by the American Institute of Physics].

The continuous research of new materials in PV and the optimization of their performances motivated us to study, in recent years, the electronic and the structural properties of the

alloys formed by IV group elements and the III-V compounds. $(GaAs)_{1-x}(Ge_2)_x$, one of the $(A^{III}B^{V})_{1-x}(C_2^{IV})_x$ nonisovalent alloys, has direct gap ranging between 0.5 and 1.4 eV, ideal for visible-IR light absorption, and lattice constants matching with that of GaAs (5.66 Å (Wang & Ye, 2002)), representing the best candidate in some technologically oriented applications (extra junction of two junction $Ga_{0.52}In_{0.48}P$/GaAs solar cell (Norman et al., 1999)).
In this chapter we review recent results from modelling the structural and electronic properties of such non-isovalent alloys widely utilized in MJ solar cells. The relevance of Density Functional Theory (DFT) and *GW* (*G*: Green function, *W*: screened Coulomb interaction) calculations in this field is discussed and the importance of large modellization in accounting the clusterization phenomena is shown. We also review the basic concepts associated with the stabilizing self-compensation mechanism, both in defective supercells and in alloys.

2. Semiconductor alloys for photovoltaics: The case of $(GaAs)_{1-x}(Ge_2)_x$: A brief overview

The isovalent class of semiconductor alloys represents the most common one (Wei & Zunger, 1989; Shan et al., 1999; Wei et al., 2000; Janotti et al., 2002; Deng et al., 2010) (i.e., IV/IV, (III−V)/(III−V), and (II−VI)/(II−VI)). The alloys belonging to this class are characterized by a reduced band offsets between the constituents (< 1 eV), by a small and composition-independent bowing coefficient, and by a lattice mismatch that is usually below 5% (Zunger, 1999). Differently, the introduction of a low-valent and high-valent element into a III-V compound generates a so called "nonisovalent" alloy (Yim, 1969; Bloom, 1970; Funato et al., 1999; Wang & Zunger, 2003; Osorio et al., 1999; Greene, 1983) (i.e., (III−V)/IV, (III−V)/(II−VI)). High carrier mobilities, an enhanced presence of free electrons and holes according to the growth conditions, and the reduced donor−acceptor charge compensation are the attractive characteristics that make this class matter of deep investigation. In the case of ZnSe−GaAs, Wang *et al.* report only small change in the final alloy bandgap, E_G, as result of the introduction of larger bandgap II−VI dopants into the III−V host, while the opposite (GaAs in ZnSe) causes sharp drops in the bandgap of the alloy (Wang & Zunger, 2003).
In recent years, many other papers have been focused on the class of $(III\text{-}V)_{1-x}(IV_2)_x$ nonisovalent alloys both at theoretical (Holloway, 2002; Newman et al., 1989; Osorio & Froyen, 1993; Ito & Ohno, 1992; Ito & Ohno, 1993; Newman & Jenkins, 1985; Bowen et al., 1983) and experimental (Green & Elthouky, 1981; Barnett et al., 1982; Alferov et al., 1982; Banerjee et al., 1985; Noreika & Francombe, 1974; Baker et al., 1993; Rodriguez et al., 2000; Rodriguez et al., 2001) level. The employment of nonequilibrium grown techniques that incorporate high dopant concentrations in semiconductors has boosted the attention towards this class of alloys. In particular, despite the mutual insolubility of the constituents, the homogeneous single crystal $(GaAs)_{1-x}(Ge_2)_x$ can be synthesized as metastable alloy by sputter deposition technique (Barnett et al., 1982), metal-organic chemical vapor deposition (MOCVD)(Alferov et al., 1982), molecular beam epitaxy (MBE) (Banerjee et al., 1985), and rf magnetron sputtering (Rodriguez et al., 2000; Rodriguez et al., 2001). The direct gap "tailorability" for $(GaAs)_{1-x}(Ge)_{2x}$ alloys is observed in a final large, negative, and asymmetric, *V-shaped*, bowing of the bandgap. Barnett *et al.* (Barnett et al., 1982) report a minimum value of about 0.5 eV at Ge concentration of about 35%: the optical absorption of

homogeneous single-crystal metastable $(GaAs)_{1-x}(Ge_2)_x$ alloys, grown using ultra-high-vacuum ion-beam sputter deposition is there investigated.

Herbert Kroemer, Nobel prize in 2001 for his studies in semiconductor heterostructures used in electronics, clearly states *"...If lattice matching were the only constraint, the Ge-GaAs system would be the ideal heterosystem, as was in fact believed by some of us – including myself – in the early 1960s"*. (Kromer, 2001). Then he adds: *"...Covalent bonds between Ge on the one hand and Ga or As on the other are readily formed, but they are what I would like to call valence mismatched, meaning that the number of electrons provided by the atoms is not equal to the canonical number of exactly two electrons per covalent bond"*(Kromer, 2001). And indeed, at equilibrium condition, phase separation between GaAs-rich domains and Ge-rich domains occurs since GaAs and Ge are mutually insoluble due to the formation of what Kroemer calls valence mismatched, otherwise called "bad" bonds (octet-rule violating bonds, i.e., Ga−Ge, As−Ge) (Osorio et al., 1991a). In the ordered GaAs-rich phase, Ga and As preferentially form donor-acceptor pairs, whereas in the Ge-rich phase, they are randomly distributed in the alloy forming a mixture of n-type and p-type semiconductors. The origin of the large bowing is ascribed to this ordered → disordered transition (Newman et al., 1989).

Several other theoretical models have been developed in order to describe such zinc-blend-to-diamond phase transition. They are based on thermodynamic (Newman & Dow, 1983; Newman et al., 1989; Gu et al., 1987; Koiller et al., 1985), percolation (D'yakonov & Raikh, 1982), and stochastic growth approaches (Rodriguez et al., 2000; A Rodriguez at al., 2001; Kim & Stern, 1985; Davis & Holloway, 1987; Holloway & Davis, 1987; Preger et al., 1988; Capaz et al., 1989).

Owing to their intrinsic difficulty in taking into account possible different growth conditions, models based on the percolation method (D'yakonov & Raikh, 1982) predict a critical concentration for the transition at 0.57, quite far from the experimentally reported. In "growth models" the alloy configuration depends on the kinetics of the growth. In such methods no explicit functional minimization is performed (Osorio et al., 1991b) and once that atoms have satisfied the set of growth rules imposed by the model, they are considered as *frozen*, without any further possibility of including the influence of thermal effects. The prediction of the alloy configuration is based on Monte Carlo (MC) models and the atomic position of single layers depends on the epitaxial growth direction. Thus, Long Range Order (LRO) of the final structure depends on the growth direction. Imposed requirement in such models is the "wrong" bond formation (i.e., III−III and V−V) forbiddance. The absence of Sb-Sb bonds in $(GaSb)_{1-x}(Ge_2)_x$ alloy detected via extended X-ray absorption fine structure (EXAFS) experiments (Stern et. al., 1985) confirms the appropriateness of such imposed condition. Kim and Stern have proposed a model specifying in the set of growth rules the equivalent probability for the A^{III} and B^{V} species in the site occupancy, obtaining a critical concentration for the phase transition (x_c) that is 0.26(+0.03/-0.02) on (100) substrate (Kim & Stern, 1985). Such model only considers Short Range Order (SRO), reporting the critical composition below which LRO is present. They also report the x_c dependence by the growth morphology: a planar growth along [100], another, still planar along the [111] direction, and finally a spherical growth model are studied. In this last model the critical Ge concentration is calculated to be smaller than 0.18 (even if this last value is affected by computational limitations that lead to possible inaccuracy). Davis and Holloway have developed another model implemented by MC simulation and analyzed via an analytical approximation. In their model the formation of Ga-Ga and As-As nearest neighbour bonds is forbidden, and

additionally every gallium atom forms a bond with As atom present in excess during the growth process (Davis & Holloway, 1987; Holloway & Davis, 1987). They find a value for $x_c \approx 0.3$ on (100) oriented substrates (Davis & Holloway, 1987). No phase transition is at variance predicted along the <111> direction (Holloway & Davis, 1987), with remnant ZB phase present in all the range of composition. This prediction confirms the experimental findings for $(GaAs)_{1-x}(Ge_2)_x$ obtained via High Resolution X-Ray Diffraction (HRXRD) on several substrate orientations (Rodriguez et al., 2000; Rodriguez et al., 2001).
Rodriguez *et al.* (Rodriguez et al., 2001) compared LRO behaviour with the results of Raman scattering, the experimental procedure for evaluating alloy SRO (Salazar-Hernández, et al. 1999; Olego & Cardona, 1981). In details, they find different LRO for each growth direction. The long range order parameter S for $(GaAs)_{1-x}(Ge_2)_x$ alloys and more in general for the $(III\text{-}V)_{1-x}(IV_2)_x$ is expressed as (Shah et al., 1986):

$$S(x)=2r(x)-1 \qquad (2)$$

where r represents the probability that any Ga (As) occupies its site in the lattice. This corresponds to r=1 (0.5) in the case of perfect LRO (disordered crystal). The analysis of Rodriguez of the optical gap and Raman scattering, shows that near-neighbour correlations (SRO) extremely influence the optical properties while, at variance, there is no impact of the substrate orientation and the LRO on the optical properties. Figure 4 reports the mean cluster reciprocal length obtained by Rodriguez by the Monte Carlo simulation for different orientations of the alloy. It is evident from the plot the overall tendency of GaAs clusters in the alloys to reduce their size (fragmentation), increasing the Ge concentration, x. That kinetic models better describe the phase transition than thermodynamic ones is well assessed. The reasons stem from the fact that thermodynamic models do not include the details of the critical composition x_c as a function of kinetic growth, the phase transition critical concentration must be explicitly added as input, and additionally there are no restrictions on the formation of Ga-Ga and As-As bonds. Also a determining role in the nature of the alloy is played by the growth temperature (Banerjee et al., 1985): $(GaAs)_{1-x}Ge_{2x}$ layers epitaxially grown on GaAs (100) substrates at different temperatures analysed by TEM revealed that at T_g = 550 ° C, Ge separated from GaAs into domains of ~100 Å. Differently, single-phase alloys are still detected at T = 430 ° C.
The research described in this chapter has been motivated by the fact that so wide potential applicability of this class of alloys in PV is astonishingly not supported by a deepen knowledge both at Density Functional Theory (DFT) and post-DFT level of their electronic, structural, and optical properties. We thus decided to examine the bowing in $(GaAs)_{1-x}Ge_{2x}$ alloys, searching the microscopic origin of this intriguing and not yet clarified phenomenon. In particular, at first we have theoretically analyzed the properties of four different intermediate structured compounds that range between "pure" GaAs and "pure" Ge (x_{Ge} = 0.25, 0.50 (two samples), 0.75) (Giorgi et al., 2010). Enlarging our models to ones ranging between 8 and 64 atoms, we have investigated the impact of clustering effects and that of the cluster shape on the bandgap bowing (Kawai et al., 2011).
For the alloy electronic properties calculations two of the methods which are reported to give extremely reliable results have been employed: the Quasiparticle Self-consistent *GW* (QS*GW*) approximation approach (van Schilfgaarde et al., 2006) developed by Mark Van Schilfgaarde and his group at Arizona State University and the frequency dependant *GW*

method implemented in the VASP code (VASP) by Shishkin and co-workers. (Shishkin & Kresse, 2006; Shishkin & Kresse, 2007; Fuchs et al., 2007; Shishkin et. al., 2007).

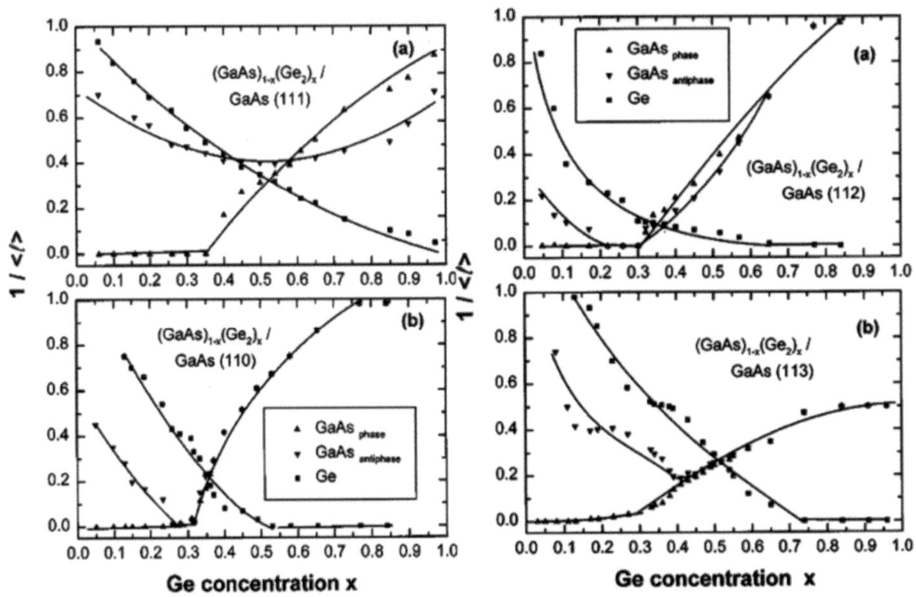

Fig. 4. Left, mean cluster lengths reciprocal (1/<*l*>) from Monte Carlo simulations for (a) (111) and (b) (110) alloys. Full up (down) triangles show the behaviour with Ge concentration for the GaAs phase (antiphase) component. The full squares give the domain size for the Ge component. Right, simulated average cluster reciprocal lengths for (a) (112) and (b) (113) alloys. Full up (down) triangles show the behaviour with Ge concentration for the GaAs phase (antiphase) component. The full squares give the domain size for the Ge component. The lines joining the points are shown for visual aid only. [Reprinted figure 4 with permission from: Rodriguez, A. G.; Navarro-Contreras, H. & and Vidal, M. A., *Physical Review B*, Vol. 63, No. 11, pp. 115328-9, ISSN 1550-235X. (2001). Copyright (2011) by the American Physical Society].

The initial analysis of the 8-atoms cell revealed that the reduction of the bandgap for intermediate x values in $(GaAs)_{1-x}(Ge_2)_x$ alloys takes place with a lattice constant increase and a symmetry reduction with the formation energies linearly related with the number of bad bonds in each model (Giorgi et al., 2010); the subsequent analysis on the shape and the clusterization effects present in these alloys have confirmed at first the main role that SRO plays on asymmetric bandgap behaviour (Rodriguez et al., 2001), further confirming experimental results, like the tendency of large clusters in the alloy to fragment (McGlinn et al., 1988). Large models have been also employed to refine the shape of the asymmetric bowing.

3. Computational details

Using Blöchl's all-electron projector-augmented wave (PAW) method (Blöchl, 1994; Kresse & Joubert, 1999), we performed spin-polarized calculations by using density functional theory (DFT), within both the local density approximation (LDA) (Perdew & Zunger, 1981; Ceperley & Alder, 1980) and the generalized gradient approximation (GGA) of Perdew and Wang (Perdew, 1991; Perdew et al., 1992) and of Perdew, Burke, and Ernzerhof (PBE, Perdew et al., 1996). In details, *d* electrons in the semicore for both Ga and Ge have been considered. Cutoff energies of 287 and 581 eV were set as the expansion and augmentation charge of the plane wave basis. The force convergence criterion for these models was 0.01 eV/Å.

3.1 The initial case of eight atom unit cells

The initial $(GaAs)_{1-x}Ge_{2x}$ models consisting of 8 atoms were optimized with a 10 x 10 x 10 Γ-centered *k*-points sampling scheme. The reliability of our initial structures has been checked by re-calculating all the total energies with the generalized LMTO method scheme developed by Methfessel *et al.* (Methfessel et al., 1996). We found almost identical results for what regards structural properties and heats of reaction, indicating that the results are well converged. The thermodynamic stability of these alloys was calculated as

$$\Delta E_{form} = E_{alloy} - \{(1-x)E_{GaAs} + xE_{Ge}\} \quad (3)$$

derived from the product-reactant equation:

$$GaAs + 2x\, Ge \rightarrow (GaAs)_{1-x}Ge_{2x} + x\, GaAs \quad (4)$$

Owing to the methodological derived large cancellation of errors, both LDA and GGA are expected to predict reasonable heats of reaction like for that of Eq. (3). At variance with structural properties, optical ones are much less well described, with a well documented underestimation of semiconductor bandgaps. Also dispersion in the conduction band is affected by this DFT shortcoming: for Ge, the LDA gap is negative and Γ_{1c} is lower than L_{1c} in contradiction to experiment. Also the Γ–X dispersion is often strongly affected: in GaAs $X_{1c}-\Gamma_{1c}$ is about twice the experimental value; underestimations that are generated by the self-interaction error (Perdew & Zunger, 1981) and the discontinuities in the derivatives of exchange-correlation energy (Perdew & Levy, 1983; Sham & Schlüter, 1983).
In the prediction of semiconductor optical properties one of the best method is based on the *GW* approximation of Hedin (Hedin, 1965). This approximation is a perturbation theory around some noninteracting Hamiltonian, H_0. In particular, the quality of this Hamiltonian highly impacts on the quality of the final *GW* result. In conjunction with a "safe" choice of the Hamiltonian, it must be stressed that for reliable results the use of an all-electron method is highly recommended (Gomez-Abal et al., 2008). To satisfy both the requirements, in this initial stage of our calculations, we have adopted here an all-electron method, where not only the eigenfunctions are expanded in an augmented wave scheme, but the screened coulomb interaction *W* and the self-energy $\Sigma = iGW$ are represented in a mixed plane-wave and local-function basis (Kotani & van Schilfgaarde, 2002; Kotani et al., 2007). In addition, all core states are treated at the Hartree-Fock level. In the following we briefly describe the methodology key points.
Usually, in literature the initial Hamiltonian (H_0) for *GW* calculations is an LDA derived guess; thus usual *GW* method may be named $G^{LDA}W^{LDA}$ approximation. Many limitations characterize this $G^{LDA}W^{LDA}$ approaches as reported in previous literature (van Schilfgaarde et al., 2006b). The QuasiParticle Self-Consistent *GW* (QSGW) approximation (van

Schilfgaarde et al., 2006a), overcomes most of these limitations. Semiconductor energy band structures are well described with uniform reliability. Discrepancies with experimental semiconductor bandgaps are small and highly systematic and the origin of the error can be explained in terms of ladder diagrams missing in the random phase approximation (RPA) to the polarizability (Shishkin et al., 2007). The RPA results in a systematic tendency for the dielectric constant, ε_∞, to be underestimated. The error is very systematic: ε_∞ is too small by a factor of approximately 0.8, for a wide range of semiconductors and insulators. This fact and also the fact that the static limit of W mainly controls the quasiparticle (QP) excitations, provides a simple and approximate remedy to correct this error: $\Sigma - V_{xc}^{LDA}$ is scaled by 0.8.

3.2 The extension up to 64 atoms

The Short Range Order effects (i.e., those involving shape and the clusterization of GaAs/Ge regions) on asymmetric bandgap behaviour was confirmed in the alloys synthesized by ion-beam sputtering techniques (McGlinn et al., 1988) and *rf* sputtering techniques (Rodriguez et al., 2001). In order to model alloy structures reproducing the SRO effect and deeply understand their effect on the bandgaps, larger supercells (ranging from 8 to 64 atoms) have to be mandatorily investigated to make reliable the comparison with experiments.

In the case of the optimization of larger supercells we made use of similar settings of those of the initial 8-atom cells, being the force convergence criterion still 0.01eV/Å. At variance with the initial case, the number of *k*-points was still 10×10×10 for models constituted by eight atoms, but in this case we used the Monkhorst-Pack (MP) scheme (Monkhorst & Pack, 1976), checking in this way the possible impact on the final results: our calculations revealed that MP scheme and Γ–centred sampling schemes gave identical results. For the $n_x \times n_y \times n_z$ multiplied supercell models derived from the eight atom unit cell we thus used a $8/n_x \times 8/n_y \times 8/n_z$ *k*-points sampling scheme. For such larger supercell E_G calculation, we employed a GGA+GW_0 (Fuchs et al., 2007) scheme, using the eigenvalues and wave functions obtained at GGA level as initial guess for GW_0 calculations (eigenvalues only updated, screened potential kept fixed). For the GGA calculations, we used the Perdew-Burke-Ernzerhof (PBE) functional (Perdew et al., 1996). Cut-off energy for response function is 90 eV, and the number of frequency points for dielectric function is 48. The number of unoccupied bands was increased up to 200. A 6×6×6 Γ-centred sampling scheme was used for eight atom models. For the $n_x \times n_y \times n_z$ supercell models of the initial eight atom unit cell we used a Γ-centered $4/n_x \times 4/n_y \times 4/n_z$ *k*-point sampling scheme.

4. Discussion

4.1 The initial case of eight atom unit cells

A common starting point for both approaches is represented by the calculations at the DFT level of the structural optimized parameters of the two most stable polymorphs of GaAs, zincblende (ZB, group 216, *F-43m*, Z=4) and wurtzite (WZ, group 186, *P63mc*, Z=2) and of Ge in its cubic form (group 227, *Fd-3m*, Z=8) Ge and GaAs, reported in Table 1.

The choice of using LDA in all the subsequent calculations stems from the fact that in this context LDA reproduces structural properties closer to experiment than GGA. Both Ga–As and Ge–Ge bond lengths are 2.43 Å in their most stable polymorph.

ZB–GaAs is constituted by interpenetrating *fcc* sublattices of cations (Ga) and anions (As). The diamond lattice of Ge may be through of as the ZB structure with Ge occupying both cation and anion sites. In this section we consider 8-atom $(GaAs)_{1-x}Ge_{2x}$ alloy models that vary the Ge composition, including pure GaAs (x=0) to x=0.25 (Ge dimers), x=0.50 (4 Ge atoms), x=0.75 (6 Ge atoms) and finally pure Ge (x=1). Figure 5 reports the structures of the four intermediate alloys. At first we performed an analysis of the Ge dimer in bulk GaAs, at site positions (0.25, 0.25, 0.5) and (0., 0.25, 0.75) , i.e. the alloy model **I**.

	GaAs (ZB) 216, F-43m, Z=4	**GaAs (WZ)** 186, P_{63mc}, Z=2	**Ge (cubic)** 227, *Fd-3m*, Z=8
Our analysis, PAW/LDA			
ΔE	---	+0.06	----
B	66.14		71.8
Lattice constant (Å)	*a*=5.605	*a*=3.917, *b*=3.886, *c*=6.505	*a*=5.612
Our analysis, PAW/PW91			
ΔE	---	+0.03	----
B	79.01		71.0
Lattice constant (Å)	*a*=5.739	*a*=4.040, *c*=6.668	*a*=5.747
Our analysis, PAW/PBE			
ΔE	---	+0.023	----
B	65.94		74.8
Lattice constant (Å)	*a*=5.744	*a*=4.045, *c*=6.670	*a*=5.741
Previous study (GGA)			
B	59.96[a]		55.9[d]
Lattice constant (Å)	*a*=5.74[a],5.722[b]	*a*=3.540, *c*=6.308[c]	*a*=5.78[d]
Previous study (LDA)			
ΔE	---	+0.0120[g]	
B	75.7[e] , 77.1[f]		73.3[d], 79.4[d]
Lattice constant (Å)	*a*=5.654[g],5.53[e] 5.508[f], 5.644[h]	*a*=3.912,*c*=6.441[g] *a*=3.912,*c*=6.407[e]	*a*=5.58[d], 5.53[d]
Experimentally			
ΔE	-----	+0.0117[h]	
B	77.[i]		75.[l]
Lattice constant (Å)	*a*=5.649[i],5.65[j]		*a*=5.678[k] , 5.66[l]

Table 1. The energy difference ($\Delta E_{per\ unit}$, eV) between ZB and WZ polymorphs of GaAs, lattice constant, *a*, and bulk moduli *B* (GPa) of GaAs (ZB) and Ge (diamond). ([a]Arabi et al., 2006; [b]A. Wronka, 2006; [c]Bautista-Hernandez et al., 2003;[d] Wang & Ye, 2003;[e]Wang & Ye, 2002; [f]Kalvoda et al.,1997;[g] Yeh et al.,1992; [h]Murayama & Nakayama, 1994;[i]Hellwege & Madelung, 1982; [j]Singh, 1993;[k]CRC, 1997-1998; [l]Levinstein, 1999.)

This model can be considered a highly concentrated molecular substitutional Ge_2 defect in GaAs, for which we predict stability owing to the donor–acceptor self-passivation mechanism (Giorgi & Yamashita, 2011). For a better understanding of this last aspect concerning self-compensation mechanism, we invite the reader to take a look at Section 4.3, where the stability of Ge substitutional defects (donor, acceptor, and donor-acceptor pairs) in GaAs matrix and its relationship with alloy self-compensation stabilizing mechanism is deeply discussed.

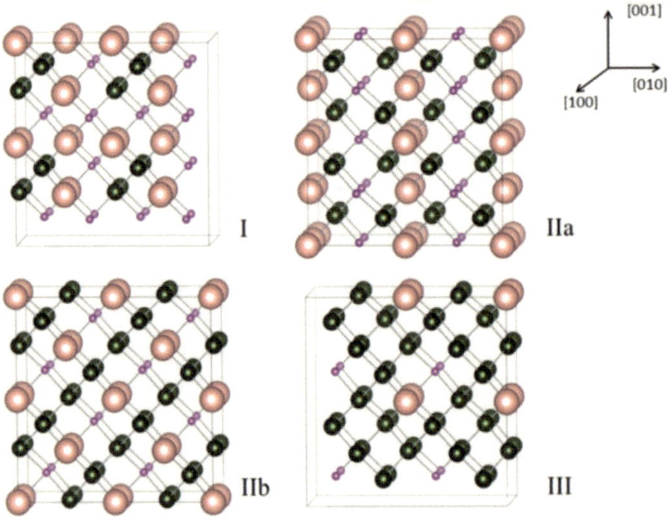

Fig. 5. The four $(GaAs)_{1-x}Ge_{2x}$ alloy models investigated. For visual aid, we here report the 2 x 2 x 2 enlarged cells [Ga, small purple; As, large pink; Ge, intermediate green].

The overall sequence of **I** is a repeated "sandwich-like" structure, ···/As/Ge–Ga/Ge–As/Ga/···along the (001) direction. The bond lengths are 2.38 (Ga–Ge), 2.42 (Ge–Ge), 2.44 (Ga–As), and 2.47 Å (Ge–As) – only slightly differing from the calculated values in bulk Ge and GaAs (2.43 Å). The small electronegativity variation ($\Delta\chi$) is the explanation of this reduced difference in the bondlength, being Ga-Ge and Ge-As nearly covalent two-center bonds (χ_{Ga}=1.81, χ_{Ge} =2.01, and χ_{As}=2.18).

In the alloy **I** the total number of III-IV and IV-V "bad bonds" (Osorio & Froyen, 1993; Kroemer, 2001; Rodriguez et al., 2001) is 12, or 37.5% of the total. According to the Bader analysis (Henkelmann et al., 2006; Tang et al., 2009; Sanville et al., 2007), in the pure host, the difference in electronegativity is responsible for charge transfer from cation to anion.

In the alloy formation process, the introduction of Ge reduces the ionic character of the GaAs bond, while increasing the ionic character of the Ge–Ge bond. When a Ge dimer is inserted in GaAs, 0.32 electrons are transferred away from Ge_{Ga} site, while Ge_{As} gains 0.21 electrons. The charge deficit on Ga, is reduced from 0.6 electrons in bulk GaAs to 0.47e , while the charge excess on As is reduced from 0.6e to 0.5e. The heat of reaction according to Eq. (3) was 0.55 eV, and the optimized lattice parameter was a=5.621 Å. We have also considered Ge donors (Ge_{Ga}) and acceptors (Ge_{As}) in the pure 8-atom GaAs host cell, separately. The formation energy has been computed according to the Zhang-Northrup formalism (Zhang & Northrup, 1991). In particular, we calculate ΔE to be 1.03 eV for Ge_{Ga} and 0.84 for Ge_{As}. The sum of the single contributions (1.87 eV) is larger than the heat of formation of the dimer, structure **I** (0.55 eV). This is ascribable to the fact that in the model alloy at least one correct bond III-V is formed while in the separate Ge_{Ga} (IV-V) and Ge_{As} (IV-III) cases only bad bonds are formed. The isolated Ge_{Ga} is a donor; the isolated Ge_{As} is an acceptor, thus both of them are unstable in their neutral charged state. We have tested it in another work (Giorgi & Yamashita, 2011) where we calculated +1 and -1 as the most stable charged state for Ge_{Ga} and Ge_{As} (both isoelectronic with GaAs), for most of the range of the electronic chemical potential. That the stabilization energy 1.32 eV (i.e., 1.87-0.55) is only slightly smaller than the host GaAs bandgap establishes that the self-passivating donor-acceptor mechanism is the stabilizing mechanism of this Ge dimer. As previously stated, for a further and deeper analysis of Ge substitutions in GaAs the reader take a look on Section 4.3.

We considered two alternative structures for the x=0.50 case. In the **IIa** structure Ge atoms are substituted for host atoms at (0.5, 0., 0.5), (0.5, 0.5, 0), (0.75, 0.75, 0.25), and (0.75, 0.25, 0.75). The initial cubic symmetry lowers toward a simple tetragonal one: the optimized lattice parameters were found to be a = 5.590 Å, b = c = 5.643 Å. The 4 intralayer bond lengths were calculated to be Ga–Ge (2.39 Å), Ge–As (2.48 Å), Ge–Ge (2.42 Å), and Ga–As (2.44). Because of the increased amount of Ge, structure **IIa** was less polarized than **I**, as confirmed by the slightly more uniform bond lengths. In **IIa** alloy the number of "bad bonds" is 16 (i.e., 50%) and ΔE rises to 0.72 eV. In the **IIb** structure Ge atoms are substituted for host atoms at (0.25, 0.25, 0.25), (0.25, 0.75, 0.75), (0.75, 0.75, 0.25), and (0.75, 0.25, 0.75). This structure consists of a stack of pure atomic layers, ···/Ga/Ge/As/Ge···, and thus it contains *only* nearest neighbors of the (Ga-Ge) and (Ge-As) type: thus *all* bonds are "bad bonds" in this **IIb** compound. Bond lengths were calculated to be 2.40 Å and 2.49 Å, respectively, and optimized lattice parameters were a = c = 5.682, b = 5.560 Å. In this structure, ΔE = 1.40 eV, almost double that of **IIa** with identical composition. According to phase transition theory, the symmetry lowering for the two intermediate systems is the fingerprint of an ordered–disordered phase transition. The calculated deviation from the ideal cubic case (c/a=1) is 0.94% and 2.15% for **IIa** and **IIb** models, respectively, confirming energetic instability for the **IIb** alloy. The last model, **III**, [Ge] = 0.75, consists of pure Ge except that Ga at (0., 0. ,0.) and As at (0.25, 0.25, 0.25). The calculated bond lengths were 2.39, 2.43, 2.45, and 2.48 Å for Ga–Ge, Ge–Ge, Ga–As, and Ge–As, respectively. Cubic symmetry is restored: the optimized lattice parameter (a = 5.624 Å) is nearly identical to structure **I**. The formation energy as from Eq.(3) is almost the same as **I** (~0.54 eV). Indeed, **I** and **III** are formally the same model with the same concentration (37.5%) of Ge in GaAs (**I**) and GaAs in Ge (**III**) and the same number of wrong bonds, 12. For sake of comparison between these two cases, we have also calculated the formation energy of a single substituted Ge in the cell.

We have also made a preliminary calculation of the stability of isolated Ga acceptors (Ga_{Ge}) and As donors (As_{Ge}) *vs* that of the substitutional molecular $GaAs_{Ge2}$ in Ge pure host supercell consisting of 64 atoms; for such concentrations (0.0312=1Ge/32GaAs unit and 0.0156=1/64GaAs), the molecular substitutional $GaAs_{Ge2}$ is only stabilized by 0.057eV with respect to the separate couple acceptor-donor. This small stabilization for $GaAs_{Ge2}$ compared to isolated Ga_{Ge} and As_{Ge} confirms the expected similar probability of finding a mixture of n-type and p-type semiconductors in the "disordered" Ge-rich phase. We have used the most stable polymorph of the elemental compounds (orthorhombic Ga and rhombohedral As) for the chemical potential, μ, of both elements (Mattila & Nieminen, 1996).

Ga-rich ($\mu_{Ga}=\mu_{Ga}^{bulk}$) and As-rich ($\mu_{As}=\mu_{As}^{bulk}$) conditions have been considered, respectively. In the case of the 8-atom cells, the formation energy for Ga_{Ge} and As_{Ge} are 0.26 eV and 0.58 eV, respectively. The model **III** stabilizes the isolated Ga and As substitutions by 0.30 eV, i.e. the ΔE between the alloy and the isolated substitutionals.

At variance with alloy model **I**, the large energy difference (~0.4 eV) between the stabilization energy and the Ge host bandgap (0.67 eV at 300 K (Kittel, 2005)) reveals that other factors, and not only a self-compensating donor-acceptor mechanism, impact on the final stability of this **III** alloy model.
Our calculations reveal an almost exactly linear relationship between the formation energy and the number of bad bonds, as reported in Figure 6. Such relationship is verified at least for systems containing same number of Ge donors (Ge_{Ga}) and Ge acceptors (Ge_{As}).

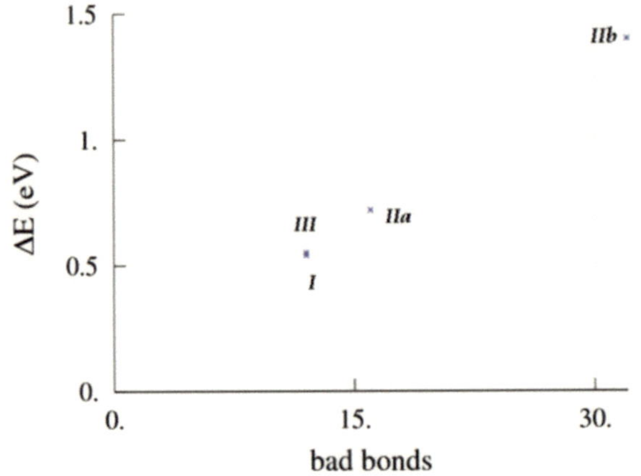

Fig. 6. Heat of formation (ΔE) of the alloy models *vs.* the number of "bad bonds".

This striking result confirms that the electronic structure of these compounds is largely described in terms of independent two center bonds. For stoichiometric compounds, it suggests an elementary model Hamiltonian for the energetics of any alloy with equal numbers of Ge cations and anions. On all the optimized structures QSG*W* calculations have been performed and also for the pure GaAs and Ge 8-atom cells. From Figure 7, where the QSG*W* bandgaps as function of [Ge] are reported, one can see the good reproduction of the asymmetric bowing both at Γ and R points. In particular, QSGW calculated bandgaps for pure GaAs and Ge are 1.66 and 1.04 eV on Γ. **IIb** model (100% of bad bonds), whose bandgap is not reported in Figure 7, has $E_G < 0$ at both the two points, confirming the tight relationship between high concentration of bad bonds and reduced values of the bandgap.

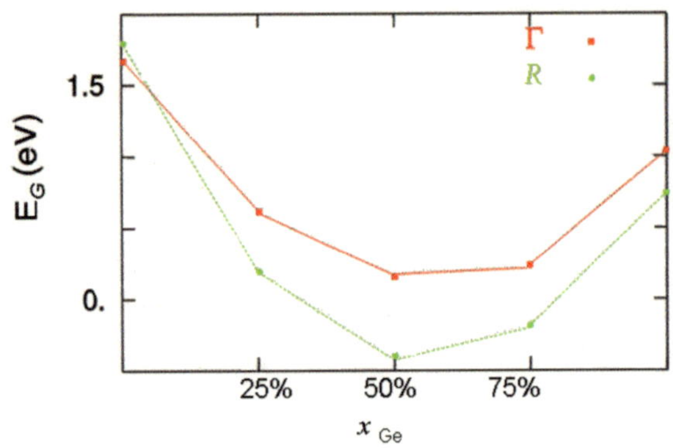

Fig. 7. QSG*W* calculated bowing of the bandgap at Γ and R *vs.* different concentrations of Ge atoms.

4.2 Extended models: The quantitative description of the asymmetric bowing minimum

The subsequent analysis focused on extended alloy models. In particular, we have taken into account here models ranging between 8 and 64 atoms.

Our choice has the two-fold target of confirming the initial results comparing different methodologies for bandgap calculations, and that of investigating the impact that cluster shape and size has in the bandgap itself. It is also supposed that enlarging the size of the models can extremely improve the reproducibility of the asymmetric V-shape of the bandgap bowing. The Special Quasi Random Structures (SQS) methodology (Zunger et al., 1990), developed to incorporate SRO and local lattice distortions in alloy systems, has been widely employed in literature for the description of alloy properties (see Fiorentini & Bernardini, 2001). We stress that the modellization we have here chosen does not lead mandatorily towards a global minimum for each concentration, stemming this choice from the metastable nature of $(GaAs)_{1-x}(Ge_2)_x$ alloys, grown only at nonequilibrium condition (Barnett et al., 1982; Rodriguez et al., 2001; Banerjee et al., 1985; Alferov et al., 1982). The appropriateness of our modellization is confirmed by the high reproducibility of the experimental results we have obtained. McGlinn. (McGlinn et al., 1988) have experimentally found that Ge regions start forming networks as the Ge concentration increased in the range between 0. and 0.3, and at [Ge]=0.3 such networks are connected with each other. This mechanism is accompanied by the GaAs region fragmentation towards size-reduced clusters. Such experimental finding is our driving force in the choice of the enlarged models: we based our study on models reproducing Ge-clusterized alloys at $0< x <0.3$ and GaAs-clusterized alloys at $0.3< x <1$. As stated, experimental results report only the formation of size reduced GaAs clusters for [Ge] > 0.3, thus we decided to compare and discuss local geometry effects and their influence on the bandgap of the two specular models at $x = 0.375$ (such Ge concentration represents the closest one in our models to the experimentally reported concentration, 0.3, where bandgap minimum is found (Barnett et al., 1982)): a Ge-clusterized (**IIIa**) and a GaAs-clusterized (**IIIb**) one. Going back to the discussion regarding our two models at x=0.375, **IIIb** and **IIIa** (whose structure is reported in Figure 8), the former alloy has larger bandgap than the latter, revealing that at low Ge concentration E_G continues decreasing as Ge concentration increases, as long as the alloy geometry is characterized by the presence of Ge clusters embedded in GaAs host, i.e., in a quantum dot-like fashion. Differently, when GaAs turns to clusterize in Ge network, the bandgap stops decreasing. The relationship between the calculated direct gaps along all the range of x is reported in Figure 9, in conjunction with the experimental values.

The close resemblance between our theoretical fitting and the experimental one reveals the extreme suitability of our models in order to reproduce the asymmetric bandgap bowing of $(GaAs)_{1-x}(Ge_2)_x$ alloys: a sharp E_G decreasing at $0< x <0.3$ accompanied by the subsequent smooth increasing at $0.3< x <1$.

Up to now we were able to explain and demonstrate that the bandgap minimum detection is due to the switching of the embedded cluster in the host from a quantum-dot-like fashion (Ge in GaAs) to an *anti*-quantum-dot like fashion (GaAs in Ge). Let us go one step further and let us try to understand the chemical origin of this minimum in the bowing.

As we mentioned, in $(GaAs)_{1-x}(Ge_2)_x$ alloys, the acceptors and donors, when nearest-neighbors, are subjected to stabilizing self-compensation mechanism, while between bad-bond pairs similar mechanism does not occur due to the presence of residual local positive and/or negative charges. However, if such charges can be delocalized, the compensation can be effective even beyond nearest-neighbor atomic sites. Thus, since the charge distribution on the bad bonds describes the nearest-neighbor atomic chemical environment, it represents a highly valuable analysis to estimate the SRO. In order to investigate the effect of the cluster type *switching* on bad bonds, we analyzed the charge distribution of **IIIa** and **IIIb** models according to the Bader charge analysis scheme (Henkelmann et al., 2006; Sanville et al., 2007; Tang et al., 2009).

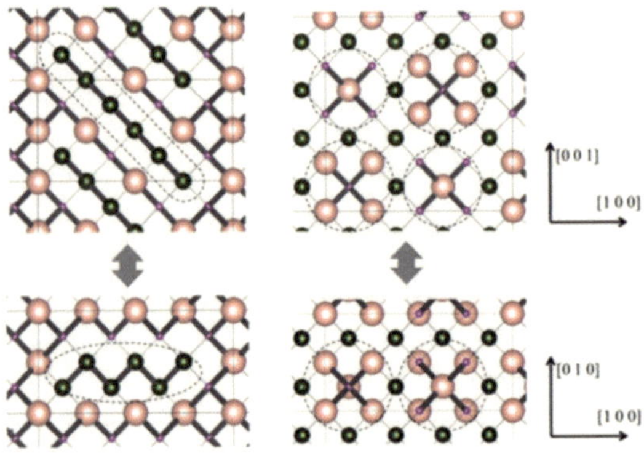

IIIa **IIIb**

Fig. 8. $(GaAs)_{1-x}(Ge_2)_x$ models at $x = 0.375$. Left: Top and lateral views of **IIIa** model. Right: Top and lateral views of **IIIb** model. [Pink, large: Ga atoms; purple, small: As atoms; green, intermediate: Ge atoms respectively].

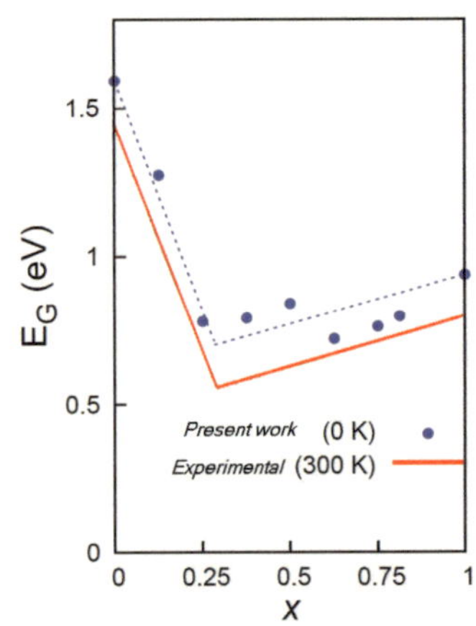

Fig. 9. Our calculated (GGA+GW_0 , (VASP)) and experimental band gap behaviour (Rodriguez et al., 2001) *vs.* Ge concentration (x). Red solid line is fitted to experimental results by two line expressions. Blue points are the calculated values. Blue dotted line is fitted to blue points by two line expressions similar to the red solid one. For $x = 0.375$, the model **IIIb** bandgap is assumed.

In optimized GaAs, charge on Ga (As) atoms is +0.55*e* (-0.55*e*). In model **IIIa**, containing Ge clusters, the charges range between +0.40 and +0.52 for Ga, -0.53 and -0.48 for As, and between -0.29 and +0.32 for Ge, respectively. The charge reduction for Ga and As with respect to those of pure GaAs are easily explained in terms of electronegativity, χ. $\Delta\chi$ are smaller on bad bonds than on correct Ga-As bonds. Charges on model **IIIb** are +0.30 on Ga, -0.41 on As, and +0.04/+0.08 on Ge, respectively, values extremely reduced if compared with those of **IIIa**. Such charge lowering is explained in terms of number of bad bonds that *single* Ga or As atom forms. In **IIIa**, each Ga and As atom forms 1 or 2 bad bonds. In **IIIb**, on the other hand, each Ga and As atom forms 3 bad bonds. The increase of bad bonds on single Ga and As atom causes partial delocalization of charges, as a consequence of the reduced $\Delta\chi$ between their constituting atoms. This result also impacts the VBM charge density distribution on Γ; indeed, in the **IIIa** model it is localized around the As and Ge, whose electronegativities are larger than Ga (strong charge localization on the formed bad bonds). Thus, similarly to the valence band forward shift ascribed to the charge density localization (non-bonding), the sharp bandgap decrease in the range $0 < [Ge] < 0.3$ is ascribed to the enhancement of the non-bonding character of the VBM. Differently, in **IIIb** the VBM is highly delocalized in the whole crystal. Here, the strong bonding character of VBM stabilizes the system, causing the backward shift of VBM and also the smooth bandgap opening at $0.3 < x < 1$, confirming that SRO effects play the main role in the asymmetric bandgap bowing of this class of alloys.

4.3 On the stability of Ge donor and acceptor defects

In previous sections we have widely taken advantage of the concept of self-passivation. We here focus on this very basic concept of semiconductor physics, showing that regardless the nature of the cell we are considering, alloys or supercells, the self-passivation stabilization mechanism between Ge donor and Ge acceptor pairs is the main stabilizing process in these IV-doped/III-V systems effective also for non nearest neighbor couples. We aim to demonstrate that alloys are super-concentrated defective cells (Giorgi & Yamashita, 2011).The formation energy of the defect is defined as the contribution deriving from the formation energy of the defect in its state of charge, plus the contribution of Ga and As potentials in GaAs, and the potential of the substituting Ge. The thermodynamic stability of the charged substitutional Ge defects is calculated as (Zhang & Northrup, 1991):

$$E_{form} = E_D^{tot} - (n_{Ga} + n_{As})\mu_{GaAs(bulk)}/2 - (n_{Ga} - n_{As})(\mu_{Ga(bulk)} - \mu_{As(bulk)} + \Delta\mu)/2 + q(\mu_e + E_{VBM}) - n_{Ge}\mu_{Ge} \quad (5)$$

where $\Delta\mu$ is the chemical potential difference between bulk GaAs ($\mu_{GaAs(bulk)}$) and bulk Ga ($\mu_{Ga(bulk)}$) and As ($\mu_{As(bulk)}$), respectively, ranges between $\pm\Delta H_{form}$. ΔH_{form} is the heat of formation of bulk GaAs. Beyond the two extreme conditions of $\Delta\mu = \pm\Delta H_{form}$ (Ga-rich and As-rich conditions, respectively) precipitation takes place. μ_e is the electronic chemical potential and E_{VBM} the energy of the top of the valence band (VBM). $n_{Ga(As,\ Ge)}$ is the number of atoms of Ga (As, Ge) in the supercell, while q is its total charge. It is straightforward that this equation is the extension (including the charged case and in the stoichiometric case) of Eq. (3) reported in Section 3.1. Similarly for the alloy case, $\mu_{Ga(As,)}$ is calculated from the orthorhombic (trigonal) polymorph for Ga (As) (Mattila & Nieminen, 1996; Giorgi & Yamashita, 2011). A method based on the combination of the Potential Alignment (PA) for the correction of supercells with a net charge and image charge correction (Lany & Zunger, 2004; Makov & Payne, 1995) has been applied.

To improve the description of the bandgap we employed an LDA + *U* scheme (Dudarev et al., 1998; Giorgi & Yamashita, 2011). VBM and CBM was also corrected and aligned with the LDA +*U* obtained bandgap. The Ge_2 *quasi*-molecule dimer defect best reproduces the high Ge-doping concentration because of the equally probable substitution of one Ga and one As atom, and represents the starting point for our analysis on the self-compensation mechanism in such systems. The neutral state is the most stable along the whole bandgap, as a consequence of the stabilization induced by the mentioned donor–acceptor self-passivating mechanism. The formation energy of the Ge molecular defect is reported in Figure 10.

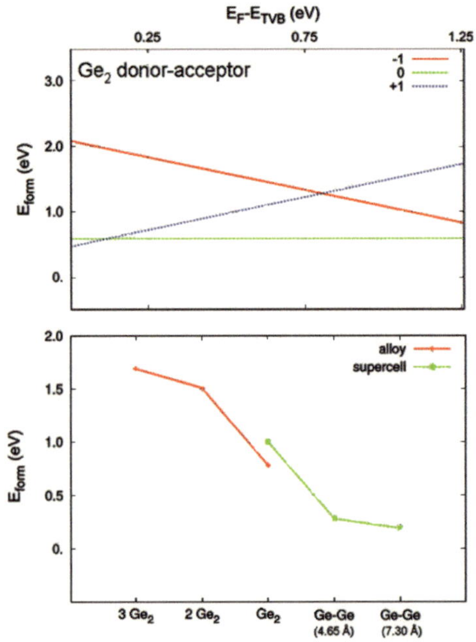

Fig. 10. Top: Formation Energy *vs.* electronic chemical potential of Ge_2 dimer defect. Bottom: Stabilization Energy *vs.* Ge concentration (red) and distance (green) in alloy and defective cells.

We also considered the case of isolated Ge_{Ga} and Ge_{As} ($Ge_{Ga}\cdots Ge_{As}$) pairs in the supercell. Comparison between the ΔE of the three structures (Ge_{2GaAs}, d=2.43 Å, $Ge_{Ga}\cdots Ge_{As}$, d=4.65, 7.30 Å, respectively) reveals the net tendency for Ge to cluster; the first configuration is indeed 0.36 eV more stable than the second and 0.41 eV more stable than the third. This energy difference is due mainly to the formation of one III–IV (Ga–Ge) and one IV–V (Ge–As) bond and the breaking of one IV–IV (Ge–Ge) bond. To evaluate the impact of distance between pairs of substitutionals on the self-compensating mechanism, we have calculated the Ge–Ge pair correlation $J(_{Ge,h\alpha,Ge,h'\beta})$ as:

$$J_{nn}(_{Ge,h\alpha,Ge,h'\beta}) = [E_{tot}(GaAs{:}Ge_{h\alpha}Ge_{h'\beta}) + E_{tot}(GaAs)] - [E_{tot}(GaAs{:}Ge_{h\alpha}) + E_{tot}(GaAs{:}Ge_{h'\beta})] \quad (6)$$

where $E_{tot}(GaAs{:}Ge_{h\alpha}Ge_{h'\beta})$ is the energy for the double Ge-substituted GaAs and $E_{tot}(GaAs{:}Ge_h)$ is that for the single substituted states (h_α and h'_β). The most correlated pair is the *quasi*-molecular Ge_{2GaAs} defect (–1.55 eV), while for the other two cases we calculate a correlation energy of –1.18 eV for $Ge_{Ga}\cdots Ge_{As}$ at d = 4.65 Å and –1.15 eV for $Ge_{Ga}\cdots Ge_{As}$ at d = 7.30 Å). The SRO included in the difference of the correlation energy between the three distances account for electronic and steric effects. Indeed, the direct formation of Ge–Ge may partly release the stress resulting from substitution of Ga and As in the host. A comparison between the defective supercells and the alloys is here straightforward (see Figure 10, self-compensation mechanism for the two systems).

As in the case of defective supercells also in alloy models, "bad bond" formation and the self-compensation mechanism were considered destabilizing/stabilizing driving forces of the final alloy; we may indeed evaluate the SRO effects on both systems and find a unified trend for alloys and defective supercells. The stabilization energy for the *quasi*-molecular defect is ($E_{Form}(Ge_{Ga}{}^0)$ + $E_{Form}(Ge_{As}{}^0)$) - $E_{Form}(Ge_{2GaAs})$ = 1.60 eV (= $-J_{nn}$, the correlation energy). Effects related to the Ge–Ge direct bond formation give a contribution of 1.01 eV to the total stability (1.60 eV - 0.59 eV). The same contribution for one Ge–Ge bond direct formation in an 8-atom cell alloy was found to be 0.78 eV. Increasing the number of Ge in the alloy results in an increase in the stabilization energy, i.e., 1.51 (= 1.87 - $E_{Form}/2$) eV for two Ge_2, and 1.69 eV (= 1.87 eV - $E_{Form}/3$) for three Ge_2. In the present case, we evaluate how the distance between pairs influences the stabilization energy, or in other words, how self-passivation increases the stability of the overall systems. The stabilization energy for the $Ge_{Ga}\cdots Ge_{As}$ at d = 4.65 Å is 1.18 eV; for $Ge_{Ga}\cdots Ge_{As}$ at d = 7.30 Å the same energy is 1.15 eV. The difference between formation energy and stabilization energy gives the self-passivation contribution, which is 0.28 eV for the former (d = 4.65 Å), 0.19 eV for the latter (d = 7.30 Å); in both cases self-passivation is reduced by the distance. A quantitative trend between the Ge atomic distance and the amount of stabilization due to the self-passivation mechanism is thus computed. Indeed, owing to the reduced difference in terms of atomic radius between the impurity and the host atoms, the stabilizing energy resulting from direct Ge–Ge formation is due purely to electronic factors. This confirms that defective supercells can be considered, regardless of the concentration of the substitutional atoms, as precursors of $(GaAs)_{1-x}Ge_{2x}$ alloys or, similarly, such alloys can be considered as extremely concentrated defective cells.

5. Conclusion

In this chapter we have reviewed the technology behind the Multi-Junction technology in solar cell assembling based on IV-doped III-V alloy showing the importance of Density Functional Theory as a tool for the prediction of the structural and electronic properties of these alloys. After an initial study focusing on eight atom cells, we have extended the analysis to systems constituted by up to 64 atoms. We detected a linear relationship between formation energy and number of bad bonds in the alloys. The relevance of this result stems by the fact that for stoichiometric compounds an elementary model Hamiltonian for the energetics of any alloy with equal numbers of Ge cations and anions as function of the number of bad bonds can be developed. The bandgap bowing for these alloys is confirmed by *GW* calculations. Increasing the cell size we were able to quantitatively reproduce the asymmetric bandgap bowing of $(GaAs)_{1-x}(Ge_2)_x$ alloys. This finding stems from an extremely suitable model choice: moving from previous experimental results, we found that Ge-clusterized alloys at $0 < x < 0.3$ and GaAs-clusterized ones at $0.3 < x < 1$ are the best in reproducing the asymmetric V-shape of the bowing. Turning from Ge-cluster to GaAs-cluster embedding alloys at concentrations close to the experimental reported for the bandgap minimum is the key-point for the interpretation of this controversial phenomenon. The last part of the Chapter has been dedicated to the discussion of the stability of the Ge donor-acceptor defects in the GaAs supercells. Regardless the distance between Ge pairs in both defective supercells and alloys, the self-passivation mechanism results the driving force of the stabilization of IV-doped III-V systems, being sensitively effective in the former case also for Ge pairs non nearest-neighbors.

6. Acknowledgment

This research was supported by a Grant from KAKENHI (#21245004) and the Global COE Program [Chemical Innovation] from the Ministry of Education, Culture, Sports, Science, and Technology of Japan. GG wants to thanks Dr. G. F. Cerofolini (University of Milano Bicocca) and Dr. A. Korkin (Arizona State University) for the fruitful and stimulating discussions and for a longstanding real friendship.

7. References

Alferov, Zh. I.; Zhingarev, M. Z.; Konnikov, S. G.; Mokan, I. I.; Ulin, V. P.; Umanskii, V. E. & Yavich, B. S. (1982). Preparation and investigation of metastable continuous solid-solutions in the Ge-GaAs system. *Soviet Physics Semiconductors*, Vol. 16, No. 5, pp. 532-537, ISSN 0038-5700.

Arabi, H.; Pourghazi, A.; Ahmadian, F. & Nourbakhsh, Z. (2006). First-principles study of structural and electronic properties of different phases of GaAs. *Physica B: Condensed Matter*, Vol. 373, No. 1, pp. 16-22, ISSN 0921-4526.

Baker, S. H.; Bayliss, S. C.; Gurman, S. J.; Elgun, N.; Bates, J. S. & Davis, E. A. (1993). The effect of varying substrate temperature on the structural and optical properties of sputtered GaAs films. *Journal of Physics: Condensed Matter*, Vol. 5, No. 5, pp. 519-534, ISSN 1361-648X.

Banerjee, I; Chung, D. W. & Kroemer, H. (1985). Properties of $(Ge_2)_x(GaAs)_{1-x}$ alloys grown by molecular beam epitaxy. *Applied Physics Letters*, Vol. 46, No. 5, pp. 494-496., ISSN 1077-3118

Barnett, S. A.; Ray, M.A.; Lastras, A.; Kramer, B.; Greene, J. E.; Raccah, P. M. & Abels, L. L. (1982). Growth and optical properties of single-crystal metastable $(GaAs)_{1-x}Ge_x$ alloys. *Electronic Letters*, Vol. 18, No. 20, pp. 891-892, ISSN 1350-911X.

Bautista-Hernandez, A.; Perez-Arrieta, L.; Pal, U. & Rivas Silva, J. F. (2003). Estudio estructural de los semiconductores AlP, GaAs y AlAs con estructura wurzita. *Revista Mexicana de Fisica*. Vol. 49, No. 001, pp. 9-14, ISSN 0035-001X.

Bernardini, F. & Fiorentini, V. (2001). Nonlinear macroscopic polarization in III-V nitride alloys. *Physical Review B*, Vol. 64, No. 8, pp. 085207-7, ISSN 1550-235X.

Blöchl, P. E. (1994). Projector augmented-wave method. *Physical Review B*, Vol. 50, No. 24, pp. 17953-17979, ISSN 1550-235X.

Bloom, S. J. (1970). Bandgap Variation in Quaternary Alloys. *Journal of Applied Physics*, Vol. 41, No. 4, pp. 1864-1865, ISSN 1089-7550.

Bowen, M. A.; Redfield, A. C.; Froelich, D. V.; Newman, K. E.; Allen, R. E. & Dow, J. D. (1983). Effects of an order–disorder transition on surface deep levels in metastable $(GaAs)_{1-x}Ge_{2x}$. *Journal of Vacuum Science & Technology B: Microelectronics and Nanometer Structures* Vol. 1, No. 3, pp. 747-750, ISSN 1071-1023.

Capaz, R. B.; Preger, G. F. & Koiller, B. (1989). Growth-driven ordering and anisotropy in semiconductor alloys. *Physical Review B*, Vol. 40, No. 12, pp. 8299-8304, ISSN 1550-235X.

Ceperley, D. M. & Alder, B. I. (1980). Ground State of the Electron Gas by a Stochastic Method. *Physical Review Letters*, Vol. 45, No. 7, pp. 566-569, ISSN 1079-7114.

Davis, L. C. & Holloway, H. (1987). Properties of $(GaAs)_{1-x}Ge_{2x}$ and $(GaSb)_{1-x}Ge_{2x}$: Consequences of a stochastic growth process. *Physical Review B*, Vol. 35, No. 6, pp. 2767-2780, ISSN 1550-235X.

Deng, H.-X.; Li, J.; Li, S.-S.; Peng, H.; Xia, J.-B.; Wang, L.-W. & Wei, S.-H. (2010). Band crossing in isovalent semiconductor alloys with large size mismatch: First-principles calculations of the electronic structure of Bi and N incorporated GaAs. *Physical Review B*, Vol. 82, pp.193204-5, ISSN 1550-235X.

Dimroth, F. (2006). High-efficiency solar cells from III-V compound semiconductors. *Physica Status solidi (c)*, Vol. 3, No. 3, pp. 373-379, ISSN 1610-1642.

D'yakonov, M. I. Raikh. M. E. (1982). *Journal Fiuzika i Tekhnika Poluprovodnikov*, Vol. 16, pp. 890. *Soviet Physics Semiconductors*, Vol. 16, pp. 570, ISSN 0038-5700.

Dudarev, S. L.; Botton, G. A.; Savrasov, S. Y.; Humphreys, C. J. & Sutton, A. P. (1998). Electron-energy-loss spectra and the structural stability of nickel oxide: An LSDA+U study. *Physical Review B*, Vol. 57, No. 3, pp. 1505-1509, ISSN 1550-235X.

Feltrin, A. & Freundlich, A. (2008). Material considerations for terawatt level deployment of photovoltaics. *Renewable Energy*, Vol. 33, No. 2, pp. 180-185. ISSN 0960-1481.

Fuchs, F.; Furthmüller, J.; Bechstedt, F.; Shishkin, M. & Kresse, G. (2007). Quasiparticle band structure based on a generalized Kohn-Sham scheme. *Physical Review B*, Vol. 76, No. 11, pp. 115109-8, ISSN 1550-235X.

Fthenakis, V. (2009). Sustainability of photovoltaics: The case for thin-film solar cells. *Renewable and Sustainable Energy Reviews*, Vol. 13, No. 9, pp. 2746-2750, ISSN1364-0321.

Funato, M; Fujita, S. & Fujita, S. (1999). Energy states in ZnSe-GaAs heterovalent quantum structures. *Physical Review B*, Vol. 60, No. 24, pp. 16652-16659, ISSN 1550-235X.

Giorgi, G.; van Schilfgaarde, M.; Korkin, A. & Yamashita, K. (2010). On the Chemical Origin of the Gap Bowing in $(GaAs)_{1-x}Ge_{2x}$ Alloys: A Combined DFT–QSGW Study. *Nanoscale Research Letters*, Vol. 5, No.3, pp. 469–477, ISSN 1556-276X.

Giorgi, G. & Yamashita, K. (2011). Amphoteric behavior of Ge in GaAs: an LDA analysis. *Modelling and Simulation in Materials Science and Engineering*. Vol. 19, No. 3, pp. 035001-14, ISSN 1361-651X.

Gomez-Abal, R.; Li, X.; Scheffler, M. & Ambrosch-Draxl, C. (2008). Influence of the Core-Valence Interaction and of the Pseudopotential Approximation on the Electron Self-Energy in Semiconductors, *Physical Review Letters*, Vol. 101, No. 10, pp. 106404-4, ISSN 1079-7114.

Greene, J. E. & Eltoukhy, A. H. (1981). Semiconductor crystal growth by sputter deposition. *Surface and Interface Analysis*. Vol 3, No. 1, pp. 34-54, ISSN 1096-9918.

Greene, J. E. (1983). A review of recent research on the growth and physical properties of single crystal metastable elemental and alloy semiconductors. *Journal of Vacuum Science & Technology B*, Vol. 1, No. 2, pp. 229-237. ISSN n.d.

Green, M. A. (1982). Solar Cells . Prentice-Hall, Englewood Cliffs, NJ,

Gu, B.-L.; Newman, K. E. & Fedders, P. A. (1987). Role of correlations in $(GaSb)_{1-x}Ge_{2x}$ alloys. *Physical Review B,* Vol. 35, No. 17, pp. 9135-9148, ISSN 1550-235X.

Guter, W.; Schöne, J.; Philipps, S. P.; Steiner, M.; Siefer, G.; Wekkeli, A.; Welser, E.; Oliva, E.; Bett, A. W. & Dimroth, F. (2009).Current-matched triple-junction solar cell reaching 41.1% conversion efficiency under concentrated sunlight. *Applied Physics Letters,* Vol. 94, No. 22, pp. 223504-3 ISSN 1077-3118.

Hedin, L. (1965). New Method for Calculating the One-Particle Green's Function with Application to the Electron-Gas Problem. *Physical Review,* Vol. 139, No. 3A, pp. A796-A823, ISSN 1943-2879.

Henkelman, G.; Arnaldsson, A. & Jónsson, H. (2006). A fast and robust algorithm for Bader decomposition of charge density. *Computational Materials Science,* Vol. 36, No. 3, pp. 354-360, ISSN 0927-0256.

Hochbaum, A. I. & Yang, P. (2010). Semiconductor Nanowires for Energy Conversion. Chemical Reviews, Vol. 110, No. 1, pp. 527-546, ISSN 1520-6890.

Holloway, H. (2002). Effect of sample size on simulations and measurements of the phase transition in $(GaAs)_{1-x}Ge_{2x}$ and related alloys. *Physical Review B,* Vol. 66, No. 7, pp. 075131-075136, ISSN 1550-235X.

Holloway, H. & Davis, L. C. (1987). Long-range order in $(GaAs)_{1-x}Ge_{2x}$ and $(GaSb)_{1-x}Ge_{2x}$: Predictions for <111> growth. *Physical Review B,* Vol. 35, No. 8, pp. 3823-3831, ISSN 1550-235X.

Ito, T. & Ohno, T. (1992). Pseudopotential approach to band structure and stability for GaAs/Ge superlattices. *Surface Science,* Vol. 267, No. 1-3, pp. 87-89, ISSN 0039-6028.

Ito, T. & Ohno, T. (1993). Electronic structure and stability of heterovalent superlattices, *Physical Review B,* Vol. 47, No. 24, pp. 16336-16342, ISSN 1550-235X.

Janotti, A.; Wei, S.-H. & Zhang, S. B. (2002). Theoretical study of the effects of isovalent coalloying of Bi and N in GaAs. *Physical Review B,* Vol. 65, pp. 115203-5, ISSN 1550-235X.

Kalvoda, S.; Paulus, B.; Fulde, P. & Stoll, H. (1997). Influence of electron correlations on ground-state properties of III-V semiconductors. *Physical Review B,* Vol. 55, No. 7, pp. 4027-4030, ISSN 1550-235X.

Kawai, H.; Giorgi, G. & Yamashita, K. (2011). Clustering and Octet Rule Violation Impact on Band Gap Bowing: Ab Initio Calculation of the Electronic Properties of $(GaAs)_{1-x}(Ge_2)_x$ Alloys. *Chemistry Letters,* Vol. 40, No. 7, pp. 770-772, ISSN 1348-0715.

Kim, K. & Stern, E. A. (1985). Model for the metastable system of type $(GaAs)_{1-x}(Ge_2)_x$. *Physical Review B,* Vol. 32, No. 2, pp. 1019–1026, ISSN 1550-235X.

King, R. R.; Law, D.C.; Edmondson, K. M.; Fetzer, C. M.; Kinsey, G. S.; Yoon, H.; Sherif, R. A. & Karam, N.H. (2007). 40% efficient metamorphic GaInP/GaInAs/Ge multijunction solar cells. *Applied Physics Letters.* Vol. 90, No. 18 pp. 183516-3, ISSN 1077-3118.

Kittel, C. (2004). Introduction to Solid State Physics, 8th edition, John Wiley & Sons (Ed.) ISBN 0-471-41526-X, New York, USA.

Koiller, B.; Davidovich, M. A. & Osorio, R. (1985). Correlation effects in metastable $(GaAs)_{1-x}Ge_{2x}$ alloys. *Solid State Communications,* Vol. 55, No. 10, pp. 861-864, ISSN 0038-1098.

Kotani, T. & van Schilfgaarde, M. (2002). All-electron GW approximation with the mixed basis expansion based on the full-potential LMTO method. *Solid State Communications*, Vol. 121, No. 9-10, pp. 461-465, ISSN 0038-1098.

Kotani, T.; van Schilfgaarde, M. & Faleev, S. V. (2007). Quasiparticle self-consistent *GW* method: A basis for the independent-particle approximation. *Physical Review B*, Vol. 76, No. 16, pp. 165106-24, ISSN 1550-235X.

Kresse, G. & Joubert, D. (1999). From ultrasoft pseudopotentials to the projector augmented-wave method. *Physical Review B*, Vol. 59, No. 3 pp. 1758-1775, ISSN 1550-235X.

Kroemer, H. (2001). Nobel Lecture: Quasielectric fields and band offsets: teaching electrons new tricks. *Reviews of modern physics*, Vol. 73, No. 3, pp. 783-793, ISSN 1539-0756.

Lany S. & Zunger, A. (2008). Assessment of correction methods for the band-gap problem and for finite-size effects in supercell defect calculations: Case studies for ZnO and GaAs. *Physical Review B*, Vol. 78, No. 23, pp. 235104-25, ISSN 1550-235X.

Makov, G. & Payne M. C. (1995). Periodic boundary conditions in *ab initio* calculations. *Physical Review B*, Vol. 51, No. 7, pp. 4014-4022, ISSN 1550-235X.

Mattila, T. & Nieminen, R. M. (1996). *Ab initio* study of oxygen point defects in GaAs, GaN, and AlN. *Physical Review B*, Vol. 54, No. 23, pp. 16676-16682, ISSN 1550-235X.

McGlinn, T. C.; Klein, M. V.; Romano, L. T. & Greene, J. E. (1988). Raman-scattering and electron-microscopy study of composition-dependent ordering in metastable $(A^{III}B^{V})_{1-x}(C_2^{III})_x$ alloys. *Physical Review B*, Vol. 38, No. 5., pp. 3362-3367, ISSN 1550-235X.

Methfessel, M.; van Schilfgaarde, M. & Casali, R. A. (2000). A full-potential LMTO method based on smooth Hankel functions, In: *Electronic Structure and Physical Properties of Solids: The Uses of the LMTO Method*, Lecture Notes in Physics, H. Dreysse, (Ed.) Vol. 535, pp. 114-147, ISBN 978-3-642-08661-8, Berlin, Germany.

Monkhorst, H. J. & Pack J. D. (1976). Special points for Brillouin-zone integrations. *Physical Review B*, Vol. 13, No. 12, pp. 5188-5192, ISSN 1550-235X.

Murayama, M. & Nakayama, T. (1994). Chemical trend of band offsets at wurtzite/zinc blende heterocrystalline semiconductor interfaces. *Physical Review B*, Vol. 49, No. 7, pp. 4710-4724, ISSN 1550-235X.

Newman, K. E. & Dow, J. D. (1983). Zinc-blende—diamond order-disorder transition in metastable crystalline $(GaAs)_{1-x}Ge_{2x}$ alloys. *Physical Review B*, Vol. 27, No. 12, pp. 7495-7508, ISSN 1550-235X.

Newman, K. E. & Jenkins, D. W. (1985). Metastable $(III–V)_{1-x}IV_{2x}$ alloys. *Superlattices and Microstructures*, Vol. 1, No. 3, pp. 275-278, ISSN 0749-6036.

Newman, K. E.; Dow, J. D.; Bunker, B.; Abels, L. L.; Raccah, P. M.; Ugur, S.; Xue, D. Z. & Kobayashi, A. (1989). Effects of a zinc blende-diamond order-disorder transition on the crystal, electronic, and vibrational structures of metastable $(GaAs)_{1-x}(Ge_2)_x$ alloys. *Physical Review B*, Vol. 39, No. 1, 657-662, ISSN 1550-235X.

Noreika, A. J. & Francombe, M. H. (1974). Preparation of nonequilibrium solid solutions of $(GaAs)_{1-x}Si_x$. *Journal of Applied Physics*, Vol. 45, No. 8, 3690-3691, ISSN 1089-7550.

Norman, A.G.; Olson, J. M.; Geisz, J. F.; Moutinho, H. R.; Mason, A.; Al-Jassim, M. M. & Vernon, S. M. (1999). Ge-related faceting and segregation during the growth of metastable $(GaAs)_{1-x}(Ge_2)_x$ alloy layers by metal-organic vapor-phase epitaxy. *Applied Physics Letters*, Vol. 74, No.10, pp.1382-1384, ISSN 1077-3118.

Olego D. & Cardona, M. (1981). Raman scattering by coupled LO-phonon—plasmon modes and forbidden TO-phonon Raman scattering in heavily doped *p*-type GaAs. *Physical Review B*, Vol. 24, No. 12, pp. 7217-7232, ISSN 1550-235X.

Olson, J.M.; Jessert, T.; Al-Jassim, M. M. (1985). GaInP/GaAs: a current- and lattice-matched tandem cell with a high theoretical efficiency. Proc. 18th IEEE Photovoltaic Specialists Conference, Las Vegas, Nevada, 1985, pp. 552-555.

O'Regan, B. & Graetzel, M. (1991). A low-cost, high-efficiency solar cell based on dye-sensitized colloidal TiO_2 films. *Nature.*Vol. 353, pp.737-740, ISSN 0028-0836.

Osório, R.; Froyen, S. & Zunger, A. (1991a). Structural phase transition in $(GaAs)_{1-x}Ge_{2x}$ and $(GaP)_{1-x}Si_{2x}$ alloys: Test of the bulk thermodynamic description. *Physical Review B*, Vol. 43, No. 17, pp. 14055-14072, ISSN 1550-235X.

Osório, R.; Froyen, S. & Zunger, A. (1991b). Superlattice energetics and alloy thermodynamics of GaAs/Ge. *Solid State Communications*, Vol. 78, No. 4, pp. 249-255, ISSN 0038-1098.

Osório, R. & Froyen, S. (1993). Interaction parameters and a quenched-disorder phase diagram for $(GaAs)_{1-x}Ge_{2x}$ alloys. *Physical Review B*, Vol. 47, No. 4, pp. 1889-1897, ISSN 1550-235X.

Perdew, J. P. & Zunger, A. (1981). Self-interaction correction to density-functional approximations for many-electron systems. *Physical Review B*, Vol. 23, No. 10, pp. 5048-5079, ISSN 1550-235X.

Perdew, J. P. & Levy, M. (1983). Physical Content of the Exact Kohn-Sham Orbital Energies: Band Gaps and Derivative Discontinuities. *Physical Review Letters*, Vol. 51, No. 20, pp. 1884-1887, ISSN 1079-7114.

Perdew, J. P. (1991). Electronic Structure of Solids 91, (Akademie Verlag, Berlin 1991). pp.11

Perdew, J. P.; Chevary, J. A.; Vosko, S. H.; Jackson, K. A.; Pederson, M. R.; Singh, D. J. & Fiolhais, C. (1992). Atoms, molecules, solids, and surfaces: Applications of the generalized gradient approximation for exchange and correlation. *Physical Review B*, Vol. 46, No. 11, pp. 6671-6687, ISSN 1550-235X.

Perdew, J. P.; Burke, K. & Ernzerhof, M. (1996). Generalized Gradient Approximation Made Simple. *Physics Review Letters*, Vol. 77, Vol. 18, pp. 3865-3868, ISSN 1079-7114.

Preger, G. F.; Chaves, C. M. & Koiller, B. (1988). Epitaxial growth of metastable semiconductor alloys: A novel simulation. *Physical Review B*, Vol. 38, No. 18, pp. 13447-13450, ISSN 1550-235X.

Rodriguez, A. G., H. Navarro-Contreras, H. & Vidal, M. A. (2000). Long-range order–disorder transition in $(GaAs)_{1-x}(Ge_2)_x$ grown on GaAs(001) and GaAs(111). *Microelectronic Journal*, Vol. 31, No. 6, pp. 439-441; Influence of growth direction on order–disorder transition in $(GaAs)_{1-x}(Ge)_{2x}$ semiconductor alloys. *Applied Physics Letters*, Vol. 77, No. 16, pp. 2497-2499, ISSN 1077-3118.

Rodriguez, A. G.; Navarro-Contreras, H. & Vidal, M. A. (2001). Physical properties of $(GaAs)_{1-x}(Ge_2)_x$: Influence of growth direction. *Physical Review B*, Vol. 63, No. 11, pp. 115328-9, ISSN 1550-235X.

Salazar-Hernández, B.; Vidal, M. A.; Constantino, M. E. & Navarro-Contreras, H. (1999) Observation of zinc-blende to diamond transition in metastable $(GaAs)_{1-x}(Ge_2)_x$ alloys by Raman scattering. *Solid State Communications*, Vol. 109, No. 5, pp. 295-300, ISSN 0038-1098.

Sanville, E.; Kenny, S. D.; Smith, R. & Henkelman, G. (2007). Improved grid-based Algorithm for Bader charge allocation. *Journal of Computational Chemistry*. Vol 28, No. 5, pp. 899-908, ISSN 1096-987X.

Shah, S. I.; Kramer, B.; Barnett, S. A. & Greene, J. E. (1986). Direct evidence for an order/disorder phase transition at $x \approx 0.3$ in single-crystal metastable $(GaSb)_{(1-x)}(Ge_2)_x$ alloys: High-resolution x-ray diffraction measurements. *Journal of Applied Physics*, Vol. 59, No. 5, pp. 1482-1487, ISSN 1089-7550.

Sham, L. J. & Schlüter, M. (1983). Density-Functional Theory of the Energy Gap. *Physical Review Letters*, Vol. 51, No. 20, pp. 1888-1891, ISSN 1079-7114.

Shan, W.; Walukiewicz, W.; Ager III, J. W.; Haller, E. E.; Geisz, J. F.; Friedmann, D. J.; Olson, J. M. & Kurtz, (1999). Band Anticrossing in GaInNAs Alloys. *Physical Review Letters*, Vol. 82, No.6, pp. 1221-1224, ISSN 1079-7114.

Shishkin, M. & Kresse, G. (2006). Implementation and performance of frequency dependent GW method within PAW framework. *Physical Review B*, Vol. 74, No. 3, pp. 035101-13, ISSN 1550-235X.

Shishkin, M. & Kresse, G. (2007). Self-consistent *GW* calculations for semiconductors and insulators. *Physical Review B*, Vol. 75, No. 23, pp. 235102-9, ISSN 1550-235X.

Shishkin, M.; Marsman, M. & Kresse, G. (2007). Accurate Quasiparticle Spectra from Self-Consistent *GW* Calculations with Vertex Corrections. *Physical Review Letters*, Vol. 99, No. 24, pp. 246403-4, ISSN 1079-7114.

Stern, E. A.; Ellis, F.; Kim, K.; Romano, L.; Shah, S. I. & Greene, J. E. (1985). Nonunique structure of metastable $(GaSb)_{1-x}(Ge_2)_x$ alloys. *Physical Review Letters*, Vol. 54, No. 9, 905-908, ISSN 1079-7114.

Tang, W.; Sanville, E. & Henkelman, G. (2009). A grid-based Bader analysis algorithm without lattice bias. *Journal of Physics: Condensed Matter*. Vol. 21, No. 8, 084204-7, ISSN 1361-648X.

van Schilfgaarde, M.; Kotani, T. & Faleev, S. V. (2006a). Quasiparticle Self-Consistent *GW* Theory, *Physical Review Letters*, Vol.96, No. 22, pp. 226402-4, ISSN 1079-7114

van Schilfgaarde, M.; Kotani, T. & Faleev, S. V. (2006b). Adequacy of approximations in GW theory. *Physical Review B*, Vol. 74, No. 24, pp. 245125-16, ISSN 1550-235X.

VASP, Vienna Ab-initio Simulation Package.

Wang, S. Q. & Ye, H. Q. (2002). A plane-wave pseudopotential study on III–V zinc- blende and wurtzite semiconductors under pressure. *Journal of Physics: Condensed Matter*, Vol. 14, No. 41, pp. 9579-9587, ISSN 1361-648X.

Wang, S. Q. & Ye, H. Q. (2003). *Ab initio* elastic constants for the lonsdaleite phases of C, Si and Ge. *Journal of Physics: Condensed Matter*, Vol. 15, No. 30, pp. L197-L202, ISSN 1361-648X.

Wang, L. G. & Zunger, A. (2003). Dilute non-isovalent (II-VI)-(III-V) semiconductor alloys: Monodoping, codoping, and cluster doping in ZnSe-GaAs. *Physical Review B*, Vol. 68, pp. 125211-8, ISSN 1550-235X.

Wei, S.-H. & Zunger, T. (1989). Band gaps and spin-orbit splitting of ordered and disordered $Al_xGa_{1-x}As$ and $GaAs_xSb_{1-x}$ alloys. *Physical Review B*, Vol. 39, No.5, pp. 3279-3304, ISSN 1550-235X.

Wei, S.-H.; Zhang, S. B. & Zunger, T. (2000). First-principles calculation of band offsets, optical bowings, and defects in CdS, CdSe, CdTe, and their alloys. *Journal of Applied Physics*, Vol. 87, No.3, pp. 1304-1311, ISSN 1089-7550.

Wronka, A. (2006). First principles calculations of zinc blende superlattices with ferromagnetic dopants. *Materials Science-Poland*, Vol. 24, No.3, pp. 726-730, ISSN 0137–1339.

Yamaguchi, M. (2003). III–V compound multi-junction solar cells: present and future. *Solar Energy Materials & Solar Cells*, Vol. 75, No. 1-2, pp. 261-269, ISSN 0927-0248.

Yamaguchi, M; Takamoto, T; Araki, K; Ekins-daukes, N (2005). Multi-junction III–V solar cells: current status and future potential. *Solar Energy*, Vol. 79, No. 1, pp. 78-85, ISSN: 0038-092X.

Yeh, C. Y.; Lu, Z. W.; Froyen, S.; Zunger, A. (1992). Zinc-blende–wurtzite polytypism in semiconductors. *Physical Review B*, Vol. 46, No. 16, pp. 10086-10097, ISSN 1550-235X.

Yim, W.M. (1969) Solid Solutions in the Pseudobinary (III-V)-(II-VI) Systems and Their Optical Energy Gaps. *Journal of Applied Physics*, Vol. 40, No.6, pp. 2617-2623, ISSN 1089-7550.

Zhang, S. & Northrup, J. (1991). Chemical potential dependence of defect formation energies in GaAs: Application to Ga self-diffusion. *Physical Review Letters*, Vol. 67, No. 17, pp. 2339-2342, ISSN 1079-7114.

Zunger, A.; Wei, S.-H.; Ferreira J. L. G. & Bernard, J. (1990). Special quasirandom structures. *Physical Review Letters*. Vol. 65, No. 3, pp. 353-356, ISSN 1079-7114.

Zunger, A. (1999). Anomalous Behavior of the Nitride Alloys. *Physica Status Solidi (b)*, Vol. 216, No. 1, pp.117-123, ISSN 1521-3951.

7

Concentrated Photovoltaics

Robert McConnell[1] and Vasilis Fthenakis[2,3]
[1]Amonix Inc.
[2]National Photovoltaic Environmental Research Center
Brookhaven National Laboratory and
[3]Columbia University
USA

1. Introduction

Concentrating sunlight by using mirrors or lenses historically is associated with generating heat. Though often discredited since the Renaissance, there are Roman records that Archimedes used mirrors and the sun's energy to attack Roman ships by setting them on fire (Archimedes, 2011). At the turn of the 19th century, several inventors and engineers used heat from solar concentrators to operate steam engines to pump water and, later, to generate electricity via rotating machinery (Backus, 2003). With the invention of modern photovoltaics, and in a quest to increase efficiencies and reduce costs, engineers in the 1970s demonstrated that concentrating sunlight and focusing the equivalent of hundreds of "suns" onto solar cells increases their efficiency (Backus, 2003). For example, 20.7% efficient mono-c-Si cells, under AM1.5 spectral conditions, reach 26.5% efficiencies under 500 suns because the efficiency of a solar cell increases as the voltage- and fill factor of the cell rise. However, further increases in solar concentration require augmented cell package designs to collect the increased electrical current and dissipate additional waste heat. Otherwise there will be a decline in efficiency as shown in Fig. 1 for a cell package optimized for 500 suns.

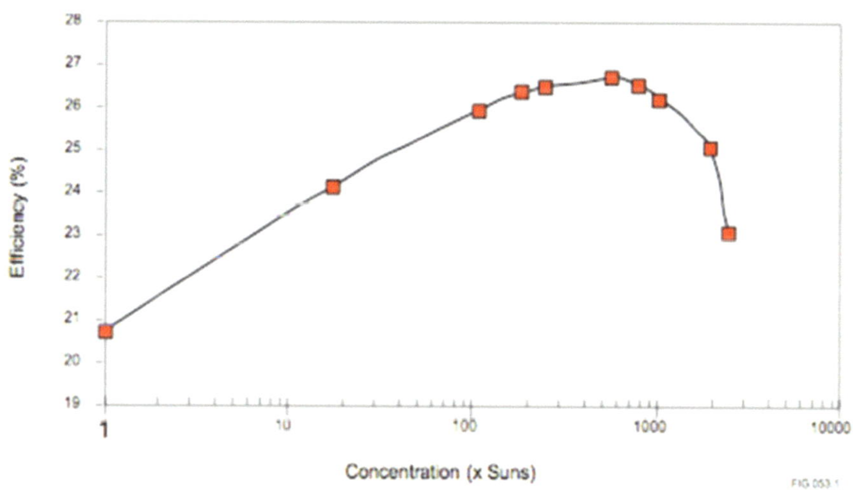

Fig. 1. Efficiency as a function of solar concentration for a mono-crystalline silicon solar cell.

Unlike the flat-plate photovoltaic systems seen on roofs, solar concentrators must track the sun to focus light on to a solar cell throughout the day. Sun tracking increases the daily energy production above that of non-tracking flat-plate PV panels. However, electrical output drops dramatically if the sun is not focused on the cell, or if clouds block the sun. A concentrator photovoltaic (CPV) system comprises of a solar concentrator using lenses (Figure 2), or mirrors (Figure 3), a tracking mechanism, solar cells, and a heat sink.

On a per-area basis, PV cells are the most expensive components of a PV system. A concentrator makes use of relatively inexpensive materials such as plastic lenses and metal housings to capture the solar energy shining on a large area and focus that energy onto a smaller area—the solar cell area. Concentrator PV systems have several advantages over flat-plate systems. First, concentrator systems reduce the size or number of cells needed and allow much higher efficiency multi-junction tandem designs that use expensive processing and materials that would otherwise be cost prohibitive. Also, concentration reduces, by the concentration ratio, the amounts needed of materials with constrained availability (e.g, Ge, Ga), enabling sustainability in very large penetration scenarios (Fthenakis, 2010). Second, a solar cell's efficiency increases under concentrated light, as shown in Fig. 1. Third, a concentrator PV module can be made of small individual cells. This is an advantage because it is harder to produce large-area, high-efficiency solar cells than it is to produce small-area cells. However, challenges exist for concentrators. First, the required concentrating optics are more expensive than the simple covers needed for flat-plate solar systems, and concentrators must track the sun throughout the day and year to be effective. Thus, achieving higher concentration ratios, via reflectors or lenses, entails complex tracking mechanisms and more precise controls. To be cost effective, the additional electricity generated by concentrator systems must outweigh the additional system component costs.

Fig. 2. Several 35-kW CPV systems built by Amonix in Torrance, California were installed at a Public Service power plant in Arizona. The system uses Fresnel lenses to concentrate sunlight onto high efficiency silicon solar cells. The pickup truck in the shade gives an idea of size of this installation.

Fig. 3. Several 25-kW CPV systems built by Solar Systems in Hawthorn, Australia, and installed on aboriginal land. These systems use mirrors for concentrating sunlight onto silicon solar cells---see www.solarsystems.com.au The people in the foreground give an idea of the system's size (photo by R. McConnell).

Fig. 4. The Amonix 7700 60 kWac System uses high efficiency multijunction solar cells.

The most promising lens for PV applications is the Fresnel lens, which uses a complex saw-tooth design to focus incoming light. When the teeth run in straight rows, the lenses act as line-focusing concentrators. When the teeth are arranged in concentric circles, light is focused at a central point. However, no lens can transmit 100% of the incident light. Typical lenses can transmit 90% to 95%, and in practice, many transmit less. Furthermore, concentrators cannot focus diffuse sunlight, which makes up about 30% of the solar radiation in some locations. High concentration ratios also introduce a heat problem. When solar radiation is concentrated, so is the amount of heat produced. Cell efficiencies decrease as temperatures increase, and higher temperatures also threaten the long-term stability of solar cells. Therefore, the solar cells must be kept cool in a concentrator system, requiring sophisticated heat transfer designs (EERE, 2011).

As these photographs show, typical CPV systems are large, and more suitable for a utility than for private customers. Some companies, however, are developing smaller CPV products for rooftop markets. In the 1980s, the U.S. Department of Energy and the Electric Power Research Institute, funded CPV projects for utility applications until both organizations curtailed their CPV studies in the early 1990s as rooftop PV began to dominate the markets. Amonix, founded in 1989, built on these early studies and continued developing its technology despite a lack of utility customers. Thus, in 2005, Amonix spearheaded the development of a 10 MW/year production facility in Spain, in a joint venture with the developer, Guascor. In 2011, Amonix completed the biggest CPV project in North America, a 5 MW plant using 84 Amonix 7700 trackers (Figure 4), in Hatch, New Mexico. For two decades, Amonix has been persistent and innovative in developing several generations of CPV designs leading to these successes in the utility market.

2. Conversion efficiencies

CPV systems are likely to be relatively low cost electricity generators because the expensive solar cells are replaced with less costly structural-steel holding mirrors or lenses. However, early CPV systems showed the importance of optical efficiencies as optical losses sometimes lowered the efficiency of CPV systems by 15 % to 20%. Today's CPV systems incorporating highly efficient III-V silicon solar-cells have system efficiencies approaching 29%. The costs of installed CPV systems today are comparable to those of utility-scale flat-plate PV systems. However, further improvement of III-V solar cells, now about 42% efficient (Figure 5), is still possible since these efficiencies are far below the physical limits for converting sunlight into electricity.

Research on multijunction solar cells began in the 1980s as part of the US Department of Energy's effort to explore new solar-cell materials and new solar-conversion processes to improve cell efficiency. A single- junction solar cell is tuned to only one wavelength of the solar spectrum, so that its maximum efficiency is attained just at that color. The term semiconductor junction refers to the interface between a p-type semiconductor material and an n-type material. P and n denote the semiconductor's charge- carriers, reminding us that solar cells behave like batteries in having positive terminals, negative terminals, and in generating direct current. Early researchers on solar cells calculated that an infinite number of junctions would be the most effective means to harvest every color in the solar spectrum, and, theoretically, such a stacked set of junctions could convert more than 80% of the sunlight into electricity. Yet the first monolithic two-junction solar cell, fabricated almost three decades after the discovery of the modern solar cell, demonstrated efficiencies less than that of a single-junction cell. Major problems with materials and difficulties in creating a tunneling diode interconnect between junctions hindered success in making the first monolithic two-junction solar cells. These multi-junction PV technologies are based on elements in columns III and V of the periodic table; accordingly, they are often referred to as III-V solar cells. These high efficiency cells found a commercial niche in space-power markets. Today, almost every commercial- and defense-satellite, as well as the Mars Rover instrumentation-packages, use multi-junction III-V solar cells as their sources of electrical power. Just before the turn of the century, collaborative research and development by the National Renewable Energy Laboratory and Spectrolab, a Boeing company, demonstrated a three-junction solar cell with a higher efficiency than that of two-junction cells. Efficiencies now are 41.6-42.4 % in the laboratory, with commercial cells showing 38% efficiency, and the promise of affordable costs is materializing. As shown in Table 1, performance pays, and module efficiencies of 32% could lead to system costs of \$2.5/W and electricity cost in the southwestern US of only 6 cents/kWh (Algora, 2004). Capturing those economic benefits involves replacing the crystalline silicon-solar cells in essentially identical solar-concentrator structures with new high-efficiency III-V multi-junction cells.

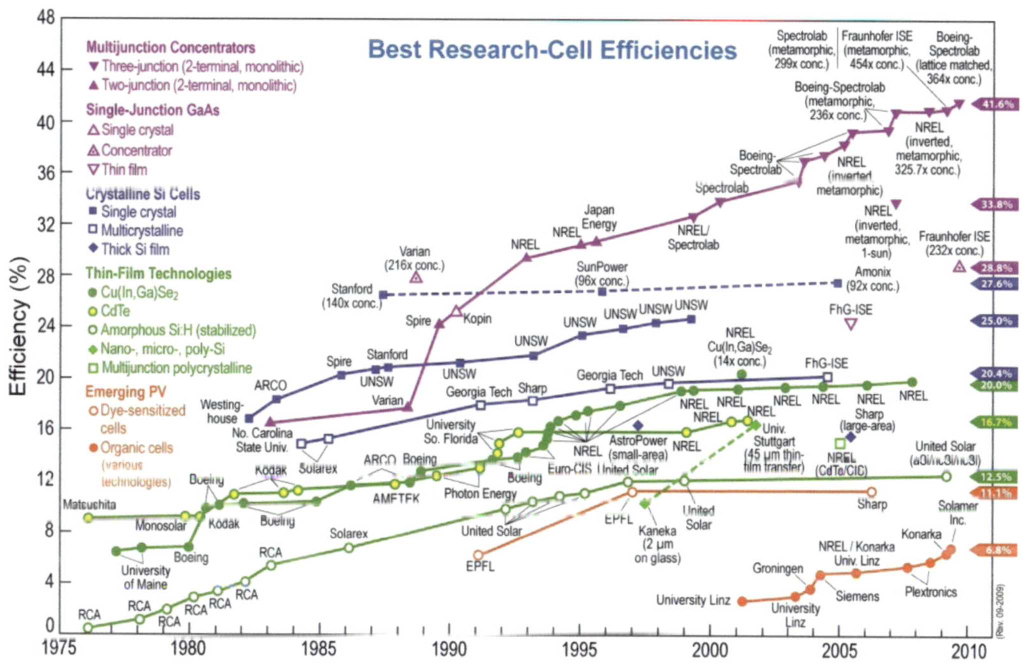

Fig. 5. Record efficiencies for different types of solar cells in the laboratory (Source: NREL).

System size	MW	10	12.5	16
Module price	$\$/W_{dc}$	4.13	3	1.56
Cell efficiency	%	26 (Si)	32 (III-V)	40 (III-V)
Module size	kW_{pdc}	40	50	64
Module efficiency	%	20	25	32
Installed system price	$\$/W_{dc}$	5.95	4.3	2.52
LCOE	$\$/kWh_{ac}$	0.15–0.27	0.10–0.15	0.06–0.11

Table 1. Benchmark (10-MW) system parameters and impact of efficiency of multijunction (III-V) solar cell on a reference solar-concentrator PV utility system (Algora, 2004).

However, companies want to be sure that these new multijunction solar cells will operate reliably in their CPV systems because they typically function at higher voltages, generate higher current, and behave differently under environmental temperature cycles and humidity than do crystalline silicon solar cells. While reliability is best confirmed by decades of field operation, accelerated aging test procedures with qualification standards provide the optimum engineering alternative to minimize risks to reliability.

3. CPV reliability

Flat-plate crystalline silicon technologies have been renowned for their reliability in generating electricity for decades. What is often forgotten is that before flat-plate PV environmental tests were established in the early 1980s, some field projects of these systems failed terribly. Standards organizations afford an important service for all technology development by providing a forum for companies, customers and independent engineers to create a set of agreed-upon qualification tests for identifying weaknesses in products before they are marketed. Qualification standards, especially military standards, were critical to the success of space solar-cells developed for defense satellites in the 1960s and 1970s. However, with the first efforts to bring space PV down to earth, project leaders noted the inadequacy of the space test standards for terrestrial PV systems. Programs begun in the early 1980s at the Jet Propulsion Laboratory successfully developed terrestrial qualification standards for crystalline silicon flat-plate PV technologies. Today, crystalline silicon solar cells are renowned for their long-term reliability, and few people are aware of the early disasters.

Many early CPV systems suffered the same fate in that reliability was a serious issue. Electrical- and electronic-engineers developing PV systems and PV standards were unaccustomed to solving mechanical engineering problems or developing standards for large mechanical structures. In the late 1980s, Sandia National Laboratories devised a set of environmental stress tests (termed accelerated environmental testing) for CPV systems based on their early field tests funded by the U.S. Department of Energy. This work was the basis for the first CPV qualification standard developed in the late 1990s and was published in 2001 by the International Electrical and Electronics Engineers (IEEE) standards organization. This first qualification standard was most suitable for concentrator PV systems using Fresnel lenses, typical of many U.S. CPV designs (Ji and McConnell, 2006). A new qualification standard for CPV was published in 2007 by The International Electrotechnical Commission (IEC), the world's leading organization in preparing and publishing international standards for all electrical-, electronic-, and related technologies. All IEC standards are fully consensus-based and represent the needs and expectations of key stakeholders from every nation participating in IEC work. Some key CPV stakeholders are the companies who design and manufacture CPV systems; another group consists of customers and project developers who specify, purchase, or install systems. As the CPV market rapidly emerged, the interest in developing IEC CPV standards increased by leaps and bounds. The IEC CPV Working Group 7 (WG7) has over 60 members from 15 countries, linking companies, test laboratories and customers in developing and publishing international CPV standards. WG7 is one of the working groups within TC82, the IEC technical committee for all solar photovoltaic energy systems. The scope of IEC 's WG7 is to formulate international standards for photovoltaic concentrator systems, including cells, receivers, modules, arrays, assemblies, trackers, power plants, and solar simulators; this effort encompasses the general areas of safety, photovoltaic performance, qualification and environmental reliability tests, as well as thermal performance, high-voltage performance, fault resistance, and fault-tolerant design (McConnell et al., 2011).

4. Life-Cycle Assessment

Life cycle assessment (LCA) is an analytical tool used to measure material- and energy-inputs and emissions, and waste-outputs throughout the entire life-cycle stages of a product or process, ultimately aiming to evaluate the system's environmental- and health-impacts. LCA is especially useful in comparing power-generation technologies.
Fthenakis and Kim (2011) undertook a life-cycle assessment of Amonix 7700 CPV based on detailed bill of materials, fuel- and electricity-usage, and operational data from Amonix Inc. The life cycle starts with the acquisition of materials, encompasses their production, the manufacturing of components, their transportation, assembly, and installation, their operation/maintenance, and then ends with their disposal (Figure 6).

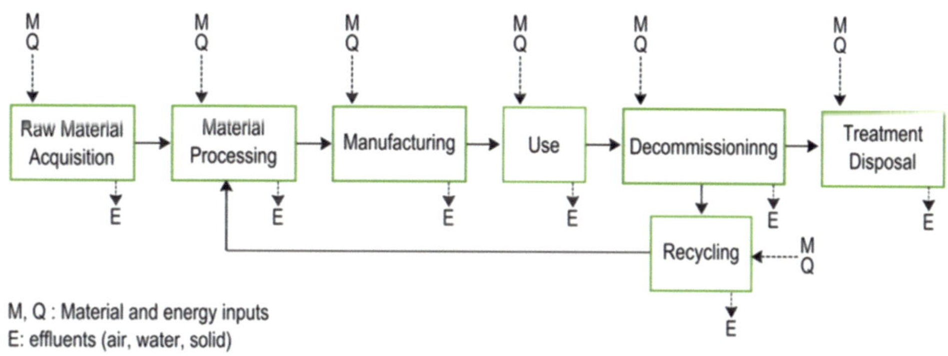

Fig. 6. The Life Cycle of Photovoltaic Systems.

The system consists of seven concentrating module units, so-called MegaModules, mounted on a two-axis tracker (Fig. 4). Sunlight is concentrated onto 7560 focal spots at a ratio of 500:1. This system uses multi-junction GaInP/GaInAs/Ge cells grown on a germanium substrate rated at 37% efficiency under the test conditions of 50W/cm^2, 25°C, and AM 1.5D. With an aperture area of 267 m^2, the capacity of this unit corresponds to 53 kW$_p$ AC power under the test conditions of the Photovoltaics for Utility Scale Applications (PVUSA), i.e., 850 W/m^2 direct normal incidence (DNI), 20°C ambient temperature, and 1 m/sec wind velocity. With multiple increases in system component efficiencies (e.g. cell, lens, heat sink, interconnect, inverter, etc.) Amonix found the same unit produced up to 63 kW$_p$ AC power in 2011 under the same test conditions.

The materials composition and mass balance of the Amonix 7700 system are detailed elsewhere (Fthenakis and Kim, 2011). The MegaModules (36%) and tracker (58%) account for most of the components, while steel (75%), concrete (11%), and aluminum (11%) dominate the material usages. The cumulative energy-demand for all the parts of the system were calculated, and contrasted with the energy generated by the system. Fthenakis and Kim (2011) determined the electricity generation of the Amonix 7700 based on the energy production of a 7500 CPV system in Las Vegas over a year, accounting for the following potential losses in field performance: Additional soiling (2%), ac wiring and transformer losses (2%), availability (1%), wind stow (0.5%), shading (0.5%) and losses from the limit on elevation angle (0.8%). Their analysis yielded 144,000 kWh/year of electricity generation for Las Vegas where the DNI with a 2-axis tracker is 2600 kWh/m^2/yr. Table 3 gives the EPBT and GHG emissions estimated for different parts of the US-SW.

The greenhouse gas (GHG) emissions during the life cycle stages of the Amonix 7700 HCPV system were estimated as an equivalent of CO_2, using an integrated time-horizon of 100 years . Unlike fixed, standard PV configurations in which the emissions mostly are evolved during the production of solar cells, the tracking- and concentrating-equipment contributes the majority of the GHG emissions from this system. After normalizing for the electricity generated, the system will generate 26-27 g CO_2-eq./kWh during its 30-year life-cycle (Table 2) under its current operating locations. Extending the system's life to 50 years by properly maintaining it, and replacing the solar cells and Fresnel lenses every 25 years, will lower its life-cycle emissions to about 16 g CO_2-eq./kWh.

Stage	**Energy (MJ)**	%	**GHG (kg CO_2-eq.)**	%
Parts	**1470633**	**88.3**	**102108**	**92.4**
Cells	14562	0.9	615	0.6
Foundation	2341	0.1	430	0.4
Frame	234218	14.1	16836	15.2
Fresnel lenses	171974	10.3	9086	8.2
Heat Sink	440089	26.4	31194	28.2
Tracker (Pedestal and tube)	427106	25.7	31420	28.4
Inverter	33395	2.0	2130	1.9
Transformer	11973	0.7	566	0.5
Hydraulic drive	117972	7.1	8912	8.1
Motor	2056	0.1	113	0.1
Cables	5278	0.3	265	0.2
Controller	8907	0.5	498	0.5
Anemometer and Sensor	762	0.05	43	0.04
Assembly/Installation	**162**	**0.01**	**12**	**0.01**
Operation/Maintenance	**111830**	**6.7**	**2463**	**2.2**
Transportation	**61364**	**3.7**	**4480**	**4.1**
End-of-Life decommissioning	**20745**	**1.2**	**1512**	**1.4**
Total	1664733	100.0	110575	100.0

Table 2. Breakdown of the Life Cycle Primary-Energy Demand (From Fthenakis and Kim, 2011).

Location	DNI with 2-axis tracker (kWh/m2/yr)	Energy generation (MWh/yr)		EPBT (yrs)		GHG emissions (g CO_2-eq/kWh)	
		2009	2011	2009	2011(est)	2009	2011(est)
Las Vegas, NV	2600	144	178	0.9	0.8	26	21
Phoenix, AZ	2480	136	170	0.9	0.8	27	22
Glendale, AZ	2570	139	176	0.9	0.8	27	21

Table 3. EPBT- and GHG-emissions of Amonix 7700 for 2009 (from Fthenakis and Kim) and estimates for 2011 based on increased energy generation.

These EPBT- and the GHG-emissions from the life-cycles of ground-mount PV modules were estimated to be to 1.8 years and 39 g CO_2-eq./kWh, respectively, when operating in south-facing latitude tilt under the optimal angle insolation of Phoenix--2370 kWh/m^2/yr. Thus, for both parameters, the Amonix 7700 HCPV has a significant advantage over the flat, fixed crystalline silicon solar-cell systems (Fthenakis and Kim, 2011).

5. CPV following wind-turbine developments

One renewable energy technology developed successfully is wind-turbine electricity generation. In the 1970s, wind systems and PV systems started out pretty much on the same footing, viz., only a few experimental systems for each were installed around the world. Today, there are roughly 10 times more installed wind-energy systems than PV systems Why did wind-energy deployment surpass that of photovoltaics? One reason was that wind developers quickly demonstrated the economies of production just as a market opportunity appeared. Fabrication facilities are relatively cheap for both wind- and CPV-systems compared with PV manufacturing facilities. Wind-turbine and CPV production facilities look more like automobile assembly lines (Figure 7) (McConnell, 2002). Early market incentives in California helped developers of wind- developers of wind-systems to move their technologies forward, reducing costs, and acquiring valuable operational experience that improved reliability. Their engineers formulated their system's qualifications standards during this time; like early PV technologies, wind systems often demonstrated poor reliability until their certification standards were established and required in the marketplace.

There are some noteworthy similarities between wind-energy systems and CPV systems (McConnell, 2002). They both employ relatively common materials, particularly steel. The costs of wind systems are typically less than $1 per watt; they are mainly dependent on the cost of steel, while flat-plate PV mostly is linked to the availability and cost of expensive semiconductor silicon. However, solar-concentrator structures also are amenable to an auto-assembly type of production, and CPV developers estimate that the costs of CPV production facilities are closer to those of wind systems than to those of flat-plate-PV production facilities. In the EPRI's early cost studies, the expense of CPV production facilities were estimated on the same costing basis as those of the crystalline- and amorphous-silicon facilities, to be about $28 million for a 100 MW per year installation, that is, about a quarter of the cost of the conventional silicon-PV facilities (Whisnant and Wright,1986). This lower investment can entail a faster scale-up of manufacturing facilities because risk to investors is relatively smaller than that entailed in investing in conventional photovoltaic-production facilities.

Fig. 7. An early Amonix production facility in Los Angeles.

Further, studies in Spain and Israel estimate that the costs of CPV installed system costs will, like those for wind systems , finish below $1 per watt when gigawatt levels of CPV production are reached (Algora 2004; Faiman et al., 2005). Both CPV- and wind-energy-technologies are modular, just like flat-plate PV modules although the sizes are different. Wind units are now megawatts in size, while CPV units are typically 10s of kilowatts. Both wind- and CPV-systems have moving parts, and moving parts have not limited the success of wind systems. Finally, wind systems first penetrated the energy marketplace in sites with very high, steady winds while CPV systems almost certainly will enter markets in locations with lots of sunlight and few clouds, similar to the world's desert climates and the southern US climates.

6. Low-cost hydrogen from hybrid solar-concentrator PV systems

In addition to generating clean, carbon-free solar electricity at low cost, concentrator PV systems potentially can produce hydrogen through the electrolysis of water. There are several reasons why the generation of electrolytic hydrogen from solar energy is critically important to the world's long-term energy needs. The feedstock (water) and solar energy both are carbon free, having no adverse impact on global warming. In addition, there is the potential to generate hydrogen near its markets, thus minimizing transportation costs. Previously, the principal criticism of photovoltaics for generating hydrogen has been the high cost of PV electricity and the inefficiencies of the conversion processes, particularly in the photovoltaic process.

As we discussed earlier, concentrator photovoltaic systems potentially can generate cheaper electricity, primarily by utilizing high-efficiency multi-junction III-V solar cells. But it is the heat boost from concentrator PV systems that will dramatically improve and enhance the electrolysis efficiency of water in a high-temperature solid-oxide electrolyzer. This heat boost, ~40%, measured by Solar Systems in Australia above 1100°C (Lasich, 1997, 1999), also was estimated in theoretical analyses (Licht, 2003). This new pathway provides significant engineering- and economic-benefits for generating electrolytic hydrogen from solar energy, thereby creating opportunities for PV directly to contribute to future transportation markets with low-cost hydrogen, or by producing liquid hydrogen-carrier fuels, such as methanol (Lewis, 2006). On-site storage of hydrogen in inexpensive steel tanks (similar to propane tanks) for on-demand electricity generation favorably will invite a fresh comparison with other technologies for storing electricity like batteries, pumped storage, flywheels, compressed air, and superconducting magnetic-energy storage. Solar-to-hydrogen conversion efficiencies of 40%, including optical losses, may be attainable in the near-term using high efficiency III-V multijunction solar cells, where efficiencies of 50% and higher are realistic targets within 5- to 10-years. These figures-of-merit are dramatically higher, by roughly a factor of 3 or 4, than those of any other previously considered methods for generating electrolytic hydrogen from solar electricity (16). Based on the long-term potential for concentrator PV systems to be mass-produced at costs of less than $1/W, these values will lead to the costs of hydrogen production being comparable with the energy costs of gasoline, recognizing that 1 kg of hydrogen has the energy equivalent of one U.S. gallon of gasoline (McConnell et al, 2005a; 2006).

6.1 System description

This approach first proposed by Solar Systems in Australia employs a dish concentrator that reflects sunlight on to a focal point (Figure 8).

The reflected infrared radiation is gathered by a fiber optics "light pipe" and conducted to the high-temperature solid-oxide electrolysis cell. The electrical output of the solar cells also powers the electrolysis cells. About 120 megajoules are needed, either in electrical- or thermal-form, or both, to electrolyze water and generate 1 kg of hydrogen. The result is that more of the solar energy is used for hydrogen production. The additional costs for the components of this hybrid solar- concentrator, the spectral splitter, and fiber-optic light pipe, are relatively small compared with the boost in hydrogen production, as we discuss later.

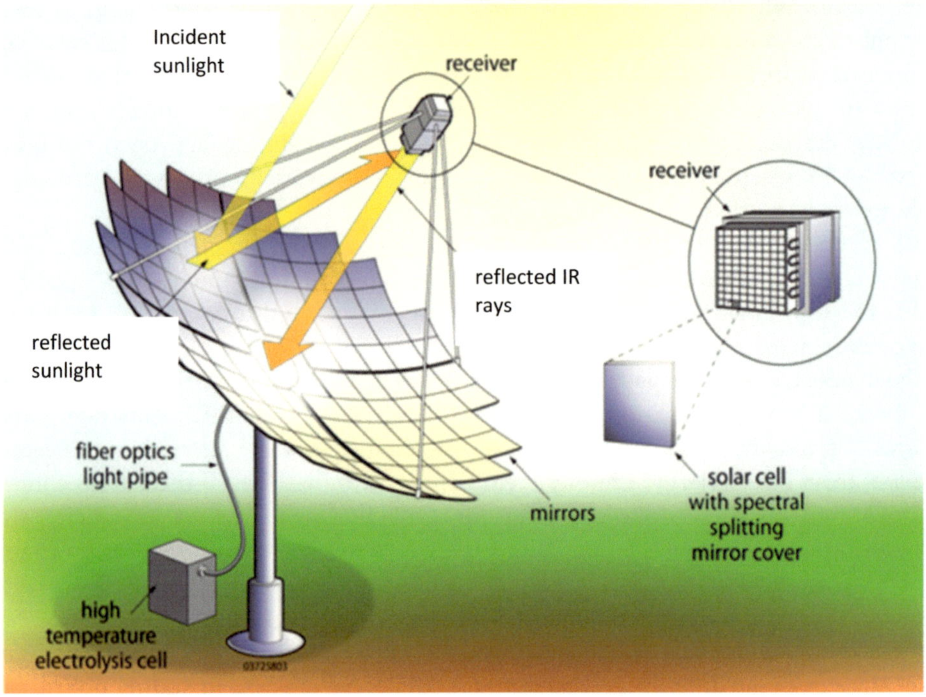

Fig. 8. Schematic of system showing sunlight reflected and focused on the receiver, with reflected infrared directed to a fiber-optics light pipe for transport to a high-temperature solid-oxide electrolysis cell. Solar electricity is sent to the same electrolysis cell that uses both heat and electricity to split water. A spectral splitter at the focal point reflects infrared solar radiation and transmits the visible sunlight to high-efficiency solar cells behind the spectral splitter. (Source: Thompson, McConnell and Mosleh, 2005).

All components shown in Figure 8 were tested in the mid-1990s as detailed previously (Faiman et al, 2005; Lasich, 1997; 1999; McConnell, 2005a, 2005b, 2008) on a small scale. The solar concentrator was a paraboloidal dish 1.5 m in diameter with two-axis tracking, and capable of more than 1000-suns concentration. The full dish was not needed and portions were appropriately shaded. At that time, the solar cell was a GaAs cell with an output voltage of 1- to 1.1-V at maximum power point with a measured efficiency of about 19%. The voltage was an excellent match for direct connection to the electrolysis cell when operating at 1000° C. The tubular solid-oxide electrolysis cell was fabricated from yttria-stabilized zirconia (YSZ); the cell had platinum electrodes since the test temperature was higher than that of typical solid-oxide cells. Figure 9 shows a schematic of the operation of solid-oxide electrolysis cells.

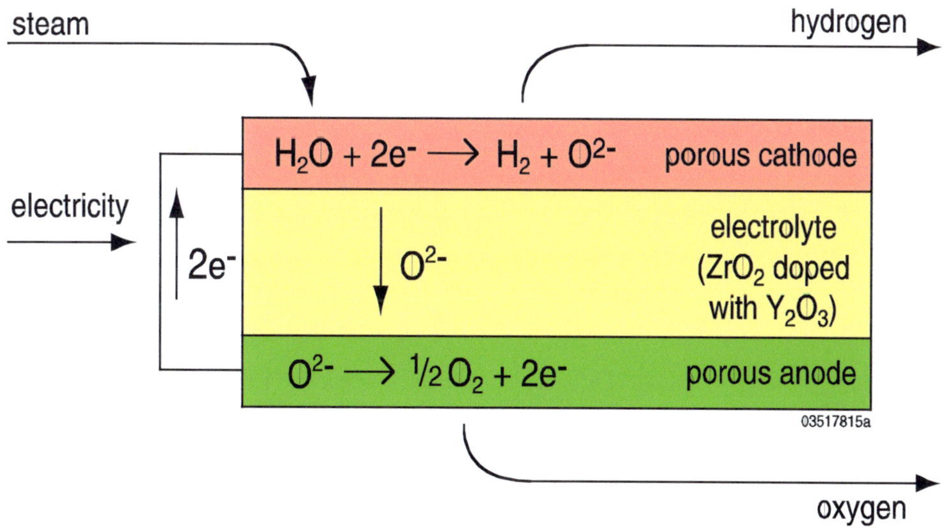

Fig. 9. Schematic of high-temperature electrolysis in a solid-oxide cell. The geometry can be planar or tubular as it was for the first demonstration of the hybrid solar-concentrator PV system. Operating the electrolysis cell in reverse corresponds to electricity- and heat-production in solid-oxide fuel cell operation.

A metal tube surrounds the cell to uniformly distribute the solar flux over the cell's surface. The test occurred over two hours of operation with an excess of steam applied to the electrolysis cell. The output stream of unreacted steam and generated hydrogen was bubbled through water, and the hydrogen was collected and measured. During a definitive 17 minutes of system operation in steady state, 80 mL of hydrogen was collected. The ratio of the thermoneutral voltage of 1.47 V to the measured electrolysis cell voltage of 1.03 V was 1.43, corresponding to a boost of more than 40% in hydrogen production attributable to the input of thermal energy. This was confirmed by the energy balance. Combining the optical efficiencies of the concentrator dish (85%), solar-cell efficiency, and thermal- energy boost, the efficiency of the total system in converting solar energy to hydrogen was 22%. When these measurements were made in the mid-1990s, the efficiency was almost three times better than that recorded for any other technology converting solar energy to hydrogen.

These early tests were not conducted with the most efficient solar cells then available. The record efficiency then was about 30% for a laboratory cell (Figure 5) and they were not easily obtainable. Today's record efficiency is 42%, and while commercial cells are 37-38% efficient. Therefore, a 40% solar-to-hydrogen efficiency is anticipated in the near term, assuming a heat boost of 40%, a multi-junction solar-cell efficiency of 35%, and an optical efficiency of 85%. Such a multi-junction solar cell would yield a solar-to-hydrogen conversion efficiency of almost 50%. As discussed previously, the theoretical electrochemical findings are consistent with these predictions, based on Solar Systems' early measurements (Licht, 2003).

7. Discussion

Today, electricity providers throughout the world are considering large-scale concentrator PV projects, some in the 100s of MW. In the future, a hybrid concentrator-PV system could generate both electricity and hydrogen for electric utilities. With low-cost tank storage on utility land, the solid-oxide electrolysis cell could be designed to operate in a regenerative mode, producing electricity from hydrogen during non-solar periods, thereby greatly increasing the value of solar electricity to utilities.

Fthenakis, Zweibel and Mason (2010) published a study on the feasibility of very-large-scale photovoltaic systems in the southwestern US, and dispersed generation throughout the country. A renewable-energy electricity mixture comprising mainly of PV, CPV and CSP could, by the end of this century, supply 100% of year-round electricity demand, while a surplus of electricity during the sunny seasons can produce 260 billion kg of hydrogen by electrolysis. By adding 5.2 TW of PV dedicated to hydrogen production, a fleet of 5666 million plug-in hybrid vehicles, trucks, ships, and planes could be fueled by hydrogen.

With today's market entry of CPV systems for electricity production, the increasing efficiencies of solar cells now approaching 40% with clearer ideas for attaining 50% efficient solar cells, and the opportunity to use wasted solar heat for augmenting solar electrolysis, this is a potential "leap frog" technology that may rapidly lower the cost of clean hydrogen.

8. Conclusion

We described CPV technologies based on high-efficiency III/V solar cells that will lower the cost of electricity production in sunny areas to parity with the present expense of grid electricity. In addition, CPV offers the future potential for enhancing the hydrolytic production of hydrogen by utilizing the thermal losses from generating electricity. We discussed the following reasons supporting the argument that CPV is a technology that will prove significantly beneficial throughout the world:

1. Short energy payback times support the potential for CPV to provide inexpensive electricity, competitive in utility markets.
2. The CPV community completed the first CPV-qualification standards in time to respond to market opportunities, while additional standards on safety, performance, and tracking are well underway.
3. The early success of wind energy technologies, similar in several respects to CPV technologies, augurs rapid success for CPV systems.
4. The possibility of efficiently producing hydrogen by splitting water using low-cost solar electricity opens future, clean pathways for storing solar electricity and generating transportation fuels.

9. References

Algora C., 2004. Chapter 6 in *Next Generation Photovoltaics,* edited by A. Marti and A. Luque, Institute of Physics Physics Publishing, Bristol and Philadelphia.

Archimedes, accessed Dec 7 2011, http://www.math.nyu.edu/~crorres/Archimedes/Mirrors/Tzetzes.html

Backus, C. E., 2003. "A Historical Perspective on Concentrator Photovoltaics", Proceedings of the International Solar Concentrator Conference for the Generation of Electricity or Hydrogen, Alice Springs, Australia.

EERE, 2011. US-DOE, Energy Efficiency and Renewable Energy, Concentrator Photovoltaic Systems, http://www.eere.energy.gov/basics/renewable_energy/concentrator_pv_systems.html

Faiman, D., Raviv I. and Rosenstreich, R. 2005. "The Triple Sustainability of CPV with the Framework of the Raviv Model", Proceedings of the 20th European Photovoltaic Solar Energy Conference and Exhibition, Barcelona, Spain.

Fthenakis, V., Mason J.E., Zweibel K. 2010. "The technical, geographical, and economic feasibility for solar energy to supply the energy needs of the US", Energy Policy 27, 387-399.

Fthenakis V.M., "Sustainability of photovoltaics: The case for thin-film solar cells", Renewable and Sustainable Energy Reviews, 13, 2746-2750, 2009.

Fthenakis V.M., Kim H.C., "Life Cycle Assessment of High-Concentration PV Systems", Progress in Photovoltaics, in press.

Hartvigsen J. 2006. Private communication, also see www.ceramatec.com

Ji L. and McConnell R., 2006. "New Qualification Test Procedures for Concentrator Photovoltaic Modules and Assemblies", Proceedings of the 2006 IEEE 4th World Conference on Photovoltaic Energy Conversion, Waikoloa, Hawaii.

Lasich, J., U.S. Patent No. 5658448, August 19, 1997

Lasich, J., U.S. Patent No. 5973825, October 26, 1999

Lewis, N. http://www7.nationalacademies.org/bpa/SSSC_Presentations_Oct05_Lewis.pdf, August 2006

Licht, S. 2003. "Solar Water Splitting to Generate Hydrogen. Fuel: Photothermal Electrochemical Analysis," *J. Phys. Chem. B* 107, 4253-4260.

McConnell, R. 2002. "Large-Scale Deployment of Concentrating PV: Important Manufacturing and Reliability Issues", Proceedings of the First International Conference on Solar Electric Concentrators, New Orleans, Louisiana, NREL/EL-590-32461, May 2002.

McConnell, R., M. Symko-Davies and D. Friedman. 2006. "Multijunction Photovoltaic Technologies for High Performance Concentrators", Proceedings of the 2006 IEEE 4th World Conference on Photovoltaic Energy Conversion, Waikoloa, Hawaii.

McConnell, R., S. Kurtz and M. Symko-Davies, 2005. "Concentrating PV Technologies: Review and Market Prospects", ReFOCUS, Elsevier Ltd, p. 35, July/August 2005.

McConnell, R.D., J. B. Lasich and C. Elam, 2005. "A Hybrid Solar Concentrator PV System for the Electrolytic Production of Hydrogen", Proceedings of the 20th European Photovoltaic Solar Energy Conference and Exhibition, Barcelona, Spain, June 2005.

McConnell, R., 2008. Chapter 4 in *Solar Hydrogen Generation,* edited by K. Rajeshwar, R. McConnell and S. Licht, Springer Science+Business Media, LLC, New York, New York.

McConnell, R. et al, 2011, "The CPV Standards Nexus", in preparation for the Eighth International Conference on Concentrating Photovoltaic Systems.

National Research Council and National Academy of Engineering. 2004. "*The Hydrogen Economy*", National Academies Press, Washington DC.

Thompson, J. R., McConell, R.D. and Mosleh M., Cost Analysis of a Concentrator Photovoltaic Hydrogen Production System, International Conference of Solar Concentration for the Generation of Electricity or Hydrogen, 1-5 May 2005, Scottsdale, Arizona (NREL/CD-520-38172)

Whisnant, R., S. Wright, P. Champagne and K. Brookshire. 1986. "Photovoltaic Manufacturing Cost Analysis: A Required-Price Approach", Volume 1, 2, EPRI AP-4369, Electric Power Research Institute, Palo Alto, CA.

8

Durable Polymeric Films for Increasing the Performance of Concentrators

T. J. Hebrink
3M Company
USA

1. Introduction

Durable solar film technology can be used to create new concentrated photovoltaic (CPV) and concentrated solar power (CSP) designs which would, potentially bring the cost of generating electricity from solar energy to parity with the cost of electricity from fossil fuels. Polymeric films are typically an order of magnitude lower in cost than conventional photovoltaic cells and can be used to effectively replace relatively expensive photovoltaic cells resulting in lower overall module costs. 3M has made the world's most reflective all polymeric wavelength selective mirror films (Weber, 2000) which are thermoformable into useful low cost solar concentrator designs. These all polymeric mirror films can also be metal vapor coated to create broad band mirror films for concentrated solar thermal, CPV, and CSP designs. 3M micro-replication technology has also been shown to capture more light into a photovoltaic module by minimizing surface reflections and refracting light at low angles into the solar cell thus increasing power output. These solar films require extreme UV stability, abrasion resistance, dimensional stability, easy cleanability, and durable adhesiveness

2. CPV (Concentrated Photovoltaics)

A major obstacle in the widespread adoption of concentrated photovoltaic designs utilizing mirrors has been the accelerated degradation of photovoltaic modules due to concentration of ultra-violet light and infrared light onto the photovoltaic cell encapsulating material. Metal coated mirrors have the disadvantage of reflecting ultra-violet and far infra-red light onto the photovoltaic cell which is not converted into electricity, but does contribute to overheating of the photovoltaic cell. Metal coated mirrors are also prone to corrosion. Since photovoltaic cell efficiencies are adversely effected by higher temperatures, thermal management designs are typically incorporated into CPV designs to remove excess heat. Expensive thermal management designs can be avoided, or minimized, with the use of infra-red transmissive mirror films made with dielectric interference stacks of polymeric materials (Hebrink, 2009). Infrared transmissive mirror films can be made with reflection band edges that correspond to the absorption band edge of any photovoltaic cell material as shown for silicon in Figure 1 and thus re-direct non-useful far IR light away from the PV cell as shown in Figure 2 to avoid over heating that can reduce photovoltaic cell power output and reduce module life. Degradation of most polymers, including adhesive and encapsulants used to assemble photovoltaic modules, accelerates with increasing temperatures. Cooler photovoltaic module temperatures result in longer module life which equates to lower cost of electricity generated in $/KWhr. As can be seen in Figure 1, these infrared transmissive mirror films will reflect useful light from 400-1150nm at intended angle of use, but transmit infrared light having wavelengths greater than 1150nm. If desired for safe non-blinding reflection in non-tracking CPV designs, mirror films can be made to reflect only a portion of the visible light. Temperature benefits are limited by reflection of Infra-Red light by some types of photovoltaic cells and even some transmission of Infra-Red light by others. Some silicon and CIGS(copper indium gallium selenide) photovoltaic cells reflect 20-30% of the infra-red light wavelengths greater than 1200 nm. However, often the encapsulants, and other materials, used to construct photovoltaic modules also absorb a portion of the infra-red light beyond 1200 nm which contributes to over heating of the photovoltaic module, especially in concentrating designs.

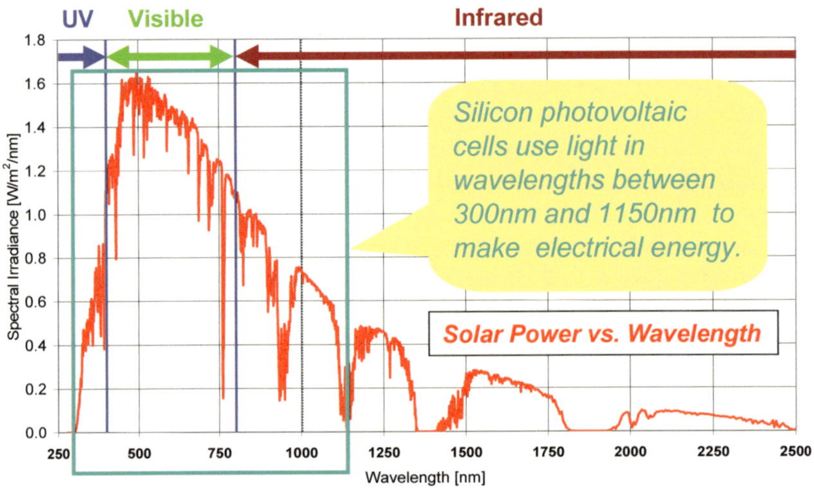

Fig. 1. Solar spectrum useful to silicon photovoltaic cell.

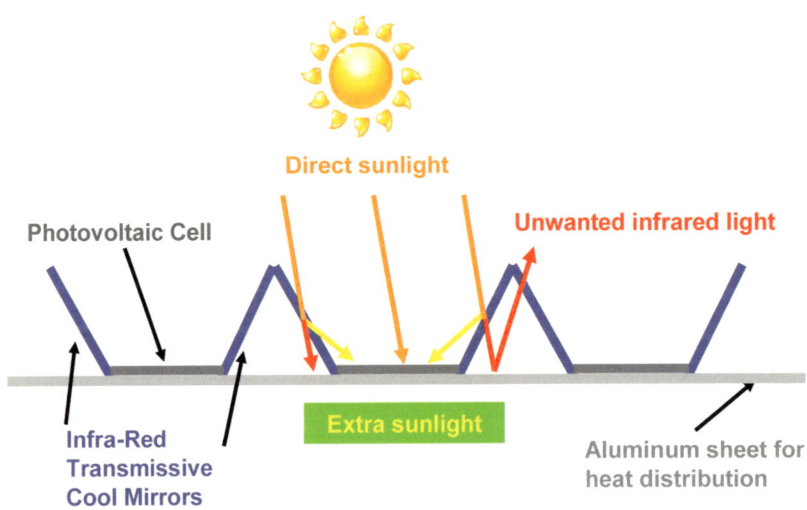

Fig. 2. LCPV design - Non-useful infrared light is redirected away from photovoltaic cell.

Ultra-violet absorbers incorporated into these mirror films will minimize concentration of UV light onto the photovoltaic cell encapsulant thereby extending the photovoltaic module life. Non-metallized polymeric mirrors are thermoformable, enabling innovative structural form factors and lightweight designs. These IR transmissive mirror films are also large-scale manufacturable at relatively low cost. Photovoltaic module costs can be reduced by replacing expensive solar cell material with relatively inexpensive mirror films. These infrared transmissive mirror films are being marketed as 3M brand Cool Mirror 330 and 3M brand Cool Mirror 550.

Infrared transmissive mirror films(3M brand Cool Mirror) have been demonstrated in multiple LCPV(low concentration ratio photovoltaic) non-tracking designs, single axis solar tracking designs, and dual axis tracking designs as shown in figures 3, 4, 5, 6, and 7. Non-tracking and single-axis solar tracking designs are particularly useful for flat commercial and sloped residential roof tops. In high irradiation regions, solar tracking devices could pay for themselves by producing 30-40% more energy per photovoltaic cell, while by increasing power production in the morning and late afternoon, as shown in Figure 3.

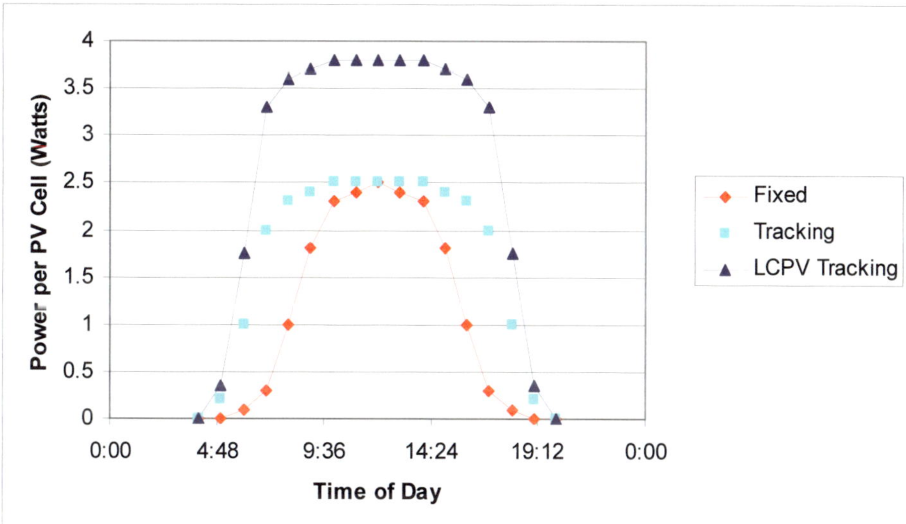

Fig. 3. Measured increase in photovoltaic cell power from solar tracking and 2x sun concentration with infrared transmissive mirror films.

Even though solar trackers are becoming more reliable every year, they are still not widely accepted due to concerns about maintenance costs. TenKsolar Inc. (www.tenksolar.com) has developed electronically smart photovoltaic modules for use in their RAIS® WAVE non-tracking LCPV design shown in Figure 4, and have increased power output by filling the already available space between rows of photovoltaic modules with reflector modules. Photovoltaic module array row spacing is used to prevent module shading during winter months at higher latitudes when the Sun's path is lower in the sky.

Fig. 4. Non-tracking LCPV design with 3M Cool Mirror (photo courtesy of Ray Colby with Sundial Solar).

Peak power output increased by up to 60% as shown in Figure 5 with the use of 3M brand Cool Mirror 330 film in the reflector modules of the RAIS WAVE reflector design.

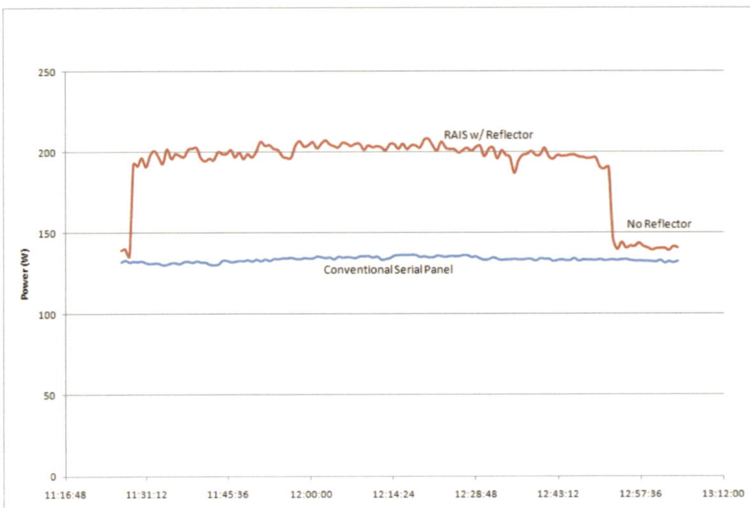

Fig. 5. Non-tracking photovoltaic peak power output increase with use of Cool Mirror reflector measured in Minneapolis, MN, USA on a clear sunny day in June, 2010.

3M brand Cool Mirror 330 reflects 650-1350nm at normal incidence and 550-1150nm at typical solar incidence for maximizing solar concentration without the blinding reflection of full visible light. RAIS® photovoltaic modules are designed to be tolerant of not only partial shading, but also non-uniform increases in solar flux provided by the non-tracking mirrors.

In latitudes where photovoltaic module row spacing is common, the value proposition for CPV with mirror films can be simply calculated by comparing the relative cost of mirror films to photovoltaic cells. For example, a 200 watt silicon photovoltaic module would cost approximately $300/square meter, or $1.5/watt. Mirror films with supporting structure can cost less than $20/square meter, and increase the peak power of the photovoltaic module by at least 100 watts. With an average annualized power output increase of 50 watts/sqmtr, the additional cost of solar mirror film modules can be less than $0.40/watt.

3M brand Cool Mirror film has also been demonstrated in an effective 3x Sun concentrator module designed by JX Crystals for use with their Solar Carousel single axis tracker(Fraas, 2008) on the flat roof of the Science & Engineering building at UNLV as shown in Figure 6. This low cost single axis solar tracker design is simple and functional for commercial flat roof tops. The 3x Sun concentrating mirrors are mounted between silicon photovoltaic cells cut into 1/3 cell size and separated by 2/3 cell size distance so that the total CPV module has a surface area that is equivalent to a conventional flat rigid photovoltaic module, however, each photovoltaic cell produces twice as much power from the area of the silicon solar cells due to the additional solar flux provided by the mirrors.

Fig. 6. Single Axis LCPV design with JX Crystal design using 3M Cool Mirror.

100% increases in power output of 1/3 cell photovoltaic modules have been measured at UNLV as shown in Figure 7 with increased solar flux from 3x sun mirror design.

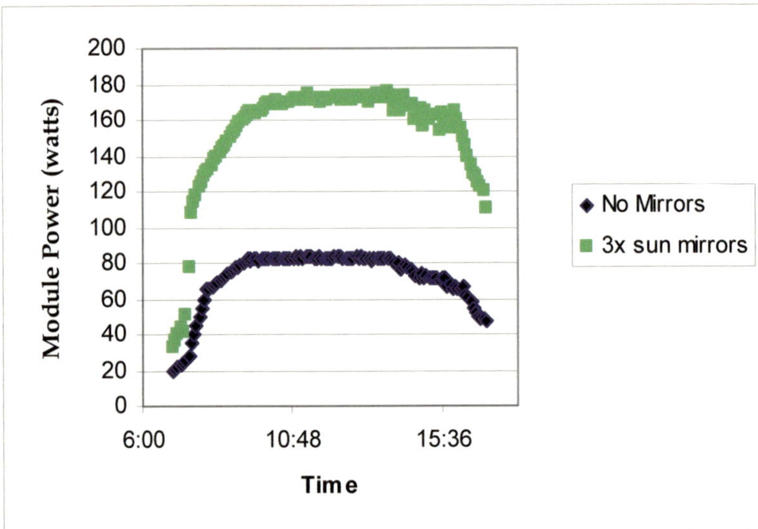

Fig. 7. Measured power output of PV modules designed with 3x sun Cool Mirror reflector measured in Las Vegas, Nevada, USA on a clear sunny day in June, 2011.

Design simplicity, and thus potential for low cost, of the JXC solar carousel tracker is shown below in Figure 8. Since racking is all ready used to mount rigid photovoltaic modules to rooftops, the additional cost entails a simple drive motor, solar sensor, and controller. Details of this design can be found in research papers (Fraas, 2008, 2009) and a book (Fraas, 2010)

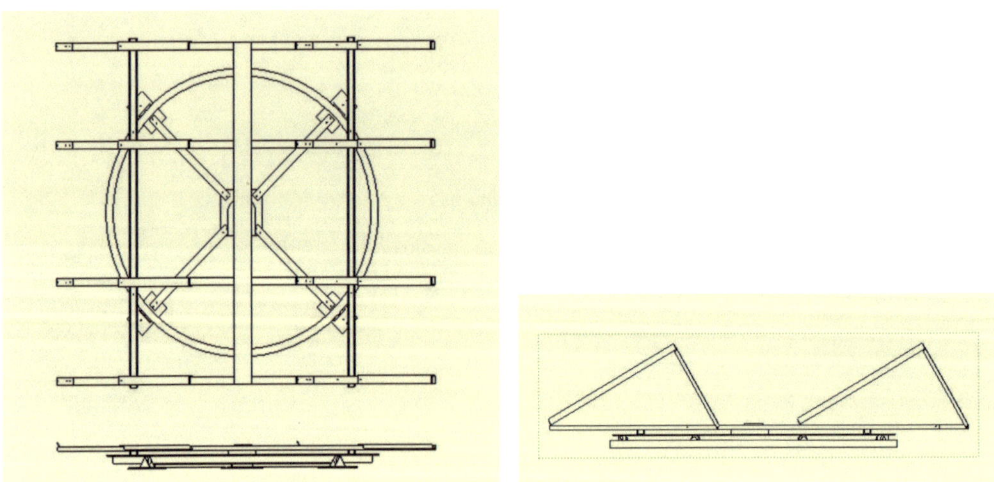

Fig. 8. Top and Side view of JX Crystal solar carousel single axis tracker.

Simple LCPV modules were also designed with 3M brand Cool Mirror linear 2x sun concentration reflectors as illustrated in Figure 2 and pictured in Figure 9 for educational and technology demonstration purposes at the Minnesota Science Museum. Cool Mirror 550 film reflecting 400-1150nm at incident sun light angles is attached to linear thermoformed prisms having 60 degree slopes and placed between rows of mono-crystalline silicon photovoltaic cells that are spaced 6" apart. Basically, every other row of photovoltaic cells is removed and replaced with less expensive mirror film prisms. A Wattsun brand AZ-125 azimuth driven 2-axis solar tracker was mounted on a pole 10 feet above the ground. Power output from each individual module is measured using Enphase M190 micro-inverters and recorded to a database. Two LCPV modules on the top of the array were made with 30% fewer photovoltaic cells than two conventional photovoltaic modules shown mounted below. On sunny days, the 2x sun LCPV modules consistently produce more power than the conventional PV modules. Increase in power output per photovoltaic cell with 2x sun mirror design is shown in Figure 3. As expected, performance is not as good on cloudy days. Concentrated photovoltaic designs require highly collimated direct sun light to be effective and during cloudy diffuse light conditions, photovoltaic power output is reduced by the inverse of the concentration ratio.

Fig. 9. Dual-axis solar tracking LCPV design with 3M brand Cool Mirror (Minnesota Science Museum).

For pole mounted solar tracking designs, the value proposition for CPV with mirror films can be simply calculated by comparing the relative cost of mirror films to photovoltaic cells. For example, a 200 watt silicon photovoltaic module would cost approximately \$300/square meter, or \$1.5/watt. Mirror films with supporting structure can cost less than \$20/square meter, and increase the peak power output of the photovoltaic modules by at least 100 watts thus producing additional electricity for only \$0.20/watt.

Real time effects of the environment (heat, cold, moisture, UV irradiation) on the polymeric mirror films are being monitored for an extended period of time in multiple climates.
Flat LCPV mirror size and angles can be calculated using geometric equations 1 and 2 below.

$$\text{Concentration Ratio} = 1 + 2^*(Wm/Wp) * \text{Cos}\ \theta \quad (1)$$

$$Wm/Wp = \tan(2^*\theta\text{-}90) / (\text{sinq} - \tan(2^*\theta\text{-}90) * \cos\theta) \quad (2)$$

Where; Wp = solar panel width, Wm = mirror width, q = mirror elevation.
For concentration ratios larger than 3:1, curved mirror designs are needed using quadratic equations. Concentration ratios larger than 3:1 will also need to incorporate additional thermal management to remove excess heat from the photovoltaic modules, and preferably, that heat is put to good use, as in hybrid photovoltaic solar thermal panel designs.
Unique solar louver tracker designs (Casperson, 2011) for sloped residential roof tops have also been created to demonstrate the efficacy of infrared transmissive mirrors (Hebrink, 2009) as shown in Figure 10. An advantage of these designs is that the more fragile photovoltaic cell strings are stationary while light weight mirror panels are driven to track the sun. 3M Company sponsored a senior project at LSSU(Lake Superior State University) where a multi-disciplinary team of electrical engineers, mechanical engineers, and computer software engineers designed, fabricated, and tested a solar louver tracker prototype utilizing 3M brand Cool Mirror film.

Fig. 10. Single-axis solar louver tracking design with 3M brand Cool Mirror (photo courtesy of Lake Superior State University).

Polymeric multilayer mirror films are also thermoformable into useful concentrating dish and other non-imaging optical designs for HCPV(high concentration ratio photovoltaic) applications. Trough, symmetrical dish, and assymetrical dish reflector designs have been thermoformed from these mirrors. Lamination of the multilayer mirror films to thicker sheets of polycarbonate or PMMA(polymethylmethacrylate) aids in thermoforming and provides structural rigidity to maintain the desired optical design in the intended environment of use. Additionally, these highly reflective mirror films can be attached to rigid metal or glass sheets formed into the desired concentrator shape.

2.1 Wavelength selective polymeric multilayer optical mirror films

Infra-red transmissive mirrors have been made utilizing constructive interference quarter wave multi-layer optical film technologies comprising hundreds of layers of transparent dielectric materials. Alternating layers of high refractive index polymers and low refractive index polymers are coextruded into optical stacks containing 100 to 1000 layers with layer thicknesses of approximately ¼ wavelength of light to be reflected (Weber, 2000). Each polymer layer pair contains polymers of differing refractive indices and works constructively with the next layer pair to create broad reflection bands as illustrated in Figure 11.

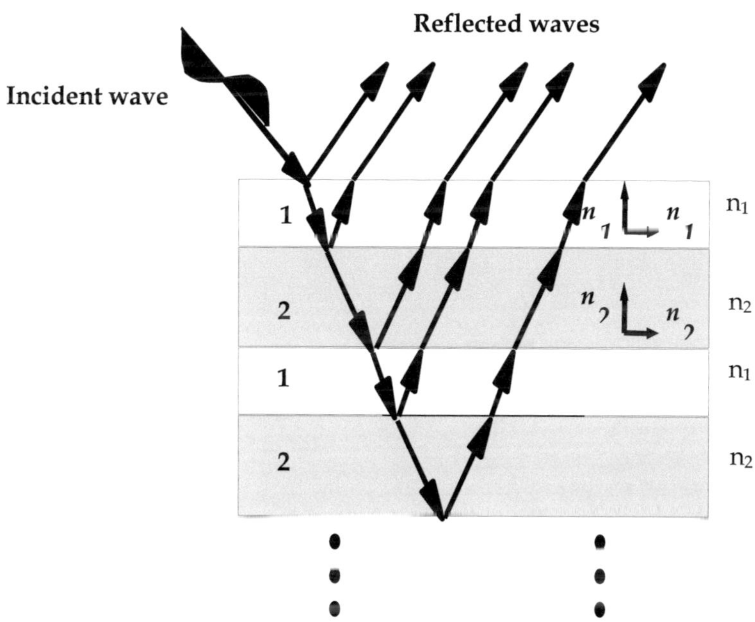

Fig. 11. Incident light reflected by polymer pairs with differing refractive indices.

Layer thickness control is critical for uniform reflection of light. 3M coextrusion and orientation process technologies have demonstrated layer thickness control at the nanometer scale as shown by the Atomic Force Analysis cross section in Figure 12.

Fig. 12. Cross section of multi-layer optical film cross section containing alternating layers of PMMA and a birefringent polyester. Layers on the left side of the image are engineered to be about 25% thicker than those on the right side.

Tunability of the reflection bandwidth and band edges of these multi-layer optical film mirrors allows matching of the reflection band edges with the absorption band edges of any photovoltaic cell as illustrated in Figure 13 for CIGS(Copper Indium Gallium Selenide) solar cell and Figure 14 for CdTe(Cadmium Telluride) solar cell. Reflectivity and reflection band is controlled by adjusting layer count, layer thicknesses, and refractive indices of polymer layers.

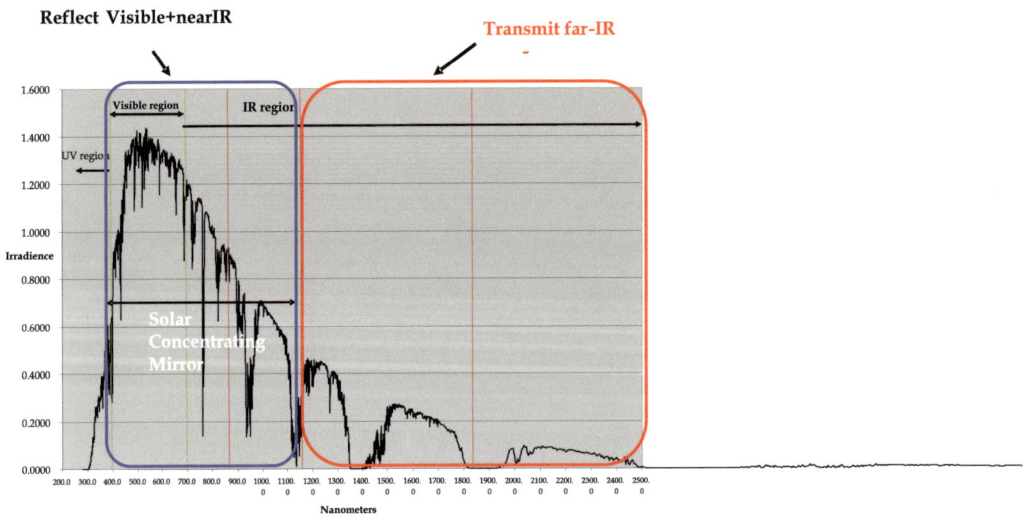

Fig. 13. IR transmissive mirror reflection band for CIGS PV cells.

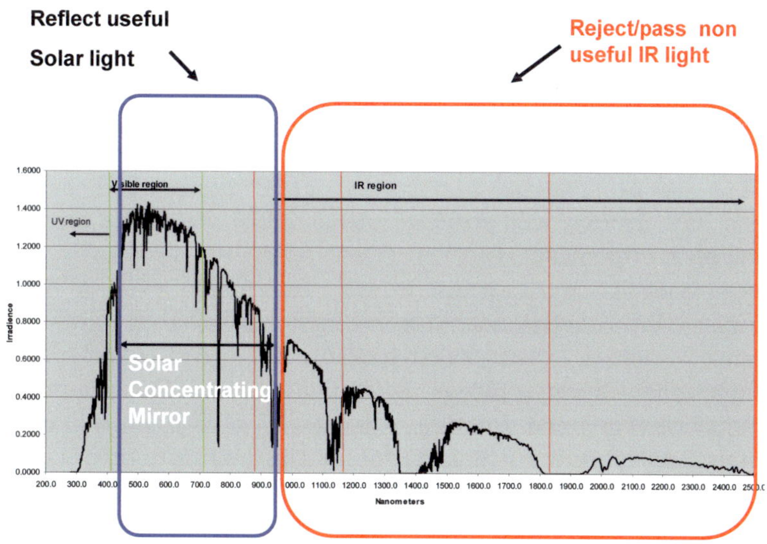

Fig. 14. IR transmissive mirror reflection band for CdTe PV cells.

Reflection band edges will shift to the left with changing incident light angle, as explained by Fresnel reflection and phase thickness equations as charted in Figure 15. Thus the exact reflection band edge will need to be optimized for every CPV design. More detailed physics of birefringent multi-layer optical films can be found in a research article titled "Giant Birefringent Optics in Multi-layer Polymer Mirrors" (Weber, 2000).

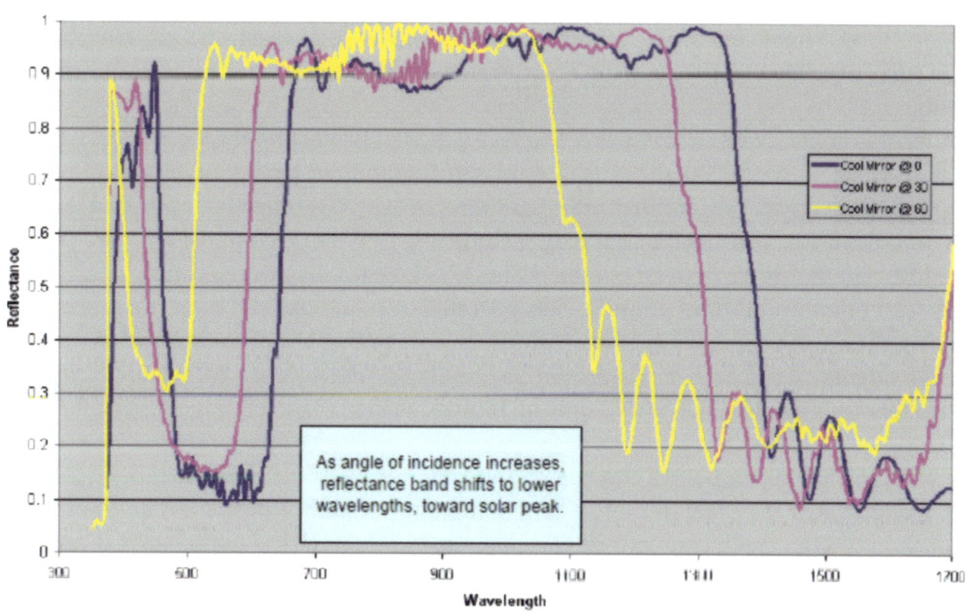

Fig. 15. Shift in reflection band with changes in incident angle of light.

2.2 UV(Ultra-Violet) light stability

Tunability of the reflection band also enables UV(ultra-violet) reflection layers to be included at or near the top surface of these solar concentrating mirrors. Incorporation of UVA (ultra-violet absorber) in the optical layers and skin layers of UV mirrors provides exceptional UV protection of any material positioned beneath them. When made with inherently UV stable materials such as PMMA(polymethylmethacrylate) and fluoropolymers, UV mirrors will provide extended UV protection even after the UV absorbers themselves extinguish.

Fluoropolymers typically are resistant to UV degradation even in the absence of stabilizers such as UVA, HALS (hindered amine light stabilizer), and anti-oxidants. Useful fluoropolymers include ethylene-tetrafluoroethylene copolymers (ETFE), tetrafluoroethylene-hexafluoropropylene copolymers (FEP), tetrafluoroethylene-hexafluoropropylene-vinylidene fluoride copolymers (THV), polyvinylidene fluoride (PVDF), and miscible blends of PVDF and PMMA. The multilayer optical film comprising fluoropolymers can also include non-fluorinated materials. Multilayer film substrates may have different fluoropolymers in different layers or may include at least one layer of fluoropolymer and at least one layer of a non-fluorinated polymer. UV protective multilayer films can comprise a few layers (e.g. 2 or 3 layers) or can comprise at least 100 layers (e.g., in a range from 100 to 2000 total layers or more). Polymer used to make the optical layers in the different multi-layer film substrates can be selected, for example, to reflect a significant portion (e.g., >50%) of UV light in a wavelength range from 350nm to 400nm or even 300nm to 400 nm.

Photo-oxidative degradation caused by UV light (e.g., in a range from 280 to 400 nm) may result in color change and deterioration of optical and mechanical properties of polymeric films. A variety of stabilizers may be added to the polymeric film substrate to improve its resistance to UV light. Examples of such stabilizers include at least one, or even better, a combination of ultra violet absorbers, hindered amine light stabilizers (HALS), and anti-oxidants.

A UV absorbing layer (e.g., a UV protective layer) aids in protecting the visible/IR-reflective optical layer stack from UV-light caused damage/degradation over time by absorbing UV-light (preferably any UV-light) that may pass through the UV-reflective optical layer stack. In general, the UV-absorbing layer(s) may include any polymeric composition (i.e., polymer plus additives) that is capable of withstanding UV-light for an extended period of time. A variety of optional additives may be incorporated into an optical layer to make it UV absorbing. Examples of such additives include at least one of UV absorbers (UVAs), HALS, or anti-oxidants. Typical UV absorbing layers have thicknesses in a range from 12 micrometers to 380 micrometers (0.5 mil to 15 mil) with a UVA loading level of 1-10% by weight.

A UVA is typically a compound capable of absorbing or blocking electromagnetic radiation at wavelengths less than 400 nm while remaining substantially transparent at wavelengths greater than 400 nm. Such compounds can intervene in the physical and chemical processes of photoinduced degradation. UVAs are typically included in a UV absorbing layer in an amount sufficient to absorb at least 80%, and even better, greater than 90% of the UV light in the wavelength region from 180 nm to 400 nm. Typically, it is desirable if the UVA is highly soluble in polymers, highly absorptive, photo-permanent and thermally stable in the temperature range from 200 °C to 300 °C for extrusion process to form the protective layer. The UVA can also be highly suitable if they can be copolymerizable with monomers to form protective coating layer by UV curing, gamma ray curing, e-beam curing, or thermal curing processes.

Red-shifted UVAs (RUVAs) typically have enhanced spectral coverage in the long-wave UV region, enabling it to block the high wavelength UV light that can cause yellowing in polyester polymers. One of the most effective RUVAs is a benzotriazole compound, 5-trifluoromethyl-2-(2-hydroxy-3-alpha-cumyl-5-tert-octylphenyl)-2H-benzotriazole. Other exemplary benzotriazoles include 2-(2-hydroxy-3,5-di-alpha-cumylphehyl)-2H-benzotriazole, 5-chloro-2-(2-hydroxy-3-tert-butyl-5-methylphenyl)-2H-benzotiazole, 5-chloro-2-(2-hydroxy-3,5-di-tert-butylphenyl)-2H-benzotriazole, 2-(2-hydroxy-3,5-di-tert-amylphenyl)-2H-benzotriazole, 2-(2-hydroxy-3-alpha-cumyl-5-tert-octylphenyl)-2H-benzotriazole, 2-(3-tert-butyl-2-hydroxy-5-methylphenyl)-5-chloro-2H-benzotriazole. Further exemplary RUVA includes 2(-4,6-diphenyl-1-3,5-triazine-2-yl)-5-hekyloxy-phenol. Some very effective UV absorbers for polymers include those available under the trade designations "TINUVIN 1577," "TINUVIN 900," and "TINUVIN 777." Some UV absorbers are available in master batch resin from Sukano Polymers Corporation, Dunkin SC, under the trade names of "TA07-07 MB" for polyesters and "TA11-10 MB" for polymethylmethacrylate. Another exemplary UV absorber available in a polycarbonate master batch from Sukano Polymers Corporation under the trade name "TA28 09 MB". In addition, the UV absorbers can be used in combination with hindered amine light stabilizers (HALS) and anti-oxidants. Exemplary HALS include those under the trade names "CHIMASSORB 944" and "TINUVIN 123." Exceptional anti-oxidants include those obtained under the trade names "IRGAFOS 126", "IRGANOX 1010" and "ULTRANOX 626".

The desired thickness of a UV protective layer is typically dependent upon an optical density target at specific wavelengths as calculated by Beers Law. When protecting polyesters from UV degradation, the UV protective layer should have an optical density greater than 3.5 at 380 nm. It is also very important that the optical densities remain constant over the extended life of the film in order to provide the intended protective function.

The UV protective layer can be a cross linkable hard coat selected to achieve the desired protective functions of UV protection and abrasion resistance. Inorganic additives that are very soluble in cross-linkable polymers may be added to the composition for improved properties. Of particular importance, is the permanence of the additives in the polymer. The additives should not degrade or migrate out of the polymer. Additionally, the thickness of the layer may be varied to achieve desired protective results. For example, thicker UV protective layers would enable the same UV absorbance level with lower concentrations of UV absorbers, and would provide more UV absorber permanence attributed to less driving force for UV absorber migration. Accelerated testing of the UV protected polymer films are being conducted in outdoor solar simulators at elevated temperatures and higher levels of UV irradiance than direct light from the sun.

3. Concentrated solar power

CSP(concentrated solar power) technology uses mirrors to direct sunlight at solar absorbing heat transfer fluid devices that heat up and whose thermal energy is then transferred for heating purposes, or turned into electricity by use of a turbine electric generator. CSP reflectors can also be used to concentrate sunlight onto photovoltaic cells, especially photovoltaic cells that make use of a wide range of the solar spectrum such as triple junction gallium arsenide indium phosphide PV cells. Conventional CSP reflectors are made with silver coated glass, which currently reflect 94% of the solar spectrum. These glass CSP reflectors are heavy, prone to breakage, and relatively expensive.

Polymeric multilayer optical mirror films have been metal vapor coated to create broadband solar mirror films for solar thermal, CPV, and CSP applications. Common reflective metals such as silver, copper, aluminum, and gold have high reflectivity in the red and infra-red wavelengths of light, but tend to absorb some visible light, especially in the blue wavelengths. Since multilayer optical mirror films can be tuned to be highly reflective of blue wavelengths of light, the combination of metal reflectors with multilayer constructive interference mirrors create a highly reflective and cost effective broadband mirror. An example of a broadband mirror is shown in Figure 16 made by vapor coating silver onto a multilayer optical film made with over 500 alternating layers of PET(polyethylene terephthalate) an CoPMMA(co-polyethylmethylmethacrylate).

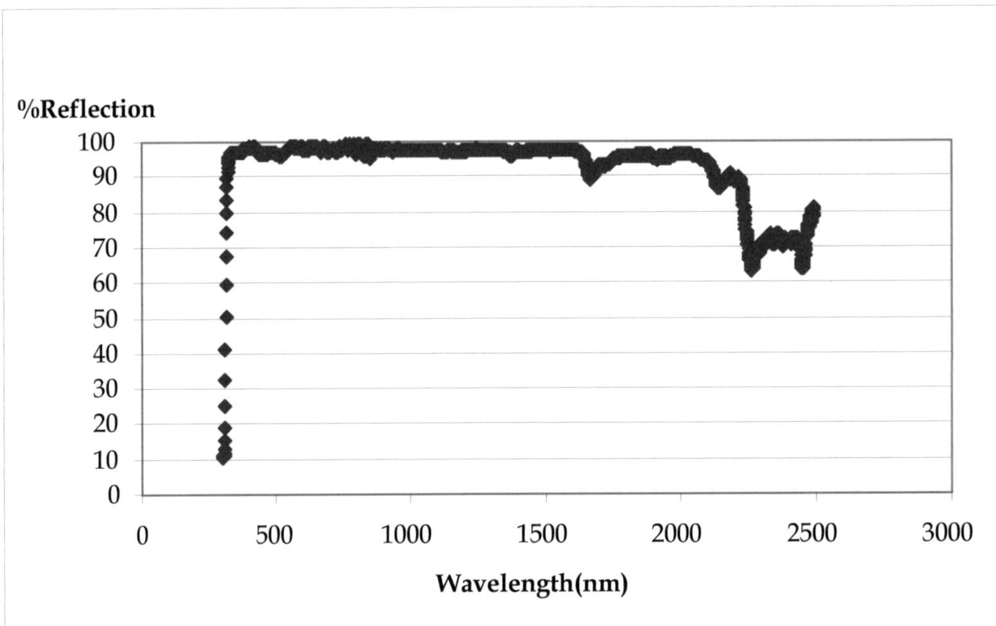

Fig. 16. Reflectivity of mirror film made by vapor coating silver onto back side of multi-layer optical film reflecting 380-800nm.

The polymeric multilayer optical mirror film reflects 97% of the light from 370 nm to 800nm while the silver reflects 96% of the available sun light from 800 nm to 2500 nm. Some losses in reflectivity are due to absorption of light in the 1600 nm to 1700 nm range by the PET and absorption of light in the 2200 nm to 2400 nm light range by the CoPMMA. Fortunately, only a small portion of the sun's available light makes it through the earth's atmosphere in these bandwidths of polymer absorption. Another 3% of the solar spectrum is wavelengths of light below 370nm which cannot be reflected by a PET based multilayer optical mirror due to absorption and degradation of the PET by UV light below 370 nm. However, UV light below 370nm can be reflected by a fluoropolymer based UV mirror with PMMA as the high refractive index optical layers. The PMMA/fluoropolymer UV mirror can be used to protect a PET/CoPMMA based visible mirror, or alternatively, a multilayer optical mirror film can be made which reflects both UV and visible light with PMMA as the high refractive index layers and fluoropolymers as the low refractive index layers. The PMMA/fluoropolymer based UV-VIS mirror can then be metal vapor coated with silver, copper, or gold to create a more reflective broadband mirror. Aluminum can also be used as the reflective metal layer. Another option for creating a broadband reflective mirror is to laminate the UV-VIS multilayer optical mirror film to a sheet of polished aluminum or stainless steel with an optically clear adhesive.

4. Structured surface anti-reflection

Front surface reflections account for 4-5% losses in conventional glass covered photovoltaic module output. These front surface reflections increase with incident light angle as illustrated in Figure 17 with a flat glass front surface having a refractive index of 1.50.

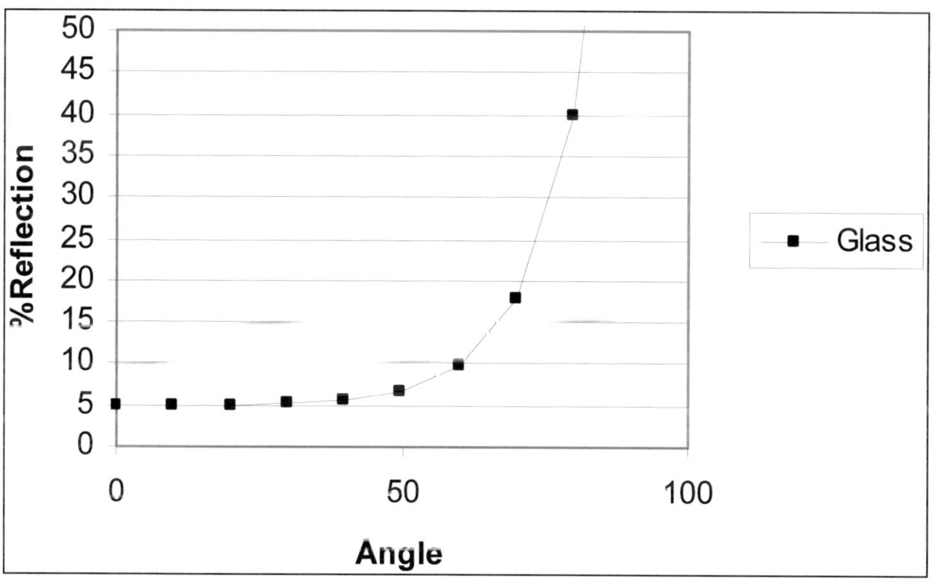

Fig. 17. Front surface reflection off flat surface.

By structuring the front surface, either with nanostructures or microstructures, the front surface reflection can be minimized. Improvements in transmission through a photovoltaic module front surface with micro-structures made with linear prisms having a range of prism apex angles were modeled with Fresnel equations and Snell's law as shown in Figure 18.

Additional increases in photovoltaic cell efficiency are expected by increasing the path length of light rays through the photovoltaic cell material itself. Initial photovoltaic module power increases of 6-8% have been measured with 53 degree apex angle prism surface structures. However, the tendency of these v-groove valley prism structures to collect dirt can reduce their anti-reflection benefits over time. Dirt attraction has prompted the investigation of additional anti-soiling surface structure geometries and anti-soiling coatings. V-groove geometries trap dirt as well as light and thus larger valley angles can be implemented to minimize dirt accumulation. Both hydrophilic and hydrophobic surface coatings have shown promising dirt resistant properties depending on the composition of dirt and climate of the photovoltaic module installation. Polymer antistat additives can also be incorporated to minimize dirt attraction.

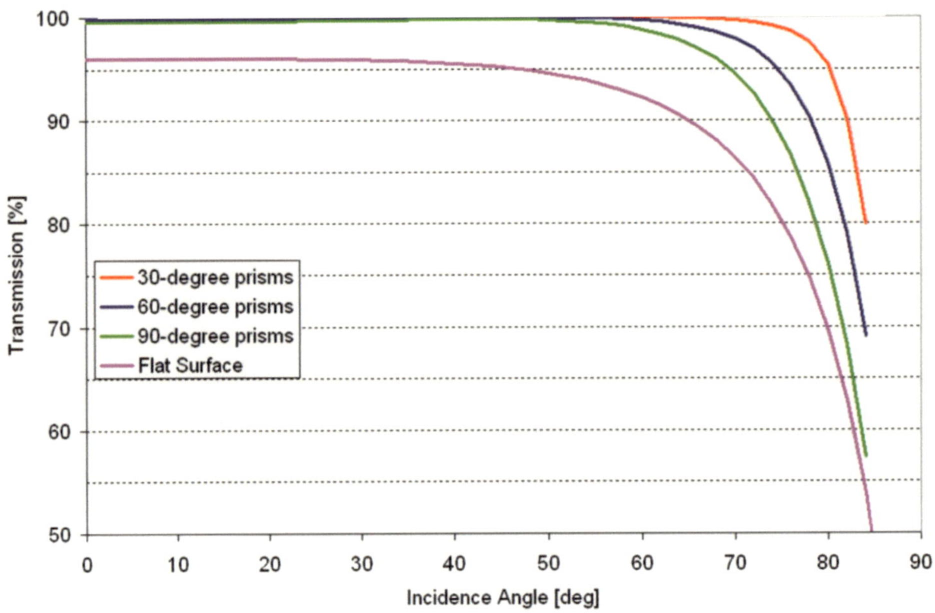

Fig. 18. Increase in light transmission dependance on prism apex angle.

Polymer films can be micro-structured at relatively high manufacturing speeds and thus low cost. Micro-structured fluoropolymers, silicone polymers, acrylate polymers, and urethane polymer films are being investigated due to their inherent UV stability. Minimizing front surface reflections off photovoltaic modules does increase the power output of a photovoltaic module. However, effective techniques for minimizing dirt attraction and soil build-up are needed for anti-reflective surface structures to truly reduce the levelized cost of generating electricity from the sun.

5. Conclusions

Durable multilayer polymeric mirror films have demonstrated significant(e.g.>50%) increases in photovoltaic cell power output in multiple CPV designs without overheating photovoltaic modules. Surface structured anti-reflection films and coatings have demonstrated 5-10% increases in photovoltaic module power output. Positive value propositions can be calculated for these polymeric mirror films and coatings when the photovoltaic power benefit they provide is greater than their additional cost per kWhr of energy produced.

6. Acknowledgments

The author wishes to thank the 3M Corporate Research Laboratory management for allocating resources to the the development of durable films and coatings. Thanks are also due to 3M Renewable Energy Division management for sponsoring and funding this work. The technology demonstration partnerships with JX Crystal Inc., Lake Superior State University, University of Nevada Las Vegas, and tenKsolar Inc. are also greatly appreciated.

7. References

Weber, M. F.; Stover, C. A.; Gilbert, L.R.; Nevitt, T. J.; Ouderkirk, A.J. (2000). Giant Birefringent Optics in Multilayer Polymer Mirrors, *Science Magazine,* Vol. 287, March 31, 2000

Hebrink, T. J. (2009). Infra-red Transmissive Mirrors for Concentrated Photovoltaics, *Proceedings of 34th IEEE Photovoltaic Specialist Conference,* June, 2009, Philadelphia, USA

Fraas, L.; Avery, J.; Minkin, L.; Huang, H. ; Maxey, C.; Gehl, A. (2008), Carousel Trackers with 1sun or 3sun Modules for Commercial Building Rooftops, *Proceedings of ASES Conference,* 2008, San Diego, California, USA

Fraas, L.; Avery, J.; Minkin, L.; Huang, H.; Schneider, H.; Larson, D. (2010), Solar PV Carousel Trackers for Building Flat Rooftops; 3 Case Studies, *Proceedings of ASES Conference,* 2010, Phoenix, Arizona, USA

Fraas, L.; Partain, L. (2010), Solar Cells and their Applications, Wiley, John & Sons, Incorporated, ISBN; 9780470446331, USA

Casperson, K.; Fill, C.; Gearin, M.; Greensky, R.; Lidbeck, K.; Weber, P.; Hebrink, T. J. (2011), Low-X Single-Axis Solar Louver Tracking System for Residential Rooftop

Applications, *Proceedings of 37th IEEE Photovoltaic Specialist Conference*, June, 2011, Seattle, Washington, USA

TenKsolar, Inc., www.tenksolar.com

9

Performance of Photovoltaics Under Actual Operating Conditions

George Makrides[1], Bastian Zinsser[2],
Matthew Norton[1] and George E. Georghiou[1]
[1]*Department of Electrical and Computer Engineering, University of Cyprus*
[2]*Institut für Physikalische Elektronik (ipe), Universität Stuttgart*
[1]*Cyprus*
[2]*Germany*

1. Introduction

Amongst the various renewable energy sources, photovoltaic (PV) technologies that convert sunlight directly to electricity have been gaining ground and popularity, especially in countries with high solar irradiation. Over the past years PV has shown rapid development and a wide variety of new technologies from different manufacturers have emerged. For each PV module type, manufacturers provide typical rated performance parameter information which includes, amongst others, the maximum power point (MPP) power, efficiency and temperature coefficients, all at standard test conditions (STC) of solar irradiance 1000 W/m^2, air mass (AM) of 1.5 and cell temperature of 25 °C. As this combination of environmental conditions rarely occurs outdoors, manufacturer data-sheet information is not sufficient to accurately predict PV operation under different climatic conditions and outdoor PV performance monitoring and evaluations are necessary.

The objective of this chapter is to provide an overview of different PV technologies ranging from crystalline silicon (c-Si) to thin-film and concentrators. Subsequently, a summary of the main outdoor evaluation performance parameters used to describe PV operation and performance is outlined. An overview of the effects of different environmental and operational factors such as solar irradiance, temperature, spectrum and degradation is also provided along with the results of previously published research efforts in this field. In the last section of the chapter, the installed PV and data acquisition infrastructure of a testing facility in Cyprus is presented and a thorough analysis of the climatic conditions and the performance of different grid-connected PV technologies that have been installed side-by-side and exposed to warm climatic conditions, typical of the Mediterranean region are given.

2. Overview of photovoltaic technologies

Over the last twenty years, the PV industry showed annual growth rates between 40 % and 80 %, proving its strength and potential to become a major worldwide power generation source (Joint Research Centre [JRC], 2010). The enormous potential of PV is also evident by the fact that the existing global energy demands could be met by over 10,000 times, had the surface area of the Earth been covered with currently available PV technologies (European

Photovoltaic Industry Association [EPIA] & Greenpeace, 2011). Nowadays, the threat of climate change and the continuous rise of oil prices have added more pressure for the integration of renewable technologies for energy production, with PV drawing considerable attention. More specifically, at the end of 2008 the cumulative worldwide installed PV capacity was approximately 16 GW (EPIA, 2011). The market growth continued throughout 2009, despite the international economic crisis and according to the European Photovoltaic Industry Association (EPIA) the installed capacity was 23 GW while in 2010, the accumulated capacity reached 40 GW worldwide with more than 50 TWh of electricity production per year (EPIA, 2011). The largest PV market was the European Union (EU) with more than 13 GW installed in 2010 and a total installed capacity of almost 30 GW as of 2010 (EPIA, 2011).

A wide range of PV technologies now exist that include mono-crystalline silicon (mono-c-Si), multi-crystalline silicon (multi-c-Si), thin-film technologies of amorphous silicon (a-Si), micromorph (microcrystalline/amorphous silicon), cadmium telluride (CdTe), copper-indium-gallium-diselenide (CIGS), concentrating PV (CPV) and other emerging PV technologies. Each technology is mainly described and classified according to the material used, manufacturing procedure, efficiency and cost.

Amongst the various existing PV technologies, c-Si is the most developed and well understood due to mainly its use in the integrated circuit industry. In addition, silicon is at present the most abundant material found in the earth's crust and its physical properties are well defined and studied. C-Si dominates the PV technology market with a share of approximately 80 % today (EPIA & Greenpeace, 2011). The type of c-Si technology depends on the wafer production and includes mono-c-Si, multi-c-Si, ribbon and sheet-defined film growth (ribbon/sheet c-Si).

The main characteristic of mono-c-Si is its ordered crystalline structure with all the atoms in a continuous crystalline lattice. Mono-c-Si technologies are highly efficient but are at the same time the most expensive amongst the flat-plate existing PV technologies mainly because of their relatively costly manufacturing processes. Over the past years, manufacturing improvements of c-Si PV technology have focused on the decrease of wafer thickness from 400 μm to 200 μm and in parallel the increase in area from 100 cm^2 to 240 cm^2. The most important limitation of this technology is the cost of the silicon feedstock which renders the material cost relatively high, particularly as the silicon substrate must have a thickness of approximately 200 μm to allow the incident light to be absorbed over a wide range of wavelengths. Despite the high material cost, this technology has remained competitive due to several manufacturing improvements such as enhancements in wire cutting techniques that have reduced the wafer thickness and also the production of kerf-less wafers. Recently, Sunpower announced an efficiency of 24.2 % for a large 155 cm^2 silicon cell fabricated on an n-type Czochralski grown wafer (Cousins et al., 2010).

The fact that mono-c-Si modules are produced with relatively expensive manufacturing techniques initiated a series of efforts for the reduction of the manufacturing cost. Such a cost improvement was accomplished with the production of multi-c-Si PV which can be produced by simpler and cheaper manufacturing processes. Multi-c-Si solar cell wafers consist of small grains of mono-c-Si and are made in a number of manufacturing processes. The substrate thickness is approximately 160 μm while attempts are being made to lower the thickness even more. In general, multi-c-Si PV cells are cheaper compared to mono-c-Si as they are produced in less elaborate manufacturing process, at the expense of slightly lower efficiencies. The lower efficiency is attributed to recombination at the grain boundaries within the multi-c-Si structure. Nonetheless, multi-c-Si currently has the largest PV market share.

Ribbon silicon is another type of multi-c-Si technology which is produced from multi-c-Si strips suitable for the photovoltaic industry. In the manufacturing process of this technology, high temperature resistant wires are pulled through molten silicon to form a ribbon which is subsequently cut and processed in the usual manner to produce PV cells. An advantage of this technology is that the production costs are lower than other c-Si technologies, while the efficiency and quality of the cells remain the same as other multi-c-Si technologies but lower than mono-c-Si.

The main incentive for the development of thin-film technologies has been their cheap production cost compared to the c-Si counterparts. Over the past years, thin-film technologies have shown very encouraging development as the global production capacity has reached around 3.5 GW in 2010 and is expected to reach between 6 - 8.5 GW in 2012 (EPIA, 2011). Amongst the many thin-film technologies some of the most promising are CdTe, a-Si, micromorph tandem cells (a-Si/µc-Si) and CIGS. The rapid growth and importance of thin-film PV is further highlighted by the fact that the world's first PV manufacturer to exceed the 1 GW/year production rate and hence to capture 13 % of the global market was First Solar, a manufacturer of thin-film CdTe modules, in 2009 (Wolden et al., 2011). Specifically, CdTe has grown from a 2 % market share in 2005 to 13 % in 2010 (EPIA & Greenpeace, 2011).

Amorphous silicon has been on the PV market longer than other thin-film technologies and this has allowed researchers and manufacturers to understand several aspects of its behavior. This technology was first commercialized in the early 1980s and since then has increased gradually in efficiency. The manufacturing of a-Si technologies is dominated by deposition processes such as plasma enhanced chemical vapor deposition (PECVD) and thus large area, flexible and cheap substrates such as stainless steel and thin foil polymer can be used (Shah et al., 1999). In comparison to mono-c-Si, a-Si PV cells have no crystalline order leading to dangling bonds which have a severe impact on the material properties and behavior. Another important material limitation arises from the fact that this technology suffers from light-induced degradation, also known as the Staebler–Wronski effect (SWE), which describes the initial performance decrease when a-Si modules are first exposed to light (Staebler & Wronski, 1977). In general, this effect has been minimized by employing double or triple-junction devices and developing micromorph tandem cells, which is a hybrid technology of c-Si and a-Si. An important advantage of a-Si is the high absorption coefficient, which is approximately 10 times higher than c-Si therefore resulting in much thinner cells.

The concept of micromorph (microcrystalline/amorphous silicon) tandem cells was introduced to improve the stability of a-Si tandem cells. The structure of a micromorph device includes an a-Si cell which is optimized with the application of a micro-crystalline silicon (µc-Si) layer of the order of 2 µm onto the substrate. The application of the µc-Si layer assists the device in increasing its absorption in the red and near infrared part of the light spectrum and hence increases the efficiency by up to 10 % (EPIA & Greenpeace, 2011). Oerlikon Solar developed and announced recently a lab cell with 11.9 % stabilized efficiency (Oerlikon Solar, 2010).

Another type of thin-film technology is CdTe, which is a II-VI semiconductor with a direct band gap of 1.45 eV. The high optical absorption coefficient of this technology further allows the absorption of light by a thin layer, as it absorbs over 90 % of available photons in a 1 µm thickness, hence films of only 1 - 3 µm are sufficient for thin-film solar cells (Ferekides & Britt, 1994). PV devices of CdTe first appeared in the 1960s (Cusano, 1963) but the technological development outbreak came in the early 1990s when efficiencies approached levels of commercial interest (Britt & Ferekides, 1993). CdTe technology is a front-runner amongst thin-film PV technologies due to the fact that it can be produced relatively cheaply and module efficiencies have reached 12.8 % (Green et al., 2011). So far, the achieved efficiency of this technology is lower compared to c-Si, but higher than triple-junction a-Si. In comparison to a-Si, the CdTe PV technology does not show initial degradation. In addition, the power is not affected to the same extent by temperature variations as c-Si based technologies (Doni et al., 2010). On the other hand, concerns have been raised related to the availability of tellurium (Te) and the environmental impact of cadmium (Cd). These concerns have been addressed by Fthenakis et al. (Fthenakis, 2004, 2009; Fthenakis et al., 2005, 2008). In order to minimize the environmental impact of this technology, a recycling process for used modules has been introduced (Meyers, 2006) and the rest of the PV industry is currently moving in this direction (PVCYCLE program).

The properties of several I-III chalcopyrite compounds are also suitable for photovoltaic applications and amongst them the most promising include copper-indium-diselenide ($CuInSe_2$) often called CIS, copper-gallium-diselenide ($CuGaSe_2$) called CGS, their mixed alloys copper-indium gallium-diselenide ($Cu(In,Ga)Se_2$) called CIGS and copper-indium-disulfide ($CuInS_2$). The first PV devices of copper chalcopyrite appeared in 1976 (Kazmerski et al., 1976) and since then it was not until the early 1990s that rapid improvements increased efficiencies to over 16 % (Gabor et al., 1994). Even though the commercial production of CIGS began in 2007, there are now a number of companies with 10 - 30 MW/year capacities (Wolden et al., 2011). Efficiencies continued to improve exceeding the 20 % threshold (Green et al., 2011) and establishing this technology as the efficiency leader amongst existing thin-film technologies. The main advantage of CIGS over other existing thin-film PV technologies is its high efficiency. In addition, CIGS modules have a performance very similar to that of c-Si technologies but have lower thermal losses as the power temperature coefficient is lower. A previous study has also shown that CIGS PV modules show an increase in power output after exposure to sunlight, a phenomenon known as light induced annealing (LIA) (Jasenek et al., 2002). On the other hand, the fabrication process of this technology is more complicated than in other technologies and as a result manufacturing costs are higher. In addition, costs may be also affected by the limited availability of indium and the difficulty in up scaling from cell to large area modules.

An emerging application of PV is in concentrator photovoltaics (CPV) systems. CPV technologies are gaining in popularity as they offer several advantages over established PV technologies. CPV make use of relatively inexpensive optical devices, such as lenses or mirrors to focus light from an aperture onto a smaller active area of solar cell. In doing so, light is 'concentrated' to higher intensities than ordinary sunlight, and less PV cell material is required for a given output. This brings several benefits: the total cost of the system can be reduced; higher system efficiencies are possible due to the increased solar flux intensities; higher efficiency cells can be used without incurring great cost; and demand for semiconductor materials can be reduced, thereby easing supply restrictions on these materials and facilitating reductions in market price. The target installation locations for CPV are predominantly in the world's sunbelts. This is because CPV systems utilize the direct normal irradiation (DNI) component of sunlight, which makes areas with high annual irradiance such as southern US states, Australia, the Middle East, North Africa and Mediterranean regions the prime target areas for this technology. Today a worldwide total of approximately 35 MW of CPV have been installed. Recent activity, particularly in the US market, has resulted in a rapid increase in projected installed capacity, which will total approximately 400 MW worldwide by the end of 2012 (Greentechmedia [GTM], 2011). Although CPV offers a promising route to lower solar electricity prices, it remains a strong technical challenge. In the last few years, the dramatic fall in the cost of conventional flat-plate PV systems has raised the bar on entry into the energy market for CPV. Systems operating above 5-fold concentration require some form of solar tracking, and most CPV systems require highly accurate tracking, which contributes significantly to the cost of the system, and reduces performance reliability. Also, as of yet there is little long-term experience of large CPV installations in operation and therefore the cost of electricity produced over the system lifetime is hard to predict. A number of CPV manufacturers are aiming to increase their competitiveness by setting a system efficiency of 30 % as a milestone to break into the solar power market, and the present trajectory of CPV cell efficiencies makes this increasingly feasible in the near future.

Table 1 summarizes the key characteristics of typical commercial PV modules.

Technology	Material thickness (μm)	Area (m^2)	Efficiency (%)	Surface area for 1 kW_p system (m^2)
Mono-c-Si	200	1.4 - 1.7 (typical)	14 - 20	~7
Multi-c-Si	160	1.4 - 1.7 (typical) 2.5 (up to)	11 - 15	~8
a-Si	1	~1.5	4 - 8	~15
a-Si/μc-Si	2	~1.4	7 - 9	~12
CdTe	~1 - 3	~0.6 - 1	10 - 11	~10
CIGS	~2	~0.6 - 1	7 - 12	~10

Table 1. Typical commercial PV module characteristics.

Costs decrease with volume of production and prices for large systems decreased as low as 2.5 €/W_p in some countries in 2010 (EPIA, 2010), while the cost of producing electricity using PV has dropped reaching an average generation cost of 15 c€/kWh in the southern parts of the EU (EPIA & Greenpeace, 2011), demonstrating clearly that PV electricity production has already reached grid-parity in some parts of the world such as southern Europe.

3. Photovoltaic performance parameters

An essential requirement in the deployment of the different existing and emerging PV technologies is the understanding of the performance exhibited by each technology, once installed outdoors. In particular, such information is necessary because the outdoor PV electrical characteristics are different from the reference STC characteristics described in manufacturer data-sheets. In this section an overview of the main outdoor performance evaluation parameters is presented and the effects of different environmental and operational factors such as solar irradiance, temperature, spectrum and degradation on PV operation and behavior are described.

3.1 Outdoor evaluated performance parameters

In general, PV manufacturers provide information about the electrical characteristics of modules at STC. Specifically, such information includes the open circuit voltage, V_{OC}, short circuit current, I_{SC}, MPP voltage, V_{MPP}, current I_{MPP}, power, P_{MPP}, efficiency, η, and temperature coefficients. As STC conditions rarely occur outdoors, these parameters are not sufficient to predict PV operation under outdoor conditions and hence the need for independent outdoor assessment of different technologies is pressing.

The main outdoor evaluated PV performance parameters include the energy yield, the outdoor efficiency and performance ratio (PR). More specifically, for grid-connected PV systems the most important parameter is the energy yield, which is closely associated with cost evaluations. In particular, the payback of a PV system and the level of investment are associated with the energy production and the feed-in-tariff scheme in place. The normalized PV system energy yield, Y_f (kWh/kW$_p$), is defined as the total energy produced by a PV system during a period with the dc energy yield, E_{dc} (kWh), further normalized to the nameplate manufacturer dc power, P_0 (kW$_p$), to allow for comparison between the different installed PV technologies (Marion et.al, 2005). The final yield, Y_f, is given by:

$$Y_f = \frac{E_{dc}}{P_0} \qquad (1)$$

Furthermore, important performance aspects are obtained by the evaluation of the outdoor efficiency, η (%), and PR (%), for each of the PV technologies installed. The efficiency is given by:

$$\eta = \frac{E_{dc}}{H \times A} \qquad (2)$$

where H (kWh/m^2) is the total plane of array irradiation and A (m^2) is the area of the PV array. From the above parameters the PR is calculated and used as a useful way of quantifying the overall effect of losses due to PV module temperature, spectrum, module mismatch and other losses such as optical reflection, soiling and downtime failures. The dc PR, PR_{dc}, is defined as the ratio between the real dc energy production, E_{real} (kWh), and the dc energy the PV array would produce, if it had no losses at STC, E_{STC} (kWh), (Zinsser et al., 2007) and is given by:

$$PR_{dc} = \frac{E_{real}}{E_{STC}} = \frac{E_{dc}}{H \times A \times \eta_{STC}} \qquad (3)$$

where η_{STC} (%) is the PV module efficiency at STC.

3.2 Environmental and operational performance effects

In the following section, a survey of previous studies on the environmental and operational effects on the performance of the above-mentioned PV technologies is given. In particular, the investigation summarizes the main findings of the effects of solar irradiance, ambient temperature and spectrum on the performance of c-Si and thin-film technologies. In addition, findings relating to the degradation of each technology are also listed.

3.2.1 Solar irradiance effects

The most important environmental parameter influencing the operation of PV technologies is the irradiance. The operating voltage of a PV device has a logarithmic dependence on irradiance while the current is linearly dependent. Many previous studies have shown that at low irradiance levels there is a decrease in efficiency and performance that also depends on the technology (Biicher, 1997; Paretta et al., 1998; Schumann, 2009; Suzuki et al., 2002; Zinsser et al., 2009).

In this section the effect of solar irradiance on the performance of PV technologies is presented along with a discussion of previously conducted indoor and outdoor investigations. The main difficulties in the assessment of solar irradiance effects arise from the fact that the irradiance is associated with other factors that also affect the performance of PV. These factors include clear sky or diffuse irradiance due to cloudy conditions, low irradiance due to early morning or late afternoon (high AM), spectral and angle of incidence (AOI) effects. In general, the effect of solar irradiance levels on PV performance has been investigated by employing indoor controlled methods. These offer the advantage that other effects such as AOI, spectrum and temperature can be controlled and excluded from the investigation. A common approach used is the acquisition of the current-voltage (I-V) curves at the cell or module level using solar flash simulators, which allow the evaluation and comparison of the efficiency at different specified irradiance levels indoors (Bunea et al., 2006; Reich et al., 2009).

Similarly, the effects of solar irradiance have been investigated in outdoor evaluations by first acquiring I-V curves at again cell or module level and secondly correcting the acquired data-sets to STC temperature, by using measured or manufacturer temperature coefficients (Merten & Andreu, 1998; Paretta et al., 1998). To minimize AOI effects, the PV devices are usually mounted on trackers while to minimize spectral effects, the investigations are usually carried out under clear sky conditions. From the acquired and corrected I-V curves the efficiency at different irradiance levels can also be evaluated and compared.

For some commercial PV technologies, the output power follows closely the irradiation level while for many commercial modules the efficiency was found to decrease by 55 – 90 % from its STC value, at irradiance levels below 200 W/m^2 (Biicher, 1997). The behavior of PV technologies at different irradiance levels has been associated with the series and shunt resistance as at high solar irradiance, high series resistance reduces the fill factor (FF) while at low solar irradiance, FF reduction occurs due to low shunt resistance (Randall & Jacot, 2003). Other investigations have further demonstrated that series resistance losses are mainly responsible for the reduction in the FF for intensities of 60 % of one sun or greater (del Cueto, 1998).

Both mono-c-Si and multi-c-Si technologies exhibit almost constant efficiencies in the irradiance range of 100 - 1000 W/m^2 with mono-c-Si found to outperform multi-c-Si in an investigation performed on commercial PV cells (Reich et al., 2009). In addition, some c-Si cells were found to have higher efficiencies at irradiance intensities in the range 100 - 1000 W/m^2 than at STC and this is attributed to series resistance effects, as a lower current leads to quadratically lower series resistance loss (Reich et al., 2009). For c-Si technologies the efficiency decreases logarithmically in the lower irradiance range of 1 - 100 W/m^2 as the open circuit voltage, V_{OC}, depends logarithmically on the short circuit current I_{SC}. Subsequently, previous work describing the low light performance based on the evaluated FF has shown that for c-Si and CIS, the FF remains approximately constant for irradiance levels above 200 W/m^2 while at lower irradiance levels the FF decreases (Mohring & Stellbogen, 2008). Furthermore, CdTe thin-film technology has been reported as having a relatively good low irradiance performance (Heesen et al., 2010;) and specifically to exhibit significant performance increase at medium irradiance levels due to the relatively high series resistance of CdTe devices (Mohring & Stellbogen, 2008). On the other hand, a-Si technology shows a constant FF over the entire range and even below 200 W/m^2 and this further implies a superior performance for sites with high diffuse light conditions (Mohring & Stellbogen, 2008). For the side-by-side irradiance dependence comparison performed for different commercial PV technologies in Nicosia, Cyprus, the a-Si and CdTe technologies have exhibited higher relative efficiencies at low light (Zinsser et al., 2009).

Because of the importance of this effect it would be very useful if all manufacturers provided, as part of their data-sheet information, the efficiencies at different irradiance levels.

3.2.2 Thermal effects

PV technologies that operate in warm climates experience module temperatures significantly above 25 °C and this is a very important performance loss factor. The parameters which describe the behavior of the electrical characteristics of PV with the operating temperature and hence the thermal effects, are the temperature coefficients (King et al., 1997; Makrides et al., 2009). Another important thermal parameter that describes the temperature of a PV module is the nominal operating cell temperature (NOCT), which is provided by PV manufacturers as an indication of how module temperature is affected by the solar irradiation, ambient temperature and thermal properties of the PV material.

Temperature coefficients of PV devices are usually evaluated using indoor laboratory techniques. A commonly used methodology is to illuminate a PV cell or module that is placed on a temperature controlled structure. Accordingly, the I-V curves of the device are acquired over a range of different cell temperatures but at controlled STC irradiance and AM. The rate of change of either the voltage, current or power with temperature is then calculated and provides the value of the temperature coefficients (King et al., 1997).

In addition, a useful technique to obtain the temperature coefficients under real operating conditions is to employ outdoor field test measurements. In outdoor investigations the PV devices are first shaded to lower the temperature close to ambient conditions and as soon as the device is uncovered and left to increase in temperature, several I-V curves are acquired at different temperatures (Akhmad et al., 1997; King et al, 1997; Makrides et al., 2009; Sutterlüti et al., 2009). As in indoor investigations, the rate of change of the investigated parameter against temperature provides the temperature coefficient. Both techniques are used by manufacturers and professionals within the field. Previous studies have shown that the power of c-Si PV modules decreases by approximately -0.45 %/K (Virtuani et al. 2010; Makrides et al., 2009). On the other hand, thin-film technologies of CdTe and CIGS show lower power temperature coefficients compared to c-Si technologies and in the case of CdTe modules the measured temperature coefficient is around -0.25 %/K (Dittmann et al., 2010). In addition, a-Si shows the lowest power temperature coefficient of up to approximately -0.20 %/K (Hegedus, 2006) while numerous studies have further shown that high module operating temperatures improve the performance of stabilized a-Si modules due to thermal annealing (Dimitrova et al., 2010; King et al., 2000; Ransome & Wohlgemuth, 2000). The thermal behavior of a-Si suggests that a unique temperature coefficient as in the case of other PV technologies cannot characterize completely the temperature behavior of this technology (Carlson et al., 2000). In general, the output power and performance of CdTe and a-Si modules is less temperature sensitive than CIS and c-Si technologies. Table 2 summarizes the MPP power, P_{MPP}, temperature coefficients of commercial PV technologies.

Technology	Approximate MPP power temperature coefficient, P_{MPP} (%/K)
Mono-c-Si	-0.40
Multi-c-Si	-0.45
a-Si	-0.20
a-Si/μc-Si	-0.26
CIGS	-0.36
CdTe	-0.25

Table 2. Typical power temperature coefficients for different technologies.

3.2.3 Spectral effects

PV devices are affected by the change and variation of the solar spectrum. In practice, the power produced by a PV cell or module can be calculated by integrating the product of the spectral response and the spectrum, at a given temperature and irradiance level, over the incident light wavelength range (Huld et al., 2009). The effect of spectrum is a technology dependent parameter as some technologies are affected more by spectral variations than others (King et al., 1997).

The spectral response of PV technologies is usually known but as the spectral irradiance at different installation locations is unknown, the spectral losses can be difficult to evaluate (Huld et al., 2009). The spectral content of a location is affected by several factors such as the AM, water vapor, clouds, aerosol particle size distribution, particulate matter and ground reflectance (Myers et al., 2002). In clear-sky conditions the spectrum can be described as a function of air mass and relative humidity (Gueymard et al., 2002). In cloudy weather the spectral effects are more complex and in general the light under these conditions is stronger in the blue region of the spectrum than the standard AM 1.5 spectrum. Conversely, the blue region of the spectrum is attenuated as the sun moves lower in the sky (Huld et al., 2009).

A number of studies have been performed both indoors and outdoors to investigate spectral effects (Gottschalg et al., 2007; Merten & Andreu, 1998; Zanesco & Krenziger, 1993). The spectral response of PV cells and modules can be determined indoors using specialized equipment such as solar simulators and special filters at controlled irradiance and temperature conditions (Cannon et al., 1993; Virtuani et al., 2011). In outdoor investigations the spectral behavior of PV devices is usually found by mounting the PV device on a tracker and acquiring measurements of the short circuit current or I-V curves in conjunction with measurements acquired using a pyranometer and a spectroradiometer (King et al., 1997).

The effect of the spectrum has been further described in different ways. Several authors have presented spectral effects by calculating the fraction of the solar irradiation that is usable by each PV technology (Gottschalg et al., 2003). Others have included the average photon energy (APE) parameter, even though this requires knowledge of the spectrum under varying conditions (Gottschalg et al., 2005; Norton et al., 2011). Empirical models have also been considered to account for the influence of the solar spectrum on the short circuit current (Huld et al., 2009).

Technologies of c-Si and CIGS have a wide spectral response and this allows a large spectral absorption. In the case of c-Si technologies an increase in efficiency at high AM and clear sky conditions has been reported (King et al., 2004; Zdanowicz et al., 2003), while other investigations performed on c-Si modules mounted on a tracker under clear sky conditions showed a slight decrease in performance with increasing AM (Kenny et al., 2006). CdTe and a-Si technologies have a narrower spectral response which ranges approximately between 350 - 800 nm and this leads to lower photon absorption. Modules of a-Si have shown higher energy yield compared to c-Si for diffuse light irradiation and high sun elevation angles (Grunow et al., 2009).

Specifically, in a previous study in Japan, the ratio of spectral solar irradiation available for solar cell utilization to global solar irradiation, was found to vary from 5 % for multi-c-Si cells, to 14 % for a-Si cells, throughout a year (Hirata & Tani, 1995). In addition, the experimental results of a study carried out in the UK, showed that on an annual basis, the usable spectral fraction of solar irradiation for a-Si varied from +6 % to -9 % with respect to the annual average, while for CdTe and CIGS it varied in the range of +4 % to -6 % and ±1.5 % (Gottschalg et al, 2003). Spectral effects on PV performance are therefore important depending on the location, climatic conditions and spectral sensitivity of each technology.

3.2.4 PV degradation

The performance of PV modules varies according to the climatic conditions and gradually deteriorates through the years (Adelstein & Sekulic, 2005; Cereghetti et al., 2003; Dunlop, 2005; Osterwald et al., 2006; Sanchez-Friera et al., 2011; Som & Al-Alawi, 1992). An important factor in the performance of PV technologies has always been their long-term reliability especially for the new emerging technologies. The most important issue in long-term performance assessments is degradation which is the outcome of a power or performance loss progression dependent on a number of factors such as degradation at the cell, module or even system level. In almost all cases the main environmental factors related to known degradation mechanisms include temperature, humidity, water ingress and ultra-violet (UV) intensity. All these factors impose significant stress, over the lifetime of a PV device and as a result detailed understanding of the relation between external factors, stability issues and module degradation is necessary. In general, degradation mechanisms describe the effects from both physical mechanisms and chemical reactions and can occur at both PV cell, module and system level.

More specifically, the degradation mechanisms at the cell level include gradual performance loss due to ageing of the material and loss of adhesion of the contacts or corrosion, which is usually the result of water vapor ingress. Other degradation mechanisms include metal mitigation through the p-n junction and antireflection coating deterioration. All the above-mentioned degradation mechanisms have been obtained from previous experience on c-Si technologies (Dunlop, 2005; Quintana et al., 2002; Som & Al-Alawi, 1992).

In the case of a-Si cells an important degradation mechanism occurs when this technology is first exposed to sunlight as the power stabilizes at a level that is approximately 70 - 80 % of the initial power. This degradation mechanism is known as the Staebler-Wronski effect (Staebler & Wronski, 1977) and is attributed to recombination-induced breaking of weak Si-Si bonds by optically excited carriers after thermalization, producing defects that decrease carrier lifetime (Stutzmann et al., 1985).

Other degradation mechanisms have also been observed for thin-film technologies of CdTe and CIGS at the cell level. For CdTe technologies the effects of cell degradation can vary with the properties of the cell and also with the applied stress factors. More specifically, in CdTe technologies as the p-type CdTe cannot be ohmically contacted with a metal, most devices use copper to dope the CdTe surface before contacting (Chin et al., 2010; Dobson et al., 2000). Copper inclusion may cause dramatic changes in the electrical properties of the CdTe thin-film (Chin et al., 2010). As copper is very mobile it can diffuse along grain boundaries of the CdTe cell and result in a recombination center situated close to the p-n junction. Very low levels of copper reduce the conductivity of CdTe and it is possible that the diffusion of copper can transform the back contact to non-ohmic. Another effect associated with CdTe degradation is due to the applied voltage either arising from the cell or the external voltage, which as a result of the electric field it can force copper ions towards the front contact. It was previously found that open-circuit conditions affected cell degradation during accelerated ageing for different CdTe cell types (Powell et al., 1996). In addition, impurity diffusion and changes in doping profiles may affect device stability (Batzner et al., 2004; Degrave et al., 2001), but the industry has resolved this problem by using special alloys.

CIGS has a flexible structure that enhances its tolerance to chemical changes and because of this it has been previously argued that copper atoms do not pose stability problems for CIGS cells (Guillemoles et al., 2000). Damp heat tests performed on unencapsulated CIGS cells have indicated that humidity degrades cell performance and is more obvious as V_{OC} and FF degradation due to the increased concentration of deep acceptor states in the CIGS absorber (Schmidt et al., 2000). Other important factors include donor-type defects (Igalson et al., 2002) and the influence of Ga-content on cell stability (Malmström et al., 2003).

At the module level, degradation occurs due to failure mechanisms of the cell and in addition, due to degradation of the packaging materials, interconnects, cell cracking, manufacturing defects, bypass diode failures, encapsulant failures and delamination (King et al., 1997; Pern at al., 1991; Wenham et al., 2007).

At the system level, degradation includes all cell and module degradation mechanisms and is further caused by module interconnects and inverter degradation. Table 3 summarizes the main thin-film failure modes and failure mechanisms (McMahon, 2004).

Indoor degradation investigations are mainly performed at the module level as the interconnection and addition of other materials to form a modular structure increases stability issues. In particular, accelerated ageing tests performed indoors and under controlled conditions can provide information about different degradation mechanisms. Degradation investigations using indoor methodologies are based on the acquisition of I-V curves and power at STC. The electrical characteristics of PV modules are initially measured at STC and then the modules are either exposed outdoors or indoors through accelerated procedures (Carr & Pryor, 2004; Meyer & van Dyk, 2004; Osterwald et al., 2002). For each investigated PV cell or module the electrical characteristics are regularly acquired using the solar simulator and the current, voltage or power differences from the initial value provide indications of the degradation rates at successive time periods.

In addition, many groups have performed outdoor monitoring of individual PV modules through the acquisition and comparison of I–V curves, as the modules are exposed to real outdoor conditions (Akhmad et al., 1997; Ikisawa et al., 1998; King et al., 2000). Another method to investigate degradation outdoors has been based on power and energy yield measurements of PV systems subjected to actual operating conditions. A common approach has been to first establish time series usually on a monthly basis, of either the PR or the maximum power normalised to Photovoltaics for Utility Scale Applications (PVUSA) Test Conditions (PTC) of solar irradiance 1000 W/m^2, air temperature of 20 °C and wind speed of 1 m/s. Time series analysis such as linear regression, classical series decomposition (CSD) and Autoregressive Integrated Moving Average (ARIMA) is then used to obtain the trend and hence the degradation rate (Jordan & Kurtz, 2010; Osterwald et al., 2002). Outdoor field tests are very important in exploring the degradation mechanisms under real conditions. These mechanisms cannot otherwise be revealed from indoor stability tests. The outcome of such outdoor investigations can provide useful feedback to improve the stability, enhance the understanding of the different technology dependent degradation mechanisms and can be used as tools for the adaptation of accelerated ageing tests so as to suit the degradation mechanisms for each technology.

Failure modes	Effect on I-V curve	Possible failure mechanisms
1. Cell degradation		
a. Main junction: increased recombination	Loss in fill factor, *Isc*, and *Voc*	Diffusion of dopants, impurities, etc. Electromigration
b. Back barrier; loss of ohmic contact (CdTe)	Roll-over, cross-over of dark and light I-V, higher R_{series}	Diffusion of dopants, impurities, etc. Corrosion, oxidation Electromigration
c. Shunting	R_{shunt} decreases	Diffusion of metals, impurities, etc.
d. Series; ZnO,Al	R_{series} increases	Corrosion, diffusion
e. De-adhesion SnO_2 from soda-lime glass	Isc decreases and R_{series} increases	Na ion migration to SnO_2/glass interface
f. De-adhesion of back metal contact	Isc decreases	Lamination stresses
2. Module degradation		
Interconnect degradation		
a. Interconnect resistance; ZnO:Al/Mo or Mo, Al interconnect	R_{series} increases	Corrosion, electromigration
b. Shunting; Mo across isolation scribe	R_{shunt} decreases	Corrosion, electromigration
Busbar degradation	R_{series} increases or open circuit	Corrosion, electromigration
Solder joint	R_{series} increases or open circuit	Fatigue, coarsening (alloy segragation)
Encapsulation failure		
a. Delamination	Loss in fill-factor, *Isc*, and possible open circuit	Surface contamination, UV degradation, hydrolysis of silane/glass bond, warped glass, 'dinged' glass edges, thermal expansion mismatch
b. Loss of hermetic seal		
c. Glass breakage		
d. Loss of high-potential isolation		

Table 3. Thin-film failure modes and failure mechanisms (McMahon, 2004).

For both indoor and outdoor evaluations a variety of degradation rates have been reported and a survey of the results of degradation studies is given below. A recent study has shown that on average the historically reported degradation rates of different PV technologies was 0.7 %/year while the reported median was 0.5 %/year (Jordan et al., 2010). More specifically, investigations performed on outdoor exposed mono and multi-c-Si PV modules

showed performance losses of approximately 0.7 %/year (Osterwald et al., 2002). Results of field tests have generally shown stable performance for CdTe devices (del Cueto, 1998; Mrig & Rummel, 1990; Ullal et al., 1997), although field results are limited for modules utilizing new cell structures (Carlsson & Brinkman, 2006). Previous studies performed on thin-film CIS modules, showed that after outdoor exposure the efficiency was found to decrease (Lam et al., 2004) and to exhibit either moderate, in the range of 2 - 4 %/year, to negligible or less than 1 %/year degradation rates due to increases in the series resistance in some of the modules (del Cueto et al., 2008).
Evaluations based on monthly PR and PVUSA values revealed degradation rates, for the PR investigation, of 1.5 %/year for a-Si, 1.2 %/year for CdTe and 0.9 %/year for mono-c-Si (Marion et al., 2005). The results were slightly different for the PVUSA investigation which showed a degradation rate of 1.1 %/year for the a-Si, 1.4%/year for the CdTe and 1.3 %/year for the mono-c-Si (Marion et al., 2005). Based on linear fits applied to the PVUSA power rating curves over the six year time period for a thin-film a-Si system, degradation rates of 0.98 %/year at the dc side and 1.09 %/year at the ac side of the system were obtained while the same investigation on PR data-sets indicated a similar degradation rate of 1.13 %/year at the ac side (Adelstein & Sekulic, 2005). Additionally, in a recent long-term performance assessment of a-Si tandem cell technologies in Germany it was demonstrated that an initial two year stabilization phase occurred and was then followed by a stable phase with a minor power decrease of maximum 0.2 %/year (Lechner et al., 2010). In a different study it was reported that thin-film modules showed somewhat higher than 1 %/year degradation rates (Osterwald et al., 2006). On the other hand, an important consideration in relation to thin-film degradation rate investigations was found to be the date of installation of the modules as it appeared that in the case of CdTe and CIGS modules manufactured after 2000 exhibited improved stability relative to older designs (Jordan et al., 2010).

4. Performance assessment of different PV technologies under outdoor conditions

In the previous section a general description of the main outdoor evaluation performance parameters and the effects of different environmental and operational parameters was given. In the following section, a discussion on the work carried out at the outdoor test facility in Cyprus, related to the performance assessment of different installed PV technologies is presented. An infrastructure was set up for continuous and simultaneous monitoring of a number of PV systems (together with weather and irradiation data) and to thereby assess their performance under the exact same field conditions. The knowledge acquired from the field testing, described in this section, is important to enhance the understanding of the underlying loss processes and to optimise the systems performance. Furthermore, it is essential to continue testing as the current PV technologies become more mature and new technologies are entering the market. The same infrastructure installed in Cyprus was also replicated in two other locations for the scope of investigating the performance of different PV technologies under different climatic conditions. The three selected locations include the Institut für Physikalische Elektronik (*ipe*) University of Stuttgart, Germany, the University of Cyprus (UCY), Nicosia, Cyprus and the German University in Cairo (GUC) Cairo, Egypt.

4.1 PV test facility description

The outdoor test facility at the University of Cyprus, Nicosia, Cyprus was commissioned in May 2006 and includes, amongst others, 12 grid-connected PV systems of different technologies. The fixed-plane PV systems installed range from mono-c-Si and multi-c-Si, Heterojunction with Intrinsic Thin layer (HIT), Edge defined Film-fed Growth (EFG), Multi-crystalline Advanced Industrial cells (MAIN) to a-Si, CdTe, CIGS and other PV technologies. Table 4 provides a brief description of the installed systems (Makrides et al., 2010).

Manufacturer	Module type	Technology	Rated module efficiency (%)
Atersa	A-170M 24V	Mono-c-Si	12.9
BP Solar	BP7185S	Mono-c-Si (Saturn-cell)	14.8
Sanyo	HIP-205NHE1	Mono-c-Si (HIT-cell)	16.4
Suntechnics	STM 200 FW	Mono-c-Si (back contact-cell)	16.1
Schott Solar	ASE-165-GT-FT/MC	Multi-c-Si (MAIN-cell)	13.0
Schott Solar	ASE-260-DG-FT	Multi-c-Si (EFG)	11.7
SolarWorld	SW165 poly	Multi-c-Si	12.7
Solon	P220/6+	Multi-c-Si	13.4
Mitsubishi Heavy Industries (MHI)	MA100T2	a-Si (single cell)	6.4
Schott Solar	ASIOPAK-30-SG	a-Si (tandem cell)	5.4
First Solar	FS60	CdTe	8.3
Würth	WS 11007/75	CIGS	10.3

Table 4. Installed PV types of modules.

The monitoring of the PV systems started at the beginning of June 2006 and both meteorological and PV system measurements are being acquired and stored through an advanced measurement platform. The platform comprises meteorological and electrical sensors connected to a central data logging system that stores data at a resolution of one measurement per second. The monitored meteorological parameters include the total irradiance in the POA, wind direction and speed as well as ambient and module temperature. The electrical parameters measured include dc current and voltage, dc and ac power at MPP as obtained at each PV system output (Makrides et al., 2009).

4.2 PV performance evaluation

The weather conditions recorded over the evaluation period in Cyprus showed that there is a high solar resource and exposure to warm conditions. The annual solar irradiation, over the period June 2006 - June 2010 is summarised in Table 5.

Period	Solar Irradiation (kWh/m^2)
June 2006 - June 2007	1988
June 2007 - June 2008	2054
June 2008 - June 2009	1997
June 2009 - June 2010	2006

Table 5. Solar irradiation over the period June 2006 - June 2010 in Nicosia, Cyprus.

A detailed analysis of the prevailing climatic conditions was performed on the acquired 15-minute average measurements, in order to obtain the fraction of solar irradiation in Cyprus, the average ambient air temperature and PV operating temperature at different solar irradiance levels over the four-year evaluation period. Table 6 shows the results of the average ambient and PV module temperature (Atersa mono-c-Si fixed-plane module temperatures presented) at different solar irradiation levels over the first, second, third and fourth year respectively. The results indicate that the PV module operating temperatures increased above the STC temperature of 25 °C at POA solar irradiance over 201 W/m^2. During the first three years, the highest amount of solar irradiation occurred within the range 801 - 900 W/m^2 while in the fourth year within the range 901 - 1000 W/m^2.

Solar Irradiance (W/m”)	Total irradiation fraction (%)				Ambient temperature (°C)				PV module temperature (°C)			
	2006 2007	2007-2008	2008-2009	2009-2010	2006-2007	2007-2008	2008-2009	2009-2010	2006-2007	2007-2008	2008-2009	2009-2010
0 - 100	2.0	1.8	1.9	2.0	15.4	16.2	16.4	16.9	14.7	15.4	15.8	16.3
101 - 200	2.8	2.4	2.6	2.7	20.2	21.0	20.3	20.1	24.1	24.6	24.0	23.9
201 - 300	4.1	4.0	4.3	4.2	21.5	22.3	22.1	22.6	27.3	27.8	27.9	28.5
301 - 400	5.3	5.1	5.8	5.6	22.3	23.3	23.0	23.4	30.7	31.4	31.4	31.8
401 - 500	7.0	7.3	7.4	7.0	22.9	23.6	23.5	23.7	33.5	34.0	34.4	34.6
501 - 600	9.7	10.0	9.5	9.2	23.4	23.7	23.9	24.5	36.8	37.0	37.4	38.1
601 - 700	12.9	13.3	12.0	11.6	24.7	24.6	25.1	25.4	41.3	40.8	41.4	41.6
701 - 800	16.9	18.1	15.7	14.5	25.7	25.5	25.9	25.9	45.2	44.3	44.9	44.8
801 - 900	20.9	21.4	20.0	18.0	26.4	28.2	27.2	27.5	48.1	49.7	48.8	48.6
901 - 1000	15.8	14.3	15.9	21.2	27.2	29.0	29.0	30.2	50.4	51.6	52.0	53.5
1001 - 1100	2.5	2.2	4.4	3.8	23.3	23.6	23.0	26.1	47.3	46.7	46.5	50.2
> 1101	0.0	0.0	0.4	0.1	24.7	24.0	20.1	17.5	47.2	51.9	46.0	40.2

Table 6. Solar irradiation fraction, average ambient and module temperature (Atersa mono-c-Si) at different irradiance levels, over the period June 2006 - June 2010 in Nicosia, Cyprus.

During the first year of operation the fixed-plane PV systems showed an average annual dc energy yield of 1738 kWh/kW_p while during the second year of operation and for the same systems the average dc energy yield was 1769 kWh/kW_p, showing an increase of 1.8 % in comparison to the first year. The average dc energy yield was lower during the third and fourth year with 1680 kWh/kW_p and 1658 kWh/kW_p respectively. The annual dc energy yield normalized to the manufacturer's rated power over the period June 2006 - June 2010 in Nicosia, Cyprus is shown in Table 7. It must be noted that partial shading affected the BP Solar mono-c-Si and Solon multi-c-Si systems specifically during the second, third and fourth year while the Schott Solar a-Si system had a broken module since October 2006.

System	Normalized DC Energy Yield (kWh/kW_p)			
	2006 - 2007	2007 - 2008	2008 - 2009	2009 - 2010
Atersa (A-170M 24V)	1753	1810	1744	1719
BP Solar (BP7185S)	1612	1593	1457	1510
Sanyo (HIP-205NHE1)	1790	1814	1731	1703
Suntechnics (STM 200 FW)	1864	1890	1800	1793
Schott Solar (ASE-165-GT-FT/MC)	1752	1810	1736	1712
Schott Solar (ASE-260-DG-FT)	1721	1783	1714	1688
SolarWorld (SW165)	1731	1772	1689	1654
Solon (P220/6+)	1715	1761	1681	1637
MHI (MA100T2)	1734	1734	1644	1617
Schott Solar (ASIOPAK-30-SG)	1599	1650	1571	1554
Würth (WS 11007/75)	1827	1863	1748	1707
First Solar (FS60)	1755	1752	1645	1605

Table 7. Annual dc energy yield normalized to the manufacturer's rated power over the period June 2006 - June 2010 in Nicosia, Cyprus.

During the first year of operation the best performing technologies in Nicosia, based on the annual dc energy yield, were the Suntechnics mono-c-Si, the Würth CIGS, the Sanyo HIT mono-c-Si and the First Solar CdTe. During the second year the mono-c-Si technologies of Sanyo, Suntechnics and the CIGS retained their high energy yield. During the third year the highest energy yield was produced by the Suntechinics mono-c-Si, Würth CIGS and Atersa mono-c-Si system. During the fourth year the first three technologies which produced the highest yield were entirely c-Si, the Suntechnics, Atersa mono-c-Si and the Schott Solar (MAIN) multi-c-Si while the Würth CIGS system followed.
The comparison of the annual dc energy yield produced by the same technology modules, Atersa mono-c-Si fixed-plane, installed in the POA of 27.5° and also mounted on a two-axis tracker is shown in figure 1. Over a four-year period, the tracker provided on average 21 % higher energy yield compared to the fixed-plane system. During the first year, the solar irradiation collected by the reference cell installed at the tracker was 2532 kWh/m^2 while during the second year it was 2606 kWh/m^2 (Makrides et al., 2010). Subsequently, during the third and fourth year the solar irradiation collected by the tracker was 2510 kWh/m^2 and 2483 kWh/m^2 respectively.

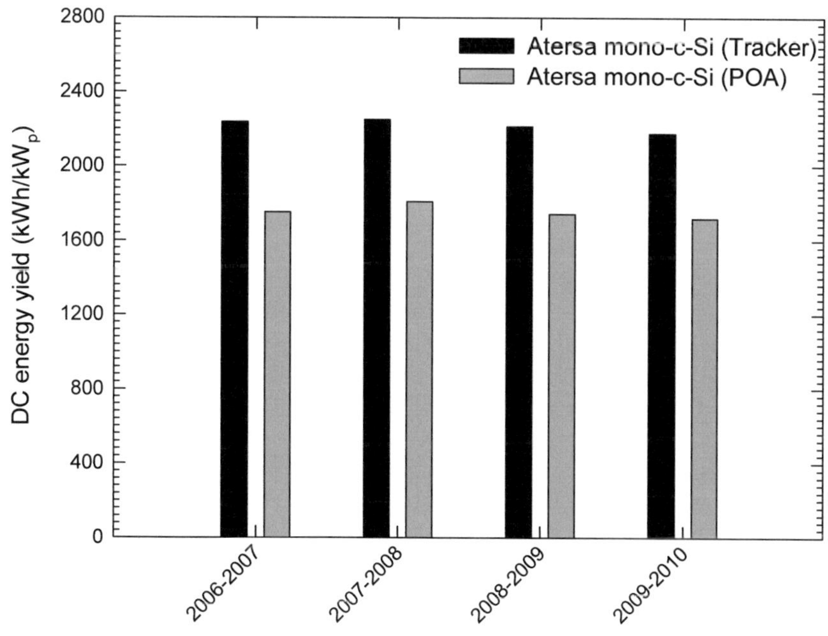

Fig. 1. Comparison of the annual dc energy yield of the tracker and fixed-plane Atersa mono-c-Si systems over the period June 2006 - June 2010.

Table 8 shows the annual ac energy yield normalized to the manufacturer's rated power.

System	Normalized AC Energy Yield (kWh/kW$_p$)			
	2006 - 2007	2007 - 2008	2008 - 2009	2009 - 2010
Atersa (A-170M 24V)	1593	1646	1583	1564
BP Solar (BP7185S)	1463	1445	1320	1370
Sanyo (HIP-205NHE1)	1630	1659	1581	1555
Suntechnics (STM 200 FW)	1692	1717	1641	1638
Schott Solar (ASE-165-GT-FT/MC)	1588	1642	1575	1552
Schott Solar (ASE-260-DG-FT)	1562	1620	1554	1532
SolarWorld (SW165)	1573	1613	1535	1500
Solon (P220/6+)	1567	1609	1533	1495
MHI (MA100T2)	1573	1575	1495	1466
Schott Solar (ASIOPAK-30-SG)	1462	1506	1433	1419
Würth (WS 11007/75)	1653	1691	1581	1543
First Solar (FS60)	1599	1600	1500	1461

Table 8. Annual ac energy yield normalized to the manufacturer's rated power over the period June 2006 - June 2010 in Nicosia, Cyprus.

4.3 Effects of environmental and operational parameters on PV performance

In the following section, a summary of the investigations and outcomes related to the seasonal performance and the effect of temperature, soiling and STC power normalization on the performance assessment of the installed technologies in Cyprus is given.

4.3.1 PV seasonal performance evaluation using outdoor measurement analysis

In order to observe the effects of environmental conditions on the outdoor performance of the installed PV technologies, a seasonal performance investigation was carried out. Specifically, a time series was constructed of the monthly average dc PR over the four-year evaluation period. The plots in figure 2 depict the constructed monthly average dc PR time series of all the PV technologies. It is evident from the plots that all technologies exhibit a seasonal behavior with peaks according to the seasons and with progressive performance loss that is more evident in some technologies than others. Both mono-c-Si and multi-c-Si technologies exhibited PR peaks during the cold winter season and performance decrease during the warm summer months as depicted in figures 2a and 2b respectively. The Suntechnics mono-c-Si exhibited high monthly PR that approached the optimum (PR of 100 %) during the winter seasons and in one case, December 2006, this value was even exceeded. This can occur because of the associated power rating and irradiation uncertainties that are present also in the calculated monthly PR value. From the PR plot of figure 2c of the a-Si technologies it was obvious that during the summer and early autumn, the performance was higher than in the winter. In addition, the high initial monthly PR of the a-Si technologies is primarily attributed to the fact that these technologies had not yet stabilized. Accordingly, the same seasonal performance pattern as the one of c-Si technologies was observed for the Würth CIGS and First Solar CdTe, shown in figure 2d. In the case of the First Solar CdTe system a narrower peak-to-peak PR variation between the seasons was observed compared to the c-Si and CIGS seasonal behavior.

4.3.2 Thermal effects

In countries such as Cyprus with a high solar resource and warm climate the extent to which PV technologies are affected by temperature is an important criterion for their selection. Investigations to evaluate the effect of temperature were performed based on an indoor and outdoor procedure for the extraction of the MPP power temperature coefficients of the installed technologies (Makrides et al., 2009).

For the outdoor procedure, the temperature coefficients at the MPP power were extracted from a series of acquired I-V curve measurements over a range of temperatures (from ambient to maximum module temperature during the period of outdoor measurements). The outdoor investigation was performed during periods of the day with conditions of stable sunshine and calm winds (lower than 2 m/s) around solar noon. All the PV systems were equipped with back surface temperature sensors that were mounted at the centre of each investigated module. At the same time, the MPP power temperature coefficients were also calculated through a filtering and analysis technique (data-evaluated technique) on acquired data over a period of a year. In this investigation 15-minute average data acquired over a period of one year were used. The MPP power data-sets that occurred when the solar irradiance was between 700 and 1100 W/m^2, were chosen in order to minimize the influence of large AM in the morning and the evening. Figure 3 summarizes the measured, calculated and manufacturer provided MPP power temperature coefficients obtained by both techniques. For most PV technologies the outdoor evaluated results showed satisfactory agreement when compared to manufacturer provided data.

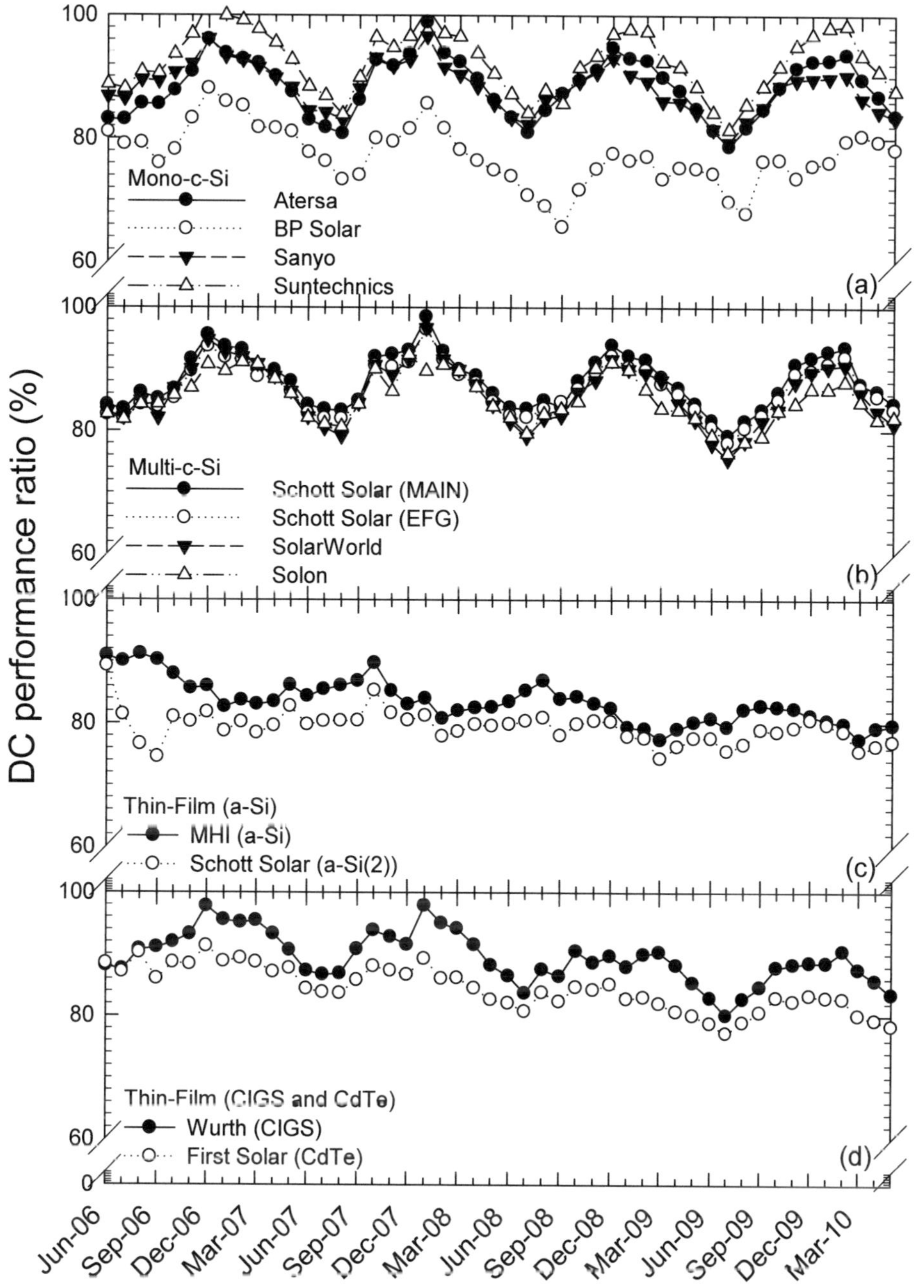

Fig. 2. Monthly average dc PR of installed PV systems over the period June 2006 - June 2010 in Nicosia, Cyprus.

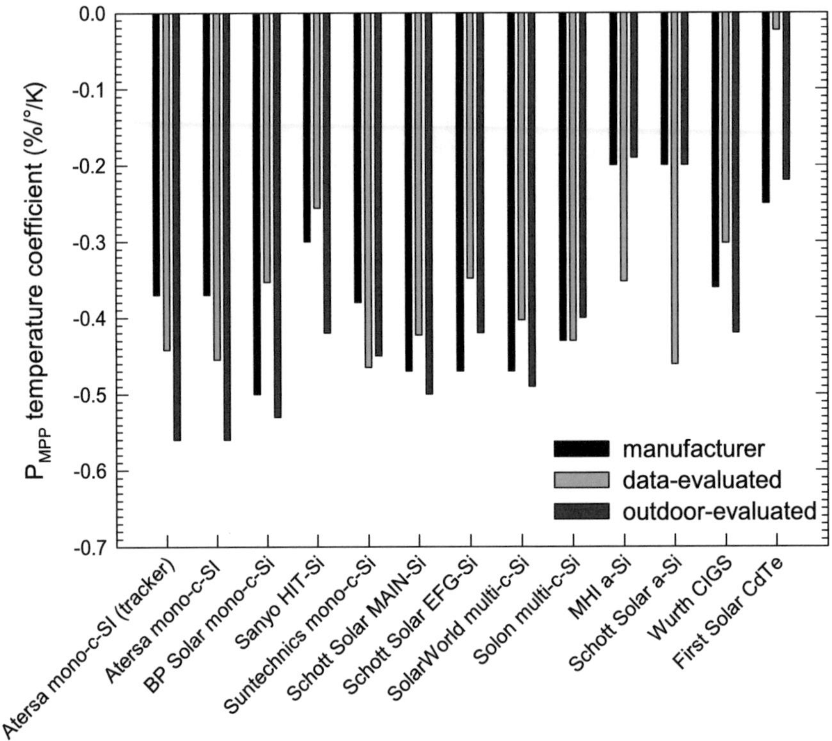

Fig. 3. Comparison of the MPP power temperature coefficients (γ_{PMPP} %/K) obtained by the two methods outlined above (outdoor measurements and data analysis) and the manufacturers' data for the installed systems.

4.3.3 Soiling effects

Soiling describes the accumulation of dirt on the front surface of PV modules and is an important loss factor particularly in locations when there is scarce rain, very dry conditions and even frequent dust or sand storms. The power loss due to soiling is therefore a function of the type of dust, the length of time since the last rainfall and the cleaning schedule (Kymakis et al., 2009). In warm climates such as the one in Cyprus, soiling losses increase as the periods between successive rainfalls increase and this is more noticeable during the summer period.

In general, the standard industry assumption of soiling losses ranges from 1 - 4 % on an annual basis (Detrick et al., 2005). In areas of frequent rainfall, it was demonstrated that the rain could clean the PV modules to an extent of restoring the performance to within 1 % of full power (Hammond et al., 1997). Accordingly, in a more recent soiling analysis performed in Crete, with climatic conditions almost identical to Cyprus, the annual soiling loss was 5.86 %, with the winter losses being 4 - 5 % and 6 - 7 % in the summer (Kymakis et al., 2009).

A soiling investigation was carried out also for the systems installed in Egypt and specifically by comparing the energy produced by a clean module, a module that has been exposed to dust for a period of one year and a module that has been exposed to dust but cleaned every two months. The energy production results showed that the 'one year dusty module' produced 35 % lower energy while the 'two month dusty module' produced 25 % lower energy compared to the clean module (Ibrahim et al., 2009). Figure 4 shows the soiling accumulation after a period of one year for the systems installed in Egypt.

Fig. 4. Dust layer accumulation on PV modules in Egypt (Ibrahim et al., 2009).

4.3.4 Energy yield normalization to rated power and uncertainties

The tolerance of the rated power provided by manufacturers is another important factor that affects the PV performance investigation as it increases the uncertainty of the results. In general, the rated power value is associated with a typical tolerance of ±3 % for c-Si PV modules, and ±5 % for thin-film modules. This uncertainty arises due to the power mismatch of PV cells during module production and the sorting which is necessary so as to avoid power mismatch at array wiring. Subsequently, manufacturers measure the power of each produced module using a flasher and then sort the modules into power classes (Zinsser et al., 2010). The uncertainty associated with the power rating is particularly important in outdoor PV performance evaluations and comparisons as in the case of the normalized annual energy yield (kWh/kW_P) to the rated power (Zinsser et al., 2010).

By considering an uncertainty of ±3 % due to the STC power and a ±2 % due to the ac energy measurement and acquisition, a difference of up to 10 % could arise for comparisons between the annual yields of two PV systems at the same location. The high power rating uncertainty value is a limiting factor in performance investigations as it is very difficult to distinguish which of those systems performed better over a period of time. Figure 5 shows the annual ac energy yield and associated power rating and measurement uncertainties over the four-year period and the average energy yield of the flat plate systems.

The uncertainty makes it difficult to accurately distinguish which technology had produced the highest energy. In addition, this uncertainty is also large enough to mask other lower order performance effects, such as degradation rate, spectral losses and other performance loss factors. In the absence of the power rating uncertainty these effects would have been important in the energy yield comparisons and the selection of the best performing technology at a particular location. Therefore, there is a high need for low uncertainties in the power tolerance of PV modules.

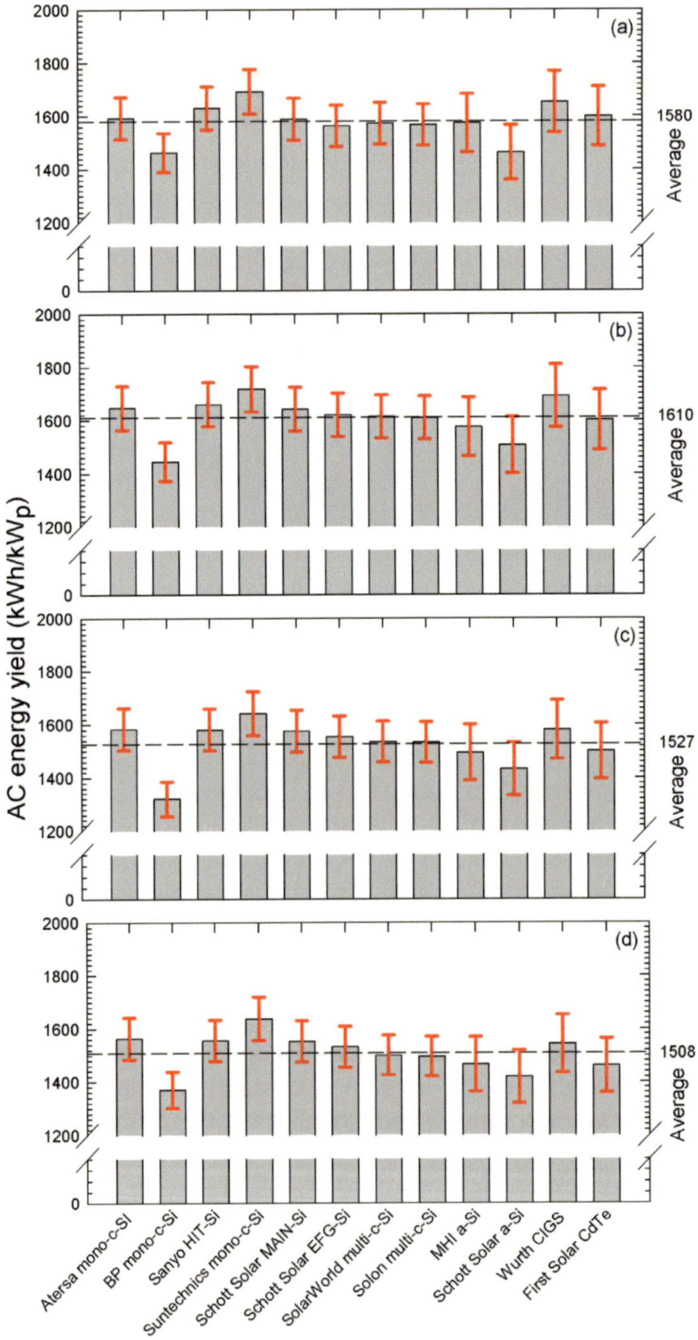

Fig. 5. Annual ac energy yield normalized to rated power over the period a) June 2006 - June 2007, b) June 2007 - June 2008, c) June 2008 - June 2009 and d) June 2009 - June 2010. The error bars represent the associated power rating and measurement uncertainty.

5. Conclusion

The emergence and continuous increase in deployment of different PV technologies such as c-Si, thin-film and CPV, provide evidence that PV can become a leading energy source in the future. The success of each technology depends mainly on the capability of meeting targets such as the enhancement of manufacturing procedures while at the same time, accomplishing efficiency increases and cost reductions.

With the vast variety of PV technologies present in the market, it is important to acquire information about their outdoor performance. The main PV performance parameters include the energy yield, the efficiency and PR. These parameters provide the basis of all performance assessments and loss factor investigations. The main environmental factors affecting PV performance include solar irradiance, ambient temperature and solar spectrum. Another important factor for consideration is degradation. Good understanding of the outdoor performance of different PV technologies is a key requirement for their successful integration under different climatic conditions.

In addition to the review of several factors affecting PV performance, the main results of the outdoor investigation carried out in Cyprus over a four-year period have been presented. In particular, useful information on the performance of different PV technologies installed side-by-side was obtained by investigating their seasonal performance and the effects of temperature, soiling and power rating. The outcome of the outdoor performance assessment also showed that these technologies have enormous potential in countries with high solar resource.

6. Acknowledgment

The authors would like to thank J. H. Werner and M. Schubert for their vision, continuous support and encouragement. The authors also gratefully acknowledge the contributions of the German Federal Ministry for the Environment, Nature Conservation and Nuclear Safety (BMU), which supported this work under contract No. 0327553. We also gratefully acknowledge the support by the companies Atersa, First Solar GmbH, Phönix Sonnenstrom AG, Q-cells AG, Schott Solar GmbH, SMA Technologies AG, SolarWorld AG, Solon AG and Würth Solar GmbH & Co.KG. Finally the authors would like to acknowledge the financial support by the Cyprus Research Promotion Foundation.

7. References

Adelstein, J. & Sekulic, B. (2005). Performance and Reliability of a 1-kW Amorphous Silicon Photovoltaic Roofing System, *Proceedings of the 31st IEEE Photovoltaics Specialists Conference*, pp. 1627-1630, ISBN 0-7803-8707-4, Lake Buena Vista, USA, January 2005

Akhmad, K.; Kitamura, A.; Yamamoto, F.; Okamoto, H.; Takakura H. & Hamakawa, Y. (1997). Outdoor Performance of Amorphous Silicon and Polycrystalline Silicon PV Modules. *Solar Energy Materials and Solar Cells*, Vol.46, No.3, (June 1997), pp. 209-218

Batzner D.; Romeo, A.; Terheggen, M.; Dobeli, M.; Zogg, H. & Tiwari, A.N. (2004). Stability Aspects in CdTe/CdS Solar Cells. *Thin Solid Films*, Vol.451-452, March 2004, pp. 536-543

Biicher, K. (1997). Site Dependence of the Energy Collection of PV Modules. *Solar Energy Materials and Solar Cells*, Vol.47, No.1-4, (October 1997), pp. 85-94

Britt, J. & Ferekides, C. (1993). Thin-Film CdS/CdTe Solar Cell With 15.8% Efficiency. *Applied Physics Letter*, Vol.62, No.22, (May 1993), pp. 2851-2852, ISSN 0003-6951

Bunea, G.E.; Wilson, K.E.; Meydbray, Y.; Campbell, M.P. & De Ceuster, D.M. (2006). Low Light Performance of Mono-Crystalline Silicon Solar Cells, *Proceedings of the 4th IEEE World Conference on Photovoltaic Energy Conversion*, pp. 1314-1314, ISBN 1-4244-0017-1, Waikoloa, USA, May 2006

Cannon, T.W.; Hulstrom, R. & Trudell, D.T. (1993). New Instrumentation for Measuring Spectral Effects During Outdoor and Indoor PV Device Testing, *Proceedings of the 23rd IEEE Photovoltaic Specialists Conference*, pp. 1176-1179, ISBN 0-7803-1220-1, Lousiville, USA, May 1993

Carlson, D.E.; Lin, G. & Ganguly, G. (2000). Temperature Dependence of Amorphous Silicon Solar Cell PV Parameters, *Proceedings of the 28th IEEE Photovoltaic Specialists Conference*, pp. 707- 712, ISBN 0-7803-5772-8, Anchorage, USA, September 2000

Carlsson, T. & Brinkman, A. (2006). Identification of Degradation Mechanisms in Field-tested CdTe Modules. *Progress in Photovoltaics: Research and Applications*, Vol.14, No.3, May 2006, pp. 213–224

Carr, A.J. & Pryor, T.L (2004). A Comparison of the Performance of Different PV Module Types in Temperate Climates. *Solar Energy*, Vol.76, No.1-3, January-March 2004, pp. 285–294

Cereghetti, N.; Bura, E.; Chianese, D.; Friesen, G.; Realini, A. & Rezzonico, S. (2003). Power and Energy Production of PV Modules Statistical Considerations of 10 Years Activity, *Proceedings of the 3rd World Conference on Photovoltaic Energy Conversion*, pp. 1919-1922, ISBN 4-9901816-0-3, Osaka, Japan, May 2003

Chin, K. K.; Gessert, T.A. & Su-Huai, W. (2010). The Roles of Cu Impurity States in CdTe Thin Film Solar Cells, *Proceedings of the 35th IEEE Photovoltaics Specialists Conference*, pp. 1915-1918, ISBN 978-1-4244-5890-5, Honolulu, USA, June 2010

Cousins, P.J.; Smith, D.D.; Hsin-Chiao, L.; Manning, J.; Dennis, T.D.; Waldhauer, A.; Wilson, K.E.; Harley, G. & Mulligan, W.P. (2010). Generation 3: Improved Performance at Lower Cost, *Proceedings of 35th IEEE Photovoltaics Specialists Conference*, pp. 278-278, ISBN 978-1-4244-5890-5, Honolulu, Hawaii, USA, June 2010

Cusano, D.A. (1963). CdTe Solar Cells and Photovoltaic Heterojunctions in II–VI Compounds. *Solid-State Electron*, Vol.6, No.3, (May-June 1963), pp. 217-218

Degrave S.; Nollet P.; Stojanoska G.; Burgelman M. & Durose K. (2001). Interpretation of Ageing Experiments on CdTe/CdS Solar Cells. *Proceedings of 17th International Photovoltaic Science and Engineering Conference*, pp. 1058-1061, Fukuoka, December 2007 del Cueto, J.A. (1998). Method for Analyzing Series Resistance and Diode Quality Factors From Field Data of Photovoltaic Modules. *Solar Energy Materials and Solar Cells*, Vol.55, No.3, (August 1998), pp. 291-297

del Cueto, J.A; Rummel, S.; Kroposki, B.; Osterwald, C. & Anderberg, A. (2008). Stability of CIS/CIGS Modules at the Outdoor Test Facility Over Two Decades, *Proceedings of the 33rd IEEE Photovoltaic Specialists Conference*, pp. 1-6, ISBN 978-1-4244-1640-0, San Diego, USA, May 2008

Detrick, A.; Kimber, A. & Mitchell, L. (2005). Performance Evaluation Standards for Photovoltaic Modules and Systems, *Proceedings of the 31st IEEE Photovoltaics Specialists Conference*, pp. 1581-1586, ISBN 0-7803-8707-4, Lake Buena Vista, USA, January 2005

Dimitrova, M.; Kenny, R.P.; Dunlop, E.D. & Pravettoni, M. (2010). Seasonal Variations on Energy Yield of A-Si, Hybrid and Crystalline Si PV Modules. *Progress in Photovoltaics: Research and Applications*, Vol.18, No.5, (May 2010), pp. 311-320

Dittmann, S.; Durisch, W.; Mayor, J.C.; Friesen, G. & Chianese, D. (2010). Comparison of Indoor and Outdoor Characterisation of a CdTe Module, *Proceedings of the 25th European Photovoltaic Solar Energy Conference*, pp. 3508-3512, ISBN 3-936338-26-4, Valencia, Spain, September 2010

Dobson, K.D.; Visoly-Fisher, I.; Hodes, G. & Cahen, D. (2000). Stability of CdTe/CdS Thin-Film Solar Cells. *Solar Energy Materials and Solar Cells*, Vol.62, No.3, May 2000, pp. 295-325

Doni, A.; Dughiero, F. & Lorenzoni, A. (2010). A Comparison Between Thin Film and C-Si PV Technologies for MW Size Applications, *Proceedings of the 35th IEEE Photovoltaics Specialists Conference*, pp. 2380-2385, ISBN 978-1-4244-5890-5, Honolulu, USA, June 2010

Dunlop, E.D. & Halton, D. (2005). The Performance of Crystalline Silicon Photovoltaic Solar Modules After 22 Years of Continuous Outdoor Exposure. *Progress in Photovoltaics: Research and Applications*, Vol.14, No.1, June 2005, pp. 53-64

EPIA & Greenpeace. (2011). Solar Photovoltaic Energy Empowering the World, In: *Solar Generation 6*, 24.06.2011, Available from: http://www.greenpeace.org/international/en/publications/reports/Solar-Generation-6/

EPIA. (2010). 2010 Market Outlook, In: *EPIA Publications*, 24.06.2011, Available from: http://www.epia.org/publications/photovoltaic-publications-global-market-outlook.html

EPIA. (2011). Global Market Outlook for Photovoltaics Until 2015, In: *EPIA Publications*, 24.06.2011, Available from: http://www.epia.org/publications/photovoltaic-publications-global-market-outlook.html

Ferekides, C. & Britt, J. (1994). CdTe Solar Cells With Efficiencies Over 15%. *Solar Energy Materials and Solar Cells*, Vol.35, (September 1994), pp. 255-262

Fthenakis, V.M. (2004). Life Cycle Impact Analysis of Cadmium in CdTe Photovoltaic Production. *Renewable and Sustainable Energy Reviews*, Vol.8, (August 2004), pp. 303-334

Fthenakis, V.M. (2009). Sustainability of photovoltaics: The Case for Thin-Film Solar Cells. *Renewable and Sustainable Energy Reviews*, Vol.13, (December 2009), pp. 2746-2750

Fthenakis, V.M.; Fuhrmann, M.; Heiser, J.; Lanzirotti, A.; Fitts, J. & Wang, W. (2005). Emissions and Encapsulation of Cadmium in CdTe PV Modules during Fires. *Progress in Photovoltaics: Research and Applications*, Vol.13, (2005), pp. 713-723

Fthenakis, V.M.; Kim, H.C. & Alsema, E. (2008). Emissions from Photovoltaic Life Cycles. *Environmental Science & Technology*, Vol.42, No.6, (2008), pp. 2168–2174

Gabor, A.M.; Tuttle, J.R.; Albin, D.S.; Contreras, M.A.; Noufi, R. & Herman, A.M. (1994). High-Efficiency CuInxGa1−xSe2 Solar Cells Made From (Inx,Ga1−x)2Se3 Precursor Films. *Applied Physics Letters*, Vol.65, No.2, (July 1994), pp. 198-200

Gottschalg, R.; Betts, T.R.; Hohl-Ebinger, J.; Herrmann, W. & Müllejans, H. (2007). Effect of Spectral Mismatch on Power Rating Measurements – A Comparison of Indoor and Outdoor Measurements for Single and Multi-Junction Devices, *Proceedings of the 22nd European Photovoltaic Solar Energy Conference*, pp. 2621-2624, Milan, Italy, September 2007

Gottschalg, R.; Betts, T.R.; Infield, D.G. & Kearney, M.J. (2005). The Effect of Spectral Variations on the Performance Parameters of Single and Double Junction Amorphous Silicon Solar Cells. *Solar Energy Materials and Solar cells*, Vol.85, No.3, January 2005, pp. 415-428

Gottschalg, R.; Infield, D.G. & Kearney, M.J. (2003). Experimental Study of Variations of the Solar Spectrum of Relevance to Thin-Film Solar Cells. *Solar Energy Materials and Solar Cells*, Vol.79, No.4, September 2003, pp. 527-537

Green, M.A.; Emery, K.; Hishikawa, Y. & Warta, W. (2011). Solar Cell Efficiency Tables (version 37). *Progress in Photovoltaic: Research and Applications*, Vol.19, No.1, (January 2011), pp. 84–92

Green, M.A.; Emery, K.; Hishikawa, Y. & Warta, W. (2011). Solar Cell Efficiency Tables (version 38). *Progress in Photovoltaic: Research and Applications*, Vol.19, No.5, (August 2011), pp. 565-572

Grunow, P.; Preiss, A.; Koch, S. & Krauter, S. (2009). Yield and Spectral Effects of A-Si Modules, *Proceedings of the 24th European Photovoltaic Solar Energy Conference*, pp. 2846-2829, ISBN 3-936338-25-6, Hamburg, Germany, September 2009

GTM Research, (2011). Concentrating Photovoltaics 2011: Technology, Costs and Markets, In: *GTM Research*, 24.06.2010, Available from:
http://www.greentechmedia.com/research/report/concentrating-photovoltaics-2011

Gueymard, C.; Myers, D. & Emery, K. (2002). Proposed Reference Irradiance Spectra for Solar Energy Systems Testing. *Solar Energy*, Vol.73, No.6, December 2002, pp. 443-467

Guillemoles, J.-F.; Kronik, L.; Cahen, D.; Rau, U.; Jasenek, A. & Schock, H.-W. (2000). Stability Issues of Cu(In,Ga)Se2-Based Solar Cells. *Journal of Physical Chemistry B*, Vol.104, No.20, April 2000, pp. 4849-4862

Hammond, R.; Srinivasan, D.; Harris, A.; Whitfield, K. & Wohlgemuth, J. (1997). Effects of Soiling on PV Module and Radiometer Performance, *Proceedings of the 26th IEEE Photovoltaics Specialists Conference*, pp. 1121-1124, ISBN 0-7803-3767-0, Anaheim, USA, September 1997

Heesen, H.; Pfatischer, R.; Herbort, V. & von Schwerin, R. (2010). Performance Evaluation of Thin-Film Technologies Based on Operating Data, *Proceedings of the 25th*

European Photovoltaic Solar Energy Conference, pp. 3766-3768, ISBN 3-936338-26-4, Valencia, Spain, September 2010

Hegedus, S. (2006). Thin Film Solar Modules: the Low Cost, High Throughput and Versatile Alternative to Si Wafers. *Progress in Photovoltaics: Research and Applications*, Vol.14, No.5, June 2006, pp. 393-411

Hirata, Y. & Tani, T. (1995). Output Variation of Photovoltaic Modules with Environmental Factors-I. The Effect of Spectral Solar Variation on Photovoltaic Module Output, *Solar Energy*, Vol.55, No.6, December 1995, pp. 463-468

Huld, T.; Sample, T. & Dunlop, E.D. (2009). A Simple Model for Estimating the Influence of Spectrum Variations on PV Performance, *Proceedings of the 24th European Photovoltaic Solar Energy Conference*, pp. 3385-3389, ISBN 3-936338-25-6, Hamburg, Germany, September 2009

Ibrahim, M.; Zinsser, B.; El-Sherif, H.; Hamouda, E.; Makrides, G.; Georghiou, G.E.; Schubert, M. & Werner, J.H. (2009). Advanced Photovoltaic Test Park in Egypt for Investigating the Performance of Different Module and Cell Technologies, *Proceedings of the 24th Symposium Photovoltaic Solar Energy*, Staffelstien, Germany, March 2009

Igalson, M.; Wimbor, M.; Wennerberg, J. (2002). The Change of the Electronic Properties of CIGS Devices Induced by the Damp Heat Treatment. *Thin Solid Films*, Vol.403-404, February 2002, pp. 320-324

Ikisawa, M.; Nakano, A.; Igari, S. & Terashima, H. (1998). Outdoor Exposure Tests of Photovoltaic Modules in Japan and Overseas. *Renewable Energy*, Vol.14, No.1-4, May-August 1998, pp. 95-100

Jasenek, A.; Rau, U.; Weinert, K.; Schock, H.W. & Warner, J.H. (2002). Illumination-Enhanced Annealing of Electron-Irradiated Cu(In,Ga)Se2 Solar Cells, *Proceedings of the 29th IEEE Photovoltaic Specialists Conference*, pp. 872-875, ISBN 0-7803-7471-1, New Orleans, USA, May 2002

Jordan, D.C. & Kurtz, S.R. (2010). Analytical Improvements in PV Degradation Rate Determination, *Proceedings of the 35th IEEE Photovoltaics Specialists Conference*, pp. 2688-2693, ISBN 978-1-4244-5890-5, Honolulu, USA, June 2010

Jordan, D.C.; Smith, R.M.; Osterwald, C.R.; Gelak, E. & Kurtz, S.R. (2010). Outdoor PV Degradation Comparison, *Proceedings of the 35th IEEE Photovoltaics Specialists Conference*, pp. 2694-2697, ISBN 978-1-4244-5890-5, Honolulu, USA, June 2010

JRC. (2010). PV Status Report 2010, In: *Scientific & Technical Reference on Renewable Energy and End-Use Energy Efficiency*, 24.06.2011, Available from:
http://re.jrc.ec.europa.eu/refsys/

Kazmerski, L.L.; White, F.R. & Morgan, G.K. (1976). Thin-Film CuInSe2/CdS Heterojunction Solar Cells. *Applied Physics Letters*, Vol.29, No.4, (August 1976), pp. 268-271

Kenny, R.P.; Ioannides, A.; Mullejans, H.; Zaaiman W. & Dunlop, E.D. (2006). Performance of Thin Film PV Modules. *Thin Solid Films*, Vol.511-512, July 2006, pp. 663-672

King, D.; Boyson, W.E. & Kratochvyl, J.A. (2004). Photovoltaic Array Performance Model, In: *SAND2004-3535*, 24.06.2010, Available from:
http://photovoltaics.sandia.gov/docs/PDF/King%20SAND.pdf

King, D.; Kratochvil, J.A. & Boyson, W.E. (1997). Measuring the Solar Spectral and Angle of Incidence Effects on Photovoltaic Modules and Solar Irradiance Sensors, *Proceedings of the 26th IEEE Photovoltaic Specialists Conference*, pp. 1113-1116, ISBN 0-7803-3767-0, Anaheim, USA, September 1997

King, D.L.; Kratochvil J.A. & Boyson, W.E. (2000). Stabilization and Performance Characteristics of Commercial Amorphous-Silicon PV Modules, *Proceedings of the 28th IEEE Photovoltaic Specialists Conference*, pp. 1446-1449, ISBN 0-7803-5772-8, Anchorage, USA, September 2000

King, D.L; Hansen, B.R.; Kratochvil, J.A. & Quintana, M.A. (1997). Dark Current-Voltage Measurements on Photovoltaic Modules as a Diagnostic or Manufacturing Tool, *Proceedings of the 26th IEEE Photovoltaic Specialists Conference*, pp. 1125-1128, ISBN 0-7803-3767-0, Anaheim, USA, September 1997

King, D.L; Kratochvil, J.A. & Boyson, W.E. (1997). Temperature coefficients for PV Modules and Arrays: Measurement Methods, Difficulties, and Results, *Proceedings of the 26th IEEE Photovoltaic Specialists Conference*, pp. 1183-1186, ISBN 0-7803-3767-0, Anaheim, USA, September 1997

Kymakis, E.; Kalykakis, S. & Papazoglou, T.M. (2009). Performance Analysis of a Grid Connected Photovoltaic Park on the Island of Crete. *Energy Conversion and Management*, Vol. 50, No. 3, March 2009, pp. 433–438

Lam, K.H.; Close, J. & Durisch, W. (2004). Modelling and Degradation Study on a Copper Indium Diselenide Module. *Solar Energy*, Vol.77, No.1, 2004, pp. 121-127

Lechner, P.; Geyer, R.; Haslauer, A.; Roehrl, T.; Lundszien, D.; Weber, C. & Walter, R. (2010). Long-Term Performance of ASI Tandem Junction Thin Film Solar Modules, *Proceedings of the 25th European Photovoltaic Solar Energy Conference*, pp. 3283-3287, ISBN 3-936338-26-4, Valencia, Spain, September 2010

Makrides, G.; Zinsser, B.; Georghiou, G.E., Schubert, M. & Werner, J.H. (2009). Error Sources in Outdoor Performance Evaluation of Photovoltaic Systems, *Proceedings of the 24th European Photovoltaic Solar Energy Conference*, pp. 3904-3909, ISBN 3-936338-25-6, Hamburg, Germany, September 2009

Makrides, G.; Zinsser, B.; Georghiou, G.E.; Schubert, M. & Werner, J.H. (2009). Temperature Behavior of Different Photovoltaic Systems Installed in Cyprus and Germany. *Solar Energy Materials & Solar Cells*, Vol.93, No.6-7, (June 2009), pp. 1095–1099

Makrides, G.; Zinsser, B.; Norton, M.; Georghiou, G.E.; Schubert, M. & Werner, J.H. (2010). Outdoor Performance Evaluation of Grid-Connected PV Technologies in Cyprus. *Journal of Energy and Power Engineering*, Vol.4, No.2, (February 2010), pp. 52–57

Makrides, G.; Zinsser, B.; Norton, M.; Georghiou, G.E.; Schubert, M. & Werner, J.H. (2010). Potential of Photovoltaic Systems in Countries with High Solar Irradiation. *Renewable and Sustainable Energy Reviews*, Vol.14, No.2, (February 2010), pp. 754-762

Malmström, J.; Wennerberg, J. & Stolt, L. (2003). A Study of the Influence of the Ga Content on the Long-Term Stability of Cu(In,Ga)Se2 Thin Film Solar Cells. *Thin Solid Films*, Vol.431-432 May 2003, pp. 436-442

Marion, B.; Adelstein, J.; Boyle K.; Hayden, H,; Hammond, B.; Fletcher, T.; Canada, B.; Narang, D.; Kimber, A.; Mitchell, L.; Rich, G. & Townsend, T. (2005).

Performance Parameters for Grid-Connected PV Systems, *Proceedings of the 31st IEEE Photovoltaics Specialists Conference*, pp. 1601-1606, ISBN 0-7803-8707-4, Lake Buena Vista, USA, January 2005

McMahon, T.J. (2004). Accelerated Testing and Failure of Thin-film PV Modules. *Progress in Photovoltaics: Research and Applications*, Vol.12, No.23, 2004, pp. 235–248

Merten, J. & Andreu, J. (1998). Clear Separation of Seasonal Effects on the Performance of Amorphous Silicon Solar Modules by Outdoor I/V-Measurements. *Solar Energy Materials and Solar Cells*, Vol.52, No.1-2, (March 1998), pp. 11-25

Meyer, E.L. & van Dyk, E.E. (2004). Assessing the Reliability and Degradation of Photovoltaic Module Performance Parameters. *IEEE Transactions on Reliability*, Vol. 53, No. 1, March 2004, pp 83-92

Meyers P.V. (2006). First Solar Polycrystalline CdTe Thin Film PV, *Proceedings of the 4th IEEE World Conference on Photovoltaic Energy Conversion*, pp. 2024-2027, ISBN 1-4244-0017-1, Waikoloa, USA, May 2006

Mohring, H.-D.M, & Stellbogen, D. (2008). Annual Energy Harvest of PV Systems– Advantages and Drawbacks of Different PV Technologies, *Proceedings of the 23rd European Photovoltaic Solar Energy Conference*, pp. 2781-2785, ISBN 3-936338-24-8, Valencia, Spain, September 2008

Mrig, L. & Rummel S. (1990). Outdoor Stability of Performance of CIS and CdTe Photovoltaic Modules at SERI. *Proceedings of the 21st IEEE Photovoltaic Specialist Conference*, pp. 1038-1039, Kissimmee, USA, May 1990

Myers, D.R.; Emery, K. & Gueymard, C. (2002). Terrestrial Solar Spectral Modeling Tools and Applications for Photovoltaic Devices, *Proceedings of the 29th IEEE Photovoltaic Specialists Conference*, pp. 1683-1686, ISBN 0-7803-7471-1, New Orleans, USA, May 2002

Norton, M.; Dobbin, A. & Georghiou, G.E. (2011). Solar Spectrum Prediction and Comparison with Measurements, *Proceedings of the International Conference on Concentrating Photovoltaic Systems*, Las Vegas, USA, April 2011

Oerlikon Solar. (2010). Oerlikon Solar Breaking Two World Records: Lowest Module Production Cost and Highest Lab Cell Efficiency, In: *Press Releases Archive 2010*, 24.06.2011, Available from: http://www.oerlikon.com/ecomaXL/index.php?site=OERLIKON_EN_archive_2010_detail&udtx_id=7720

Osterwald, C.R.; Adelstein, J.; del Cueto, J.A.; Kroposki, B.; Trudell, D. & Moriarty, T. (2006). Comparison of Degradation Rates of Individual Modules Held at Maximum Power, *Proceedings of the 4th IEEE World Conference on Photovoltaic Energy Conversion*, pp. 2085-2088, ISBN 1-4244-0017-1, Waikoloa, USA, May 2006

Osterwald, C.R.; Anderberg, A.; Rummel, S. & Ottoson, L. (2002). Degradation Analysis of Weathered Crystalline-Silicon PV Modules, *Proceedings of the 29th IEEE Photovoltaic Specialists Conference*, pp. 1392-1395, ISBN 0-7803-7471-1, New Orleans, USA, May 2002

Paretta, A.; Sarno, A. & Vicari, L. (1998). Effects of Solar Irradiation Conditions on the Outdoor Performance of Photovoltaic Modules. *Optics Communication*, Vol.153, No.1-3, (July 1998), pp. 153-163

Pern, F.J.; Czanderna, A.W.; Emery, K.A. & Dhere, R.G. (1991). Weather Degradation of EVA Encapsulant and the Effect of its Yellowing on Solar Cell Efficiency, *Proceedings of the 22nd IEEE Photovoltaic Specialists Conference*, pp. 557-561, ISBN 0-87942-636-5, Las Vegas, USA, October 1991

Powell, R.C.; Sasala, R.; Rich, G.; Steele, M.; Bihn, K.; Reiter, N.; Cox, S. & Dorer, G. (1996). Stability Testing of CdTe/CdS Thin-Film Photovoltaic Modules, *Proceedings of the 25th IEEE Photovoltaic Specialists Conference*, pp. 785-788, ISBN 0-7803-3166-4, Washington, USA, May 1996

Quintana, M.A.; King, D.L.; McMahon, T.J. & Osterwald, R.C. (2002). Commonly Observed Degradation in Field-Aged Photovoltaic Modules, *Proceedings of the 29th IEEE Photovoltaic Specialists Conference*, pp. 1436-1439, ISBN 0-7803-7471-1, New Orleans, USA, May 2002

Randall J.F. & Jacot, J. (2003). Is AM1.5 Applicable in Practice? Modelling Eight Photovoltaic Materials With Respect to Light Intensity and Two Spectra. *Renewable Energy*, Vol.28, No.12, (October 2003), pp. 1851–1864

Ransome, S.J. & Wohlgemuth, J.H. (2000). Predicting kWh/kWp Performance for Amorphous Silicon Thin Film Modules, *Proceedings of the 28th IEEE Photovoltaic Specialists Conference*, pp. 1505-1508, ISBN 0-7803-5772-8, Anchorage, September 2000

Reich, N.H.; van Sark, W.G.J.H.M.; Alsema, E.A.; Lof, R.W.; Schropp, R.E.I.; Sinke, W.C. & Turkenburg, W.C. (2009). Crystalline Silicon Cell Performance at Low Light Intensities. *Solar Energy Materials and Solar Cells*, Vol.93, No.9, (September 2009), pp. 1471–1481

Sanchez-Friera, P.; Piliougine, M.; Pelaez, J.; Carretero, J. & Sidrach M. (2011). Analysis of Degradation Mechanisms of Crystalline Silicon PV Modules After 12 Years of Operation in Southern Europe. *Progress in Photovoltaics: Research and Applications*, January 2011

Schmidt, M.; Braunger, D.; Schäffler, R.; Schock, H.W. & Rau, U. (2000). Influence of Damp Heat on the Electrical Properties of Cu(In,Ga)Se2 Solar Cells. *Thin Solid Films*, Vol.361-362, February 2000, pp. 283-287

Schumann, A. (2009). Irradiance Level Characteristics of PV Modules and the Need for Improved Data Quality, *Proceedings of the 24th European Photovoltaic Solar Energy Conference*, pp. 3468-3470, ISBN 3-936338-25-6, Hamburg, Germany, September 2009

Shah, A.; Torres, P.; Tscharner, R.; Wyrach, N. & Keppner, H. (1999). Photovoltaic Technology: The Case of Thin-Film Solar Cells. *Science*, Vol.285, No.5428, (July 1999), pp. 692-698

Som, A.K. & Al-Alawi, S.M. (1992). Evaluation of Efficiency and Degradation of Mono- and Polycrystalline PV Modules Under Outdoor Conditions. *Renewable Energy*. Vol.2, No.1, February 1992, pp. 85-91

Staebler, D.L. & Wronski, C.R. (1977). Reversible Conductivity Charges in Discharge-Produced Amorphous Si. *Applied Physics Letters*, Vol.31, No.4, (August 1977), pp. 292–294

Stutzmann, M.; Jackson, W.B. & Tsai, C.C. (1985). Light-Induced Metastable Defects in Hydrogenated Amorphous Silicon: A Systematic Study. *Physical Review B*, Vol.32, No.1, July 1985, pp. 23-47

Sutterlüti, J.; Sinicco, I.; Hügli, A.; Hälker, T. & Ransome, S. (2009). Outdoor Characterization and Modelling of Thin-Film Modules and Technology Benchmarking, *Proceedings of the 24th European Photovoltaic Solar Energy Conference*, pp. 3198-3205, ISBN 3-936338-25-6, Hamburg, Germany, September 2009

Suzuki, R.; Kawamura, H.; Yamanaka, S.; Ohno, H. & Naito, K. (2002). Loss Factors Affecting Power Generation Efficiency of a PV Module, *Proceedings of the 29th IEEE Photovoltaic Specialists Conference*, pp. 1557-1560, ISBN 0-7803-7471-1, New Orleans, Louisiana, May 2002

Ullal, H.S.; Zweibel, K. & von Roedem, B. (1997). Current Status of Polycrystalline Thin-Film PV Technologies, *Proceedings of the 26th IEEE Photovoltaic Specialist Conference*, pp. 301-305, ISBN 0-7803-3767-0, Anaheim, USA, September 1997

Virtuani, A.; Mullejans, H. & Dunlop, E.D. (2011). Comparison of Indoor and Outdoor Performance Measurements of Recent Commercially Available Solar Modules. *Progress in Photovoltaics: Research and Applications*, Vol.19, No.1, January 2011, pp. 11-20

Virtuani, A.; Pavanello, D. & Friesen, G. (2010). Overview of Temperature Coefficients of Different Thin Film Photovoltaic Technologies, *Proceedings of the 25th European Photovoltaic Solar Energy Conference*, pp. 4248-4252, ISBN 3-936338-26-4, Valencia, Spain, September 2010

Wenham, S.R.; Green, M.A. & Watt, M.E. (2007). *Applied Photovoltaics*, (Second edition), Earthscan, ISBN 978-1-84407-401-3, London, UK

Wolden, C.A.; Kurtin, J.; Baxter, J.B.; Repins, I.; Shaheen, S.E.; Torvik, J.T.; Rockett, A.A.; Fthenakis, V.M. & Aydil, E.S. (2011). Photovoltaic Manufacturing: Present Status, Future Prospects, and Research Needs. *Journal of Vacuum Science & Technology A: Vacuum, Surfaces, and Films*, Vol.29, No.3, (May 2011), pp. 030801-030801-16, ISSN 0734-2101

Zanesco, I. & Krenziger, A. (1993). The Effects of Atmospheric Parameters on the Global Solar Irradiance and on the Current of a Solar Cell. *Progress in Photovoltaics: Research and Applications*, Vol.1, No.3, July 1993, pp. 169-179

Zdanowicz, T.; Rodziewicz, T. & Waclawek, M.Z. (2003). Effect of Airmass Factor on the Performance of Different Type of PV Modules, *Proceedings of the 3rd World Conference on Photovoltaic Energy Conversion*, pp. 2019-2022, ISBN 4-9901816-0-3, Osaka, Japan, May 2003

Zinsser, B.; Makrides, G.; Schmitt, W.; Georghiou, G.E. & Werner, J.H. (2007). Annual Energy Yield of 13 Photovoltaic Technologies in Germany and Cyprus, *Proceedings of the 22nd European Photovoltaic Solar Energy Conference*, pp. 3114-3117, Milan, September 2007

Zinsser, B.; Makrides, G.; Schubert, M.; Georghiou, G.E. & Werner, J.H. (2009). Temperature and Irradiance Effects on Outdoor Field Performance, *Proceedings of the 24th European Photovoltaic Solar Energy Conference*, pp. 4083-4086, ISBN 3-936338-25-6, Hamburg, Germany, September 2009

Zinsser, B.; Makrides, G.; Schubert, M.B.; Georghiou, G.E. & Werner, J.H. (2010). Rating of Annual Energy Yield More Sensitive to Reference power Than Module Technology, *Proceedings of the 35th IEEE Photovoltaics Specialists Conference*, pp. 1095-1099, ISBN 978-1-4244-5890-5, Honolulu, USA, June 2010